T0075524

ADVANCEMENTS IN CYBERCRIME INVESTIGATION AND DIGITAL FORENSICS

ADVANCEMENTS IN CYBERCRIME INVESTIGATION AND DIGITAL FORENSICS

Edited by
A. Harisha
Amarnath Mishra, PhD
Chandra Singh

APPLE
ACADEMIC
PRESS

First edition published 2024

Apple Academic Press Inc.
1265 Goldenrod Circle, NE,
Palm Bay, FL 32905 USA

760 Laurentian Drive, Unit 19,
Burlington, ON L7N 0A4, CANADA

CRC Press
2385 NW Executive Center Drive,
Suite 320, Boca Raton FL 33431

4 Park Square, Milton Park,
Abingdon, Oxon, OX14 4RN UK

© 2024 by Apple Academic Press, Inc.

Apple Academic Press exclusively co-publishes with CRC Press, an imprint of Taylor & Francis Group, LLC

Reasonable efforts have been made to publish reliable data and information, but the authors, editors, and publisher cannot assume responsibility for the validity of all materials or the consequences of their use. The authors, editors, and publishers have attempted to trace the copyright holders of all material reproduced in this publication and apologize to copyright holders if permission to publish in this form has not been obtained. If any copyright material has not been acknowledged, please write and let us know so we may rectify in any future reprint.

Except as permitted under U.S. Copyright Law, no part of this book may be reprinted, reproduced, transmitted, or utilized in any form by any electronic, mechanical, or other means, now known or hereafter invented, including photocopying, microfilming, and recording, or in any information storage or retrieval system, without written permission from the publishers.

For permission to photocopy or use material electronically from this work, access www.copyright.com or contact the Copyright Clearance Center, Inc. (CCC), 222 Rosewood Drive, Danvers, MA 01923, 978-750-8400. For works that are not available on CCC please contact mpkbookspermissions@tandf.co.uk

Trademark notice: Product or corporate names may be trademarks or registered trademarks and are used only for identification and explanation without intent to infringe.

Library and Archives Canada Cataloguing in Publication

Title: Advancements in cybercrime investigation and digital forensics / edited by A. Harisha, PhD, Amarnath Mishra, PhD, Chandra Singh.
Names: Harisha, A., editor. | Mishra, Amarnath (Forensic scientist), editor. | Singh, Chandra (Professor of engineering), editor.
Description: First edition. | Includes bibliographical references and index.
Identifiers: Canadiana (print) 20230223524 | Canadiana (ebook) 20230223540 | ISBN 9781774913031 (hardcover) | ISBN 9781774913048 (softcover) | ISBN 9781003369479 (ebook)
Subjects: LCSH: Computer crimes—Investigation. | LCSH: Digital forensic science.
Classification: LCC HV8079.C65 A38 2023 | DDC 364.16/8—dc23

Library of Congress Cataloging-in-Publication Data

CIP data on file with US Library of Congress

ISBN: 978-1-77491-303-1 (hbk)
ISBN: 978-1-77491-304-8 (pbk)
ISBN: 978-1-00336-947-9 (ebk)

About the Editors

A. Harisha

Assistant Professor, Department of Computer Science and Engineering, SCEM, Sahyadri College of Engineering and Management, Karnataka, India

A. Harisha is an Assistant Professor in the Department of Computer Science and Engineering at Sahyadry College of Engineering and Management, Karnataka, India. He has published about nine research papers indexed in DBLP, Scopus, and SCI. He has presented his research findings at various international conferences and is an active member of several professional organizations, such as Indian Society for Technical Education (ISTE), Association for Computing Machinery (ACM), and Computer Society of India. He is also a web developer and application team member for Techno-Medicare. He is a Co-Principal Investigator of Project Digital Forensics and Intelligence, funded by the Government of Karnataka, Department of Science and Technology, under the Vision Group on Science and Technology (VGST) scheme. The Indian patent office has published his patent application on "Enhanced Detection of Attacks on Network Based on Pattern Recognition with Decision Stump." He also has two international innovation patents: "Feed Forward Neural Networks Combined with Extreme Learning Machine Approach for Large Weather Data" and "An Improved Cyber Security System with Digital Watermarking Using Combined Transformation Approach." He has served as a committee member for several international conferences. Mr. Harisha completed his master's degree with specialization in Wireless Communication and Computing from the Indian Institute of Information Technology, Allahabad, India, and his bachelor's degree with specialization in Computer Science and Engineering from Viswesvarayya Technological University Belagavi, Karnataka, India.

Amarnath Mishra, PhD
Professor (Forensic Science) & Director,
Lloyd Institute of Forensic Science,
Greater Noida, Uttar Pradesh
(affiliated to the National Forensic Sciences
University, Gandhinagar, Gujarat), India

Prof. (Dr.) Amarnath Mishra has been working as a Professor (Forensic Science) and Director at the Lloyd Institute of Forensic Science, Greater Noida, Uttar Pradesh (affiliated to the National Forensic Sciences University, Gandhinagar, Gujarat), India, since October 2022. Prior to this, he worked as an Associate Professor (Forensic Science) at the School of Forensics, Risk Management and National Security, Rashtriya Raksha University, Gandhinagar, Gujarat (An Institution of National Importance) (Pioneering National Security and Police University of India); Ministry of Home Affairs, Government of India; Amity Institute of Forensic Sciences, Amity University, Noida, Uttar Pradesh, India; Department of Forensic Science, School of Humanities, Arts & Applied Sciences, Amity University Dubai Campus, United Arab Emirates; and as an Associate Professor (Forensic Medicine) at the Department of Forensic Medicine, National Medical College, Birgunj, affiliated to Tribhuvan University; and College of Medical Sciences, Bharatpur, affiliated to Kathmandu University, recognized by Nepal Medical Council and approved by the Government of Nepal, Ministry of Education. He has over 15 years of experience in research, academics, administration, forensic practice, and casework.

Prof. Mishra has been awarded a PhD degree in Forensic Science with a specialization in Forensic and Analytical Toxicology from the Sam Higginbottom University of Agriculture, Technology & Sciences, Prayagraj (formerly SHIATS, Allahabad) in collaboration with the Central Forensic Science Laboratory, Chandigarh, DFS, MHA, Govt. of India. He completed his MPhil degree in Biochemistry with a specialization in Forensic Biochemistry from Vinayaka Missions University, Salem, Tamilnadu, India. He has completed his MSc degree in Forensic Science with a specialization in Forensic Serology & DNA Fingerprinting from the Sam Higginbottom University of Agriculture, Technology & Sciences, Prayagraj (formerly AAIDU, Allahabad) in collaboration with the Central Forensic Science Laboratory, Hyderabad, DFS, MHA, Govt. of India. He has worked as

Research Scholar for the pilot study on "Non-registration of F.I.R." at Police Stations, conducted by the Bureau of Police Research and Development (BPR&D), MHA, Govt. of India. He has qualified UGC-NET Exam. June 2007 for Lectureship in Forensic Science.

Prof. Mishra has supervised three PhD students in the area of issue and perspective of DNA typing, method development for the analysis of anaesthetic drugs, and detection of fire accelerants in arson cases. He has guided more than 150 undergraduate and postgraduate students in various programs, i.e., BSc, BTech, MSc, MBBS, and MD, for their major projects and dissertations. He has more than 100 publications and attended about 100 national and international events. He has granted/awarded four national and international patents.

Prof. Mishra is a recipient of an Academic Excellence Award. He is an NABL-empanelled Technical Assessor for ISO/IEC17025. He is a paper setter and an external examiner for various academic institutions and recruitment agencies. He is an editorial board member in different reputed national and international journals. He has organized more than 15 events.

Prof. Mishra is a Freelance Forensic Expert and Medico-legal Consultant for the Court of Law under section 45 and section 65B of the Indian Evidence Act 1872. He has also contributed his knowledge and experiences in TV and radio programs as a Forensic Expert & Subject Specialist. He is the founder of Aspire Forensics and Amar Educational and Social Welfare Trust. In addition to his above assignments, he is serving many professional bodies along with lifetime membership.

Orcid ID: https://orcid.org/0000-0001-8798-5153

Chandra Singh

Assistant Professor, Department of Electronics and Communication Engineering, Sahyadri College of Engineering & Management, Karnataka, India

Chandra Singh is an Assistant Professor at Sahyadri College of Engineering and Management, Karnataka, India. He has co-authored a textbook titled *Survivability Techniques in Optical Network*, published by Studuim Press. He has published more 15 papers indexed in Scopus & Web of Science journals. He has received two best paper awards given by the Malviya National Institute of Technology, Jaipur, and by Droncharya College of Engineering, Gurgoan, India. He twice received a Hackerank Topper for Python & Advanced CPP held at VIT, Chennai, FDP, India. He was also the topper of a java quiz held at FDP, organized by VIT, Chennai, India. He has received a Certification of Recognition as a SPOC during GUVi's RPA SKILL-A-THON 2020. He has also served as the Indian Institute of Remote Sensing (IIRS) Co-ordinator for Department of Electronics and Communication. He has received other awards for his work as well. Mr. Singh served as a TPC/PC member for several conferences, including ICSC 2019, MRCN 2020, SPTM 2020, and ICCCIS 2021. His areas of interest are optical networking and communication, wireless communication, IoT machine learning, and image processing. He earned his MTech in Digital Electronics & Communication from NMAM Institute of Technology Nitte and his BE in Electronics and Communication Engineering at Srinivas School of Engineering, Mukka, Mangaluru, India.

Contents

Contributors

Shiji Abraham
Department of Computer Science and Engineering, Sahyadri College of Engineering and Management, Mangalore, Karnataka, India

Akanksha
MSc Forensic Science Student, Amity Institute of Forensic Sciences, Amity University, Noida, Uttar Pradesh, India

K. Alakananda
Department of Computer Science and Engineering, Sahyadri College of Engineering and Management, Mangalore, Karnataka, India

Namita A. Amdalli
Department of Computer Science and Engineering, Sahyadri College of Engineering and Management, Mangalore, Karnataka, India

Vinay Aseri
Department of Forensic Science, Vivekanand Global University, Jaipur, Rajasthan, India

Rushikesh Chopade
Department of Forensic Science, Vivekanand Global University, Jaipur, Rajasthan, India

Deepika Dubey
Research Scholar, Department of Forensic Science, Amity School of Applied Sciences, Amity University, Haryana, India

A. Harisha
Assistant Professor, Department of Computer Science and Engineering, Sahyadri College of Engineering and Management, Mangaluru, Karnataka, India

M. R. Ebenezar Jebarani
Department of ECE, Sathyabama Institute of Science and Technology, Chennai, Tamil Nadu, India

Sunidhi Joshi
MSc Forensic Science Student, Amity Institute of Forensic Sciences, Amity University, Noida, Uttar Pradesh, India

P. Kavipriya
Department of ECE, Sathyabama Institute of Science and Technology, Chennai, Tamil Nadu, India

Akash Kumar
Amity School of Engineering and Technology, Amity University, Noida, Uttar Pradesh, India

S. Lakshmi
Department of ECE, Sathyabama Institute of Science and Technology, Chennai, Tamil Nadu, India

Varun Malik
Chitkara University Institute of Engineering and Technology, Chitkara University, Punjab, India

Amarnath Mishra
Professor (Forensic Science) & Director, Lloyd Institute of Forensic Science, Greater Noida, Uttar Pradesh (affiliated to the National Forensic Sciences University, Gandhinagar, Gujarat), India

Mandeep Mittal
Department of Mathematics, Amity Institute of Applied Sciences, Amity University, Noida, Uttar Pradesh, India

Ruchi Mittal
Chitkara University Institute of Engineering and Technology, Chitkara University, Punjab, India

Arpit Nirvan
Department of Computer Science and Engineering, I.T.S. Engineering College, Greater Noida, Uttar Pradesh, India

Pandit Pritam
Department of Forensic Science, Vivekanand Global University, Jaipur, Rajasthan, India

E. Fantin Irudaya Raj
Assistant Professor, Department of Electrical and Electronics Engineering, Dr. Sivanthi Aditanar College of Engineering, Tamil Nadu, India

Raja Rajendran
Department of IT, Canara Bank, India

Richa Rohatgi
Assistant Professor, Forensic Science, LNJN NICFS, National Forensic Science University, Delhi Campus, Rohini, New Delhi, India

Shipra Rohatgi
Assistant Professor, Amity Institute of Forensic Sciences, Amity University, Noida, India

Golda Sahoo
Assistant Professor–Law, FIC Centre for Study in Victimology, Tamil Nadu National Law University, Trichy, Tamil Nadu, India

Likhith Salian
Student, Computer Science and Engineering, Sahyadri College of Engineering and Management, Mangaluru, Karnataka, India

Mahipal Singh Sankla
Department of Forensic Science, University Centre for Research & Development (UCRD), Chandigarh University, Mohali, Punjab, India.

Priya R. Sankpal
Department of Electronics and Communication Engineering, BNM Institute of Technology, Bangalore, Karnataka, India

Priyanka Sharma
Rashtriya Raksha University, Gandhinagar, Gujarat, India

C. G. Ajay Shastry
PG-Student, Computer Science and Engineering, Sahyadri College of Engineering and Management, Mangaluru, Karnataka, India

Sakshi Shrivastava
MSc Student, Information and Cyber Security, NSHM College of Management and Technology, Kolkata, West Bengal, India

Chintan Singh
Research Scholar, Amity Institute of Forensic Sciences, Amity University, Noida,
Uttar Pradesh, India

Leena Singh
Amity School of Engineering and Technology, Amity University, Noida, Uttar Pradesh, India

Shubham Singh
Amity School of Engineering and Technology, Amity University, Noida, Uttar Pradesh, India

Sourabh Kumar Singh
Research Scholar, Amity Institute of Forensic Sciences, Amity University, Noida,
Uttar Pradesh, India

Susmita Singh
Assistant Professor, Department of Chemistry, Amity Institute of Applied Science, Amity University,
Kolkata, West Bengal, India

Vikram Singh
Department of Computer Science and Engineering, Chaudhary Devi Lal University, Sirsa,
Haryana, India

Vineet Kumar Singh
Marine Faculty (Marine Engineer), Sensea Maritime Academy, Kolkata, West Bengal, India

Swaroop S. Sonone
Department of Forensic Science, Dr. Babasaheb Ambedkar Marathwada University, Aurangabad,
Maharashtra, India

Hepi Suthar
Rashtriya Raksha University, Gandhinagar, Gujarat, India

Jitender Tanwar
Amity Institute of Information Technology, Amity University, Noida, Uttar Pradesh, India

Sudhanshu Tripathi
Amity University in Tashkent, Labzak Tashkent City, Uzbekistan

Vaishali Tyagi
MSc Forensic Science Student, Amity Institute of Forensic Sciences, Amity University, Noida,
Uttar Pradesh, India

Vaishali
MSc Forensic Science Student, Amity Institute of Forensic Sciences, Amity University, Noida,
Uttar Pradesh, India

Manisha Verma
Assistant Professor, Department of Computer Science and Engineering, Hindustan College of Science
& Technology, Mathura, Uttar Pradesh, India

P. A. Vijaya
Department of Electronics and Communication Engineering, BNM Institute of Technology, Bangalore,
Karnataka, India

Adish Yermal
Student, Information Science and Engineering, Sahyadri College of Engineering and Management,
Mangaluru, Karnataka, India

Abbreviations

ADF	abstract digital forensics
ADS	alternate data streams
AF	anti-forensics
AI	artificial intelligence
AML	anti-money laundering
ANN	artificial neural network
AoI	area of interest
API	application programmable interface
ArTHIR	ATT&CK remote threat hunting incident response
ATM	automated teller machine
BiGRU	bi-directional gated repeated unit
BiLSTM	long bi-directional memory
CA	certification authority
CAINE	computer-aided investigative environment
CAPTCHA	Completely Automated Public Turing Test to Tell Computers and Humans Apart
CBR	case-based thinking
CC	correlation coefficient
CDR	call detail records
CEE	convention on electronic evidence
CFS	correlation feature selection
CI	cover image
CIA	Central Intelligence Agency
CNN	convolutional neural network
COFEE	computer online forensic evidence extractor
CP	child pornography
CSRF	cross-site request forgery
DBMS	database management system
DCT	discrete cosine transform
DFF	digital forensics framework
DFIR	digital forensics and incident response
DFT	discrete Fourier transform
DFTD	digital forensic terminology database
DICOM	digital imaging and communications in medicine

DM	data mining
DNN	deep neural network
DWT	discrete wavelet transform
EDA	electrodermal activity
EDD	encrypted disk detector
EDI	electronic data interchange
EEPROM	erasable programmable read-only memory
EFA	e-mail forensics analyst
EFS	Encrypted File System
FAU	forensic acquisition utilities
FBI	Federal Bureau of Investigation
FDA	forensic data analyst
FTK	forensic toolkit
GC	garbage clear
GCS	ground control station
GDPR	general data protection regulation
GPS	global positioning system
GUI	graphical user interface
HSI	Homeland Security Investigations
HPA	hardware protected area
IC3	internet crime complaint center
ICT	information communication technology
IDES	identity ecosystem steering group
IDI	integrated digital investigation
IDS	intrusion detection system
IEA	Indian Evidence Act
IMEI	International Mobile Equipment Identity
IOCE	International Organization for Computer Evidence
IoT	internet of things
IPC	Indian Penal Code
IPsec	internet protocol security
ISP	Internet Service Provider
KDD	knowledge discovery in database
K-NN	K-nearest neighbor
LFW	labeled face in the wild
MCC	Matthew's correlation coefficient
MF	mobile forensics
ML	machine learning
MLC	multi-level cell

MSE	mean square error
NCC	normalized correlation coefficient
NDNAD	National Desoxyribonucleic Acid Data
NFC	near-field communication
NIJ	National Institute of Justice
NMap	network mapper
NPCR	number of pixels change rate
NSTIC	National Strategy for Trusted Identities in Cyberspace
NTFS	New Technology File System
NW3C	National Center for Combating Crime
OS	operating system
OSINT	open-source intelligence
PCB	printed circuit board
PGP	pretty good privacy
PSNR	peak signal-to-noise ratio
QR	quick response
RIM	research in motion
RIR	regional internet registry
ROC	receiver operator characteristic curve
ROM	read-only memory
RPA	robot process automation
RRIQA	reduced reference image quality assessment
SAS	Statistical Analysis System
SDK	software development kit
SIFT	SANS investigative forensic toolkit
SIM	Subscriber Identity Module
SIM	Subscriber Information Module
SLC	single-level memory cell
SMR	soft max regression
SSD	solid-state drives
SSIM	structural similarity index
SSL	secure socket layer
STL	self-taught learning
SVD	singular value decomposition
SVM	support vector machine
SWGDE	scientific working group on digital evidence
TLC	three-layer memory cell
TLS	transport layer security
TOR	the onion router

TPB	theory of planned behavior
UACI	unified average changing intensity
UAV	unmanned aerial vehicle
UQI	Universal Quality Index
UTAUT	unified technology acceptance and use theory
VPN	virtual private network
VTU	Visvesvaraya Technological Institute
WEP	Wired Equivalent Privacy
WHO	World Health Organization
WinRM	windows remote management
WM	watermark image

Acknowledgment

First of all, we are grateful for the incredible opportunity provided to us by the publisher, Apple Academic Press of Taylor & Francis Group, and for supporting us in realizing this book. We are also indebted to the support of our authors, who continuously supported us by sending the required information in a timely manner and coordinating with us. We would also like to thank our reviewers for their selfless service in conducting the reviews and providing us with sound advice to improve the accepted book chapters. Lastly, we would like to thank our students, family and friends, management, principal, research director, and faculty fraternity of our respective organizations; Sahyadri College of Engineering & Management, Mangaluru, and Lloyd Institute of Forensic Science, Greater Noida, Uttar Pradesh, India, for their full-fledged support and motivation whenever required.

We acknowledge that the editing work for this book has been upheld by the Vision Group on Science and Technology, Government of Karnataka, India.

Preface

The tech world today is exceptionally inclined to digital assaults due to the extensive use of advanced gadgets and organizational innovation. In recent years, many cybersecurity-related incidents have been reported. Various tools have been developed to extract digital evidence and mitigate cyber-attacks as a countermeasure to cybercrime.

For any cyber incident or crime, digital evidence plays a major role in investigation. The digital forensic investigation provides a way to recover lost or purposefully deleted files from a suspect's device. Through various surveys, it is understood that current manpower and resources need to be upgraded to investigate cybercrimes. Advanced technology in digital forensics with programming techniques reduces human errors. At the same time, machine learning combined with automation in the digital investigation process at different stages of investigation has significant potential to aid digital investigators.

A comprehensive study of such advancements is highlighted in this book. Along with highlighting recent advancements, the book aims to present research in machine learning-based digital forensic investigation, identify the gaps, and address the challenges and open issues in this field.

Key Features of the Book

This book brings together state-of-the-art data about the innovation and practices that are important for specialists to comprehend so that they can successfully direct and perform sufficient super-advanced examinations and investigations. It serves as a great source of information for crime investigations. The book will also help specialists and novices gain insight in leading assessments. The book helps to understand advancements in digital forensics in terms of methodologies used, key tactical concepts, and the tools needed to perform examinations. The proposed book starts with familiar technologies, cyber security, and digital forensics and then transcends towards their convergence.

Taking off from recognizable ideas to new ones will make the reader very comfortable and convinced about the practicality of achieving future digital forensics for computers, networks, cell phones, GPS, the cloud, and the internet.

Organization of the Book

Chapters are organized under various categories including, challenges in cybercrime reporting, review on digital forensics, security issues in transforming information on the network, digital forensics tools, and authentication techniques.

Chapter 1 discusses the challenges faced at different levels of administration and governance of cyberspace.

Chapter 2 discusses the different types of forensics used to investigate and solve different kinds of cases, such as intellectual property theft, industrial espionage, fraud investigations, forgeries-related matters, and bankruptcy investigations.

Chapter 3 discusses different security mechanisms, such as encryption and watermarking, which are explored and investigated for information protection and transmission over open networks.

Chapter 4 discusses different wearable device technologies used to track, analyze, and transmit personal data. These devices are physically carried as accessories, fixed somewhere in the body wear, implanted in the users' body, or sometimes may even be tattooed on the skin. Such smart technology can track heart rates, sleep patterns, calls, emails, messages, etc.

Chapter 5 discusses the concept of digital forensics. Digital forensics or cyber forensics may "collect and investigate information from computer framework systems, telecommunication streams (remote) and capacity media in a court-approved manner."

Chapter 6 discusses the concept of clustering and classification of digital forensic data using machine learning and data mining approaches.

Chapter 7 discusses analytical and scientific style pertaining to methodology and technological advancements in digital forensic investigation.

Chapter 8 discusses the different types of e-trade threats. Some are unintended, while others occur due to human blunders. The most common safety threats are digital payments, e-cash, statistics misuse, credit/debit card fraud, and many others.

Chapter 9 discusses and analyses relevant principle structure and storage characteristics of file data of SSD hard disk and proposes the recovery possibility and existing challenges of SSD hard disk from three aspects: logic layer, physical layer, and firmware layer.

Chapter 10 discusses and examines the various factors responsible for increase of the dark web. It also critically analyzes the effectiveness of various legal measures in developed countries relating to the dark web. It identifies the legal challenges and suggests measures to combat the issue of the dark web.

Chapter 11 discusses and analyses the different roles of mobile phones and social media in digital forensics.

Chapter 12 discusses and examines credit card fraud detection using machine learning techniques.

Chapter 13 discusses the different financial cyber crimes, the types of criminals, and how one can be a victim, and also how individuals can protect themselves from this crime along with some case studies within the category of financial cyber-crime.

Chapter 14 discusses automating the structure of a systematic model. Its major framework is that it can learn from data, recognize various examples, and make decisions with minimal or no human input.

Chapter 15 discusses artificial intelligence applied to computer forensics, the use of modern digital investment methods, research, and the opportunity to address the challenges and high-level sectors in which cybercrime occurs.

Chapter 16 discusses the use of digital signatures and details how the concept of OAuth 2.0 is more efficient in terms of security measures.

Chapter 17 discusses child pornography: internet sex crimes, ways to combat this and other online sex crimes. It details how countries collaborate internationally to amend the anti-pornography civil rights ordinance, which can minimize these threats.

Finally, Chapter 18 discusses digital forensics and its cybersecurity tools.

CHAPTER 1

Challenges to Cybercrime Reporting, Investigation, and Adjudication in India

VIKRAM SINGH,[1] VARUN MALIK,[2] and RUCHI MITTAL[2]

[1]*Department of Computer Science and Engineering, Chaudhary Devi Lal University, Sirsa, Haryana, India*

[2]*Chitkara University Institute of Engineering and Technology, Chitkara University, Punjab, India*

ABSTRACT

The Internet infrastructure is not 'secure by design.' And in recent times, fueled by cheap internet data and affordable smart mobile devices, digital technologies have been adopted at an exponential rate. Given the above facts, lately, the world has seen an unprecedented rise in cybercrime cases. And, owing to the novelty of tools, means, targets, and modi operandi of cybercrime, an altogether new approach to investigation and prosecution is required on this front. A radically different approach is required to discover (and preserve) evidence from the crime scene, analysis of evidence, and mapping them to judicial procedure. Cybercrime investigators, prosecutors, and adjudicators face a multitude of challenges in performing their respective duties and deliverance of justice in cybercriminal cases. In this background, the present chapter discusses the challenges of different types as faced at different levels of administration and governance of cyberspace.

Advancements in Cybercrime Investigation and Digital Forensics. A. Harisha, Amarnath Mishra, & Chandra Singh (Eds.)
© 2024 Apple Academic Press, Inc. Co-published with CRC Press (Taylor & Francis)

1.1 INTRODUCTION

With each passing day, the world is hooking itself more and more on the Internet. Today, entities in large numbers–be they individuals, or institutions, or transnational corporates, or governments, or NGOs, or even the crime cartels–depend on computer networks for their daily chores. This makes them vulnerable to criminal attacks–popularly called cybercrime or cyberattack–which are perpetrated using network infrastructure. A cyber geek sitting in the safety of his home in a far-off place can cripple, paralyze or even destroy the infrastructure being controlled by computer networks. As much as that, the inimical use of this potential technology has added a new (fifth) dimension of warfare, namely, cyber warfare; the other four dimensions of warfare being land, sea, air, and space.

A globally accepted definition of cybercrime is elusive; it has changed with time; and means differently in different domains. This definitional and taxonomical heterogeneity arises from the viewpoint-heterogeneity of understanding a cybercrime. Four different and popular viewpoints identified by Wall (2007) are: (i) the technical discourse–where experts work to understand the dynamics of cybercrimes to find solutions to precarious situations in the cyberspace; (ii) the academic discussion–where the cybercrime phenomenon is studied through research techniques of different disciplines like computer science, sociology, and economics; (iii) the legislative perspective–where legal framework defining the boundaries for legit conduct in the cyberspace is discussed; and (iv) the common man's perspective–where the common man's understanding of cybercrime is pondered upon.

The terms 'computer crime,' 'ICT crime,' and 'cybercrime' convey a similar meaning in that they describe the use of computer and communication networks that know no geographical bounds and share data–an intangible and volatile artifact. Victims of cybercrime and law enforcing framework (police and judiciary) experience an all-new turf in terms of: (i) plundering of assets through manipulation of intangible entities like data; (ii) falling victim to unconventional weaponry like hacking; (iii) a virtual crime scene that may encompass machines physically located thousands of miles apart and in different legal jurisdictions; (iv) the divergent criminal behavior where theft of asset (read data) may not mean the disappearance of an asset from the owner's possession; (v) radically different (from traditional crime) forms of crime where the brute processing power of modern computing machine is used to perpetrate misdeeds; and (vi) an investigative approach and judicial trial synchronized to the novelty of the crime.

Having emphasized the absence of a universal definition of cybercrime, Aiyengar (2010) has presented a somewhat largely agreed-upon description of cybercrime in the legal domain an accumulation of computer-related crimes, conventional crimes committed using computers, and content-related crimes.

Roshan (2008) has presented a general definition of cybercrime as any illegal activity in which the computer is used either as a tool in committing the crime or becomes the target of the crime or both. Govil (2007) has observed that sometimes flimsy and many a time very real anonymity of the Internet provided by 'virtual private network' (VPN) and 'the onion router' (TOR) provides an ideal playfield for unsuspecting net users falling prey to net hawks.

Although, often cybercrime and computer crime are used interchangeably, yet Ajayi (2016) has attempted to distinguish cybercrime from computer crime. He has argued that although the two are similar but different. Computer crimes comprise the crimes perpetrated against the computer system, including hardware and non-hardware components (data and software). Cybercrime, on the other hand, encompasses criminal acts committed in cyberspace–a virtual space created by computer networks. McGuire & Dowling (2013) has described two types of cybercrimes: (i) cyber-enabled crimes; and (ii) cyber-dependent crimes. Whereas cyber-enabled crime comprises traditional crimes whose potency and scale are increased by employing the computing and communication power of modern computer networks, cyber-dependent crime, however, can happen only by using computers and their networks.

The tenth UN Congress' Report on the Prevention of Crime has placed cybercrime in two categories, namely, computer crime and computer-related crime. The former type of crime compromises the security of a computer system to gain unlawful access to computing resources, including data, software, and hardware. A computer-related crime, on the other hand, is an otherwise illegal activity carried out with the help of or using a computer system/network.

Indian cyber law or for that matter any other Indian law does not define 'cybercrime.' But, the legal literature mentions cyber-terrorism as a kind of premeditated crime that is politically motivated against the information technology infrastructure of a nation-state resulting in violence against public property and people (Kumar et al., 2015). Swaathi & Kannappan (2018) have defined cybercrime as an offense wherein a computer or a network is either the target, the intermediary, or the means for commissioning of an act that is otherwise criminal under law.

1.2   CYBER FORENSICS

Etymologically, the word 'forensic' relates to the Latin word 'forensis' which means: "public, to the forum or public discussion; argumentative, rhetorical, belonging to debate or discussion." Merriam-Webster gives a topical and modern definition of 'forensic' as: *"relating to, used in, or suitable to a court of law."* Accordingly, a science used in the court of law to push a point of law is called 'forensic science.'

Digital or cyber forensics can be understood as the extension of 'traditional' forensics in cyberspace. Clancy (2011) has defined cyber forensics as the scientific discipline of identification, collection, preservation, transportation, analysis, and reporting of digital evidence in a format that is admissible in a court of law. As a corollary, it can be used in the detection, trial, and adjudication of cybercrime. It involves scientific techniques for the recovery, authentication, analysis, and interpretation of residual or reconstructed digital data.

And, in cybercrime cases, the signature and trail of the crime could be traced back to data–called 'digital evidence'–that is stored/residual on digital devices like computers, routers, smartphones, gateways, servers, tablets, smart televisions, peripheral storage devices like hard drives, USB flash sticks, and other such devices that have digital memory capacity. Rishi (2018) has bifurcated the cyber evidence into two forms: (i) active content–written text, audio, files, videos, e-mail texts, text messages, instant messages, and social media posts; and (ii) passive metadata–data about the active content, e.g., user credentials, geolocation of user, transactional data, etc.

Cyber forensics is the use of computer science and related technologies in the investigation of cybercrime cases. It may involve the collection, preservation, transportation, and analysis of digital evidence for legal purposes. Cyber forensics work upon the 'digital evidence'–the footprints left behind when somebody consumes information and communication technology resources (Albert & Venter, 2017). The digital evidence may reveal varied but important information about the person using the ICT resource.

Digital evidence is the digital footprints of users' online behavior. Internet architecture allows for (and keeps a record of) many user actions, such as posts (e.g., emoticons, text comments, video clips, images), likes, forwards, dislikes, favorites, followings, search queries, browsing histories, geolocation information, and purchase preferences. All these pieces of users' activity-related information contribute to digital evidence. The evidence is bifurcated into two categories, namely: active and passive (Micheli et al., 2018):

1. **Active Digital Footprints:** Digital footprints created (and left behind) by the users, such as comments, videos, and images posted on social media platforms, weblogs, RSS, and BBS, etc., are termed as 'active digital footprints.'
2. **Passive Digital Footprints:** The ancillary data that are created (and left behind) by the Internet and social media users are called 'passive digital footprints.' This may include the users' internet browsing trail and weblog data.

Data can be acquired and used as digital evidence in a court of law. Digital evidence is paramount in the adjudication of cybercrime cases and can be used as evidence in a courtroom. These data can be used to: (i) refute or support the testimony of a victim, witness, or suspect; (ii) prove or disprove a matter being asserted; and (iii) implicate or exculpate a suspect of a crime.

But acceptance of a digital device by a court of law as either direct or circumstantial evidence requires it to be authenticated. Authentic evidence includes: (i) content created by a user (e.g., a word processing document created by any person); (ii) content created by a digital device without any kind of user intervention (e.g., internet browsing history, Google search history, weblogs); and (iii) content created by a combination of (i) and (ii) above (e.g., a spreadsheet wherein some data were input by the users and others were generated through calculations made by spreadsheet software).

User-created content is admissible as evidence in the court of law if it can be attributed to a person and device-created content shall be admissible as evidence if: (i) its creation can be demonstrated; (ii) possibility of alteration of data so created can be ruled out; and (iii) for the content created both by device and user, its fidelity and reliability needs to be established.

Contrary to traditional evidence like weapons and paper documents, digital evidence poses singular challenges, for its following characteristics: volume, velocity, volatility, and fragility. Some jurisdictions have created specific digital evidence authentication rules while others treat the digital evidence at par with traditional evidence in respect of authentication requirements. Several cyber forensics models have been proposed and used in different jurisdictions.

1.2.1 DIGITAL FORENSIC RESEARCH WORKSHOP MODEL 2001

In 2001, Digital Forensic Research Workshop (DFRW2001) has worked out a model based on US FBI's physical crime scene search protocol, wherein the following seven phases have been listed–Table 1.1 (Palmer, 2001).

TABLE 1.1　Investigation Phases in DFRW2001 Model

SL. No.	Phase	Activity Carried Out/Artefact Used
1.	Identification	It involves the detection of cybercrime and the creation of the profile of the cybercrime/cybercriminals.
2.	Preservation	It involves the preservation of digital evidence using imaging technologies.
3.	Collection	The collection involves a mechanism of recovery and lossless compression of digital evidence from the crime scene. It is required that the evidence is preserved during collection.
4.	Examination	The examination involves trace-back, pattern matching, and validation of evidence. Preservation of evidence is required.
5.	Analysis	Analysis of digital evidence involves trace-back, data mining, statistical testing, and evidence preservation. A timeline of events is established.
6.	Presentation	The presentation involves documentation of findings and clarification by domain experts followed by the testimony by legal experts in courtrooms.
7.	Decision	Pronounce the decision.

Source: Palmer (2001).

1.2.2　*NATIONAL INSTITUTE OF JUSTICE (NIJ) 2001/2008 MODEL*

The National Institute of Justice (NIJ) of the US Department of Justice proposed an investigative model of cybercrimes in 2011 (revised in 2008). NIJ's First Responder Guide to Electronic Crime Scene Investigation focuses on counterpart actions of physical crime scene procedures, namely:

- Identification of digital and other devices possibly holding the digital evidence;
- Documentation of the crime scene;
- Collection of relevant digital and other related devices; and
- Transporting the devices possibly holding the digital evidence. It requires packaging and safeguarding the devices involved.

1.2.3　*THE ABSTRACT DIGITAL FORENSICS (ADF) MODEL 2002*

In the year 2002, the Digital Forensic Research Workshop extended its 2001 model (Reith et al., 2002) and named it the abstract digital forensics model (ADF). The model has nine phases as in Table 1.2.

TABLE 1.2   Investigation Phases in ADF2002 Model

SL. No.	Phase	Activity
1.	Identification	Observe the indicators to identify the incident and its category.
2.	Preparation	Suitable tools and requisite search warrants and authorization.
3.	Approach strategy	Case-specific strategy is prepared to maximize the collection of 'pure' evidence with minimum impact to the victim.
4.	Preservation	Isolation of digital evidence followed by its secure preservation.
5.	Collection	The crime scene is recorded, and a copy of digital evidence is created for further examination and analysis.
6.	Examination	Evidence is examined to establish relationships with the crime.
7.	Analysis	The significance of evidence is determined, and the conclusion is drawn.
8.	Presentation	Summary of investigation with explanatory notes is presented.
9.	Returning evidence	Confiscated devices are returned to the owner.

Source: Reith et al. (2002).

1.2.4   *INTEGRATED DIGITAL INVESTIGATION (IDI) MODEL 2003*

Carrier & Spafford (2004) have suggested the integrated digital investigation model (IDI) comprising five stages with each stage having its phases (Table 1.3).

1.2.5   *NATIONAL INSTITUTE OF STANDARDS AND TECHNOLOGY 2006*

In 2006, the United States National Institute of Standards and Technology suggested a digital forensics model (Grance et al., 2006). The model comprises four phases listed in Table 1.4.

1.2.6   *CYBER FORENSIC FIELD TRIAGE PROCESS MODEL 2009*

Rogers et al. (2006) has pointed out a common drawback of the aforementioned five models. They have pointed out that all these models assume that all the steps of the respective model are carried out for each crime investigation. Nowadays, practically every criminal investigation requires digital device(s) as evidence. Count of digital devices figuring in criminal cases and the volume of data are increasing at an exponential rate, thereby making it increasingly impractical to deeply examine all the concerned digital devices.

TABLE 1.3   Stages in IDI2003 Model

SL. No.	Stage	Phase/Activity	
1.	Readiness	Readiness of investigative tools assessed	
2.	Deployment	a.	Incident detected;
		b.	Requisite personnel notified;
		c.	Authority to investigate obtained;
		d.	Search warrants obtained.
3.	Investigation of physical crime scene	a.	The physical crime scene is secured;
		b.	Relevant physical evidence are identified;
		c.	The crime scene is documented;
		d.	Physical evidence is collected from the scene;
		e.	Physical evidence is examined;
		f.	Crime scene events are reconstructed;
		g.	Findings are presented in court proceedings.
4.	Investigation of digital crime scene	a.	Digital evidence secured;
		b.	Relevant digital evidence identified;
		c.	Digital evidence documented, acquired, and analyzed;
		d.	Events reconstructed;
		e.	Findings presented in court.
5.	Review	a.	Investigation concluded;
		b.	Assessment made to identify lessons learnt.

Source: Carrier & Spafford (2004).

TABLE 1.4   Phases in NIST Model (2006)

SL. No.	Phase	Activity
1.	Collection	Evidence at the scene identified, labeled, collected, and documented.
2.	Examination	Forensic tools/techniques to extract relevant digital evidence determined.
3.	Analysis	Digital evidence extracted and evaluated, and their applicability in the instant case determined.
4.	Reporting	Cyber forensic activities documented and findings presented.

Source: Grance et al. (2006).

Rogers et al. (2006) have proposed 'an onsite' digital forensics process model for quick identification, analysis, and interpretation of digital evidence. This model has no requirement of transporting the digital equipment to the cyber forensic facility for the examination of digital evidence. Casey et al. (2009) has argued that some cyber forensics establishment can afford to create a copy of every piece of digital evidence and analyze them deeply, but for most instances, examination and analysis of only a handful of digital evidence can provide sufficient support to the prosecutors. Authors have proposed trifurcation of forensic examination for use in both settings–field and lab:

1. **Survey Forensics Inspection:** Conducted to quickly identify potential sources of evidence and prioritize them for examination based on their volatility and relative importance.
2. **Preliminary Forensic Examination:** Conducted on the digital evidence identified during the survey forensics inspection phase to search the information that provides direct, or circumstantial evidence. For some cases, no further examination is required.
3. **In-Depth Forensic Examination:** Conducted on the whole body of digital evidence. This kind of examination might be necessitated when: (a) evidence destruction is not ruled out; (b) earlier examination has raised new questions; or (c) case trial is about to start.

1.3 CYBERLAW IN INDIA

In India, the Information Technology Act (2000) as amended in 2008 along with the statutes, rules, and guidelines invoked under its proviso provides for primary legal infrastructure and safeguard against cybercrimes. 2008 amendment in the IT Act 2000 has provided for the digital signature, among other things. Other additional provisions made in the amended act included Sections 66A and 69, for dealing with offensive messages and interception of any information through any computer resource. 2008 amendments also took note of pedophilia (child porn), cyber terrorism, and voyeurism. But, in a 2015 landmark decision delivered by India's apex court in Shreya Singhal v. Union of India case, Section 66A of the IT Act was struck down as unconstitutional.

In this communication, the unlawful cyber activities punishable under the IT Act (2000) have been bifurcated on the ground whether the act draws a fine or a jail term plus a fine. Accordingly, the former kind of unlawful acts has been termed as the 'cyber offense' and the latter type the 'Cybercrime.'

Quantum of limiting penalty and/or jail term under various sections of ITA2000 has been listed in Tables 1.5–1.8.

TABLE 1.5   Cyber Offenses Covered u/s 43, 44, 45 of IT Act (2000)

SL. No.	Cyber Offense Description	Relevant Section	Nature of Offense	Compensation/ Penalty up to
1.	Damage to computer and communication system.	43	Civil offense	₹10 m
2.	Failure to protect data.	43A	Civil offense	₹50 m
3.	Failure to furnish a document/ report to authority.	44(a)	Civil offense	₹0.15 m per instance
4.	Failure to file return to authority.	44(b)	Civil offense	₹5 k per day
5.	Failure to maintain account/ books/record.	44(c)	Civil offense	₹10 k per day
6.	Contravention not specified in ITAA2008.	45	Civil offense	₹25 k

TABLE 1.6   Cybercrime Covered u/s 65, 68, 69, 70 of IT Act (2000)

SL. No.	Cybercrime Description	Relevant Section	Nature of Crime	Punishment up to
1.	Tampering with computer source documents.	65	Cognizable, bailable	3 years or ₹0.2 m or both
2.	Failure to comply with the orders of the controller.	68	Noncognizable, bailable	2 years or ₹0.1 m or both
3.	Failure to assist the agency.	69	Cognizable, nonbailable	7 years and fine
4.	Failure to block public access when so directed.	69A	Cognizable, nonbailable	7 years and fine
5.	Failure to decrypt data for law enforcing agency.	69B	Cognizable, bailable	3 years and fine
6.	Unauthorized access to protected system.	70	Cognizable, nonbailable	10 years and fine

1.4   INDIA SPECIFIC AND GENERAL CHALLENGES TO CYBERCRIME INVESTIGATION

Yanbo et al. (2019) have mentioned three types of requirements for an effective investigation into cybercrime, namely, technology, mode, and means of investigation:

1. **Technological Requirement:** Investigation agencies needs to be manned with technically competent officers and equip themselves with the tools, and technology to investigate into high-tech cybercrime. Further, there exists a need for continuous upgradation of investigative equipment to keep pace with the evolution of technology.

TABLE 1.7 Cybercrimes Covered u/s 66 of IT Act (2000)

SL. No.	Cybercrime Description	Relevant Section	Nature of Crime	Punishment/Fine up to
1.	Offensive or false messages.	U/S 66A	Cognizable, bailable	3 years and fine
2.	Receiving stolen computer.	U/S 66B	Cognizable, bailable	3 years or ₹0.1 m or both
3.	Identity theft.	U/S 66C	Cognizable, bailable	3 years and ₹0.01 m
4.	Cheating by personation.	U/S 66D	Cognizable, Bailable	3 years and ₹0.1 m
5.	Violation of privacy.	U/S 66E	Cognizable, bailable	3 years or ₹0.2 m or both
6.	Cyberterrorism.	U/S 66F	Cognizable, nonbailable	Life term

TABLE 1.8 Cybercrimes Covered u/s 67 of IT Act (2000)

SL. No.	Cybercrime Description	Relevant Section	Nature of Crime	Punishment/Fine up to
1.	Obscenity in electronic form.	67	Cognizable, bailable[1]	3 years and ₹0.5 m (1st conviction)
				5 years and ₹1 m (after 1st conviction)
2.	Pornography in electronic form.	67A	Cognizable, nonbailable	5 years and ₹1 m (1st conviction)
				7 years and ₹1 m (after 1st conviction)
3.	Pedophilic in electronic form.	67B	Cognizable, nonbailable	5 years and ₹1 m (1st conviction)
				7 years and ₹1 m (after 1st conviction)
4.	Intermediary's failure to preserve information.	67C	Cognizable, bailable	3 years or ₹0.2 m or both

[1]Obscenity in electronic form u/s 67 becomes a nonbailable crime after 1st conviction

2. **Modus Operandi:** The procedure to investigate cybercrime cases needs to be radically different, for cybercrime is potentially different from a traditional crime. Accordingly, the examination and analysis of evidence and its presentation in the courtroom need to be in sync with the law and nature of the crime.
3. **Means of Investigation:** An altogether different set of artifacts is required to obtain evidence, justify arrest warrants for the accused, and bring the case to a just end.

Cybercrime Investigation Handbook for Police Officers prepared by the Ministry of Home Affairs, Government of India, has presented the following generic list of challenges faced during cybercrime investigation:

- Prompt, hassle-free, and victim-friendly reporting of cybercrimes;
- Adequate and proper 'first response' when cybercrime is reported;
- Properly trained human resources are required to investigate cybercrime;
- Capacity building education and training programs and institutes;
- Understanding and appreciation of cyber forensics by judicial officers;
- An inadequate and up-to-date legal framework to try and punish cybercrime;
- Availability of state of art hardware and software tools and technologies required for cybercrime investigation;
- Standardization of cybercrime investigative framework wherein required and desirable skill sets are mentioned.

Poonia (2014) has compiled the following list of challenges in the fight against cybercrimes in the Indian context:

- There is a stark lack of awareness regarding the security of cyberspace at an individual, societal, organizational, and government level.
- There is a huge deficit of human resources with proper education and training to counter cybercrimes and cyberattacks.
- Absence of e-mail account policy for government officials in general and security forces in particular.
- State-sponsored cyber terror attacks are a major concern for securing national cyberspace.
- An overwhelming majority of police personnel are cyber-illiterate, for there is no cyber-proficiency requirement for recruitment in the police force.
- Cyber technology changes at a very fast pace and cybercriminals outpace the cybersecurity agencies in the innovative use of modern technology.

- Almost always the cybersecurity agencies reactively approach cybercrime. Analysis of internet traffic to unearth a would-be cybercrime is a matter of research and fiction, but it is quite unheard of in actual practice.
- State and public-sponsored research and development projects in cybersecurity are not up to the mark.
- Law enforcement personnel including public prosecutors and judiciary are not well trained to address cybercrimes.
- Prevailing cybercrime investigation protocols are quite up to mark to investigate transnational cybercrimes.
- Budgetary allocation for cyberspace security at local, provincial, and union levels are dismally low.

Ajayi (2016) has discussed several challenges faced by mankind while dealing with cybercrimes:

- The identity of cybercriminals is rather difficult to establish, for there is no prerequisite of disclosing the real identity and location of the user for using the Internet.
- Jurisdictional challenges are faced in the trial and adjudication of cybercrimes, for the principles of sovereignty and territorial integrity restrict the scope of laws to *lands*.
- Extradition process-related challenges are not unique to the deliverance of justice for criminal acts culpable in cyberspace only, but unlike traditional crimes, cybercrimes could be transnational even the criminal without knowing it.
- The nature of evidence applicable to cyberspace is volatile, easy to alter, and un-established acceptability in courtrooms.
- Owing to the complexity of the investigation process and the degree of technical expertise required, the cost, time, and efforts spent in cybercrime investigation/prosecution are exorbitant.
- In several jurisdictions, the legislative framework to investigate and adjudicate the cybercrimes are inadequate, ineffective, and out of sync with new developments in cyberspace.
- International legal mechanisms to investigate and try the cybercrimes are truly global and lack commitment and enforcement.
- Law enforcement agencies' personnel are not properly trained, are not paid handsomely, and are protected from retaliation.
- There is a huge deficit of competent private and public prosecutors and adjudicators for cybercrimes.

Dhupdale (2010) has mentioned the following challenges by cybersecurity teams during investigation and prosecution:

- A global dearth of cyber experts;
- Securing cyberspace goes against privacy requirements;
- Cybersecurity teams are required to keep pace with fast-changing digital businesses;
- Cyberspace is becoming more complex with each passing day.

Iqbal & Beigh (2017) have underlined the following challenges to cybercrime investigation:

i. **No Physical Boundaries:** Cybercrimes know no physical boundaries and their investigation requires international cooperation–an unlikely proposition.

ii. **Lack of Cyber-Criminal Statutes:** Several countries either don't have a cybercrime law in place or lack effective and updated cyber laws.

iii. **Lack of Tools and Technology for Investigation:** In general, there is a dearth of tools for the conduct of a proper investigation into cybercrimes in India.

iv. **Lack of Enforceable Mutual Assistance Provisions:** Cybercrimes, possibly being transnational, may require several jurisdictions to assist in trying and prosecuting the cybercriminals. India's being out of the Budapest convention may be seen in this light apart from the fact that the convention seeks to undermine the sovereignty of its signatories.

MEITY (2021) report has listed a comprehensive but incomplete list of challenges to the cybercrime investigation:

a. Analysis of fake information/content is a singular type of challenge in the fight against the spread of disinformation on the Internet in general and social media platforms in particular. Time and again it has come to light that *social bots*[2] are increasingly used for manipulating the trends on social media.

b. Organized terror outfits and state actors make incremental use of cyber technology to pose advanced threats to national security.

---

[2]Social bots (aka socialbots or socbots) are automated software agents that communicate almost independently on social media networks with a designated task of determining and influencing the direction of a discussion thread as also the opinions of social media users. Social bots accomplish their task by convincing the social media users that a social bot is a real person.

These attacks could be pretty harmful, for there is no dearth of funds for such terror groups and *the darknet*[3] offers the best of perilous services for big money.

c. Of late, the world has seen online avatars of social influencers. Organic measures like the number of subscribers, followers, connections, likes, retweets, forwards, shares, etc., are the metrics for the clout of these influencers. Apart from these organic measures of popularity, there are services like Fiverr which can help increase the parameters such as likes, shares, followers, etc., inorganically.

d. Analysis of cybercrime to understand their social, financial, cultural, and political dynamics can be very helpful in deterrence and better control cybercrimes. But, owing to the fast-changing profile of cybercrimes and their modi operandi, building tools, and techniques for cybercrime analysis are very challenging.

e. Security in cyberspace and privacy often have conflicting goals. The human race has evolved to share its emotions and activities through socialization in the real world or cyberspace. Sharing of a user's what, when, where, etc., on social media has been used by cybercriminals to their benefit. Striking a balance between these two very important aspects of the user profile is a very daunting and challenging task.

Having reviewed the cybersecurity-related challenges discussed in various research articles and technical reports, a comprehensive discussion on the cybercrime investigation challenges in the Indian context follows.

1.4.1 CHALLENGES RELATED TO CYBERCRIME REPORTING

In a study based on 630 adolescent respondents of Delhi and NCR published on the portal 'Business Insider India' on March 16, 2020, Prerna Sindwani has reported that 9.2% of respondents have been victims of online harassment, but half of them have never reported it.

On August 30, 2019, the Ministry of Home Affairs, GoI, has launched the National Cybercrime Reporting Portal (https://cybercrime.gov.in) to report cybercrimes incidents online. Law enforcement agencies of the concerned

---

[3]The darknet (aka dark net or dark web) is the encrypted part of the internet that is not indexed by commonplace search engines such as Google, Yahoo, or Bing. It comprises the internet/websites that are only accessible to a select group of users and not to the general internet users. Accessing these sites requires authorization, special software, and configurations. Further, the dark net spans harmless sites such as academic databases as well as the shadier places in cyberspace such as pedophilic communities, black markets, cyber pirates, and hackers.

State/UT shall investigate the crimes reported on the portal. The portal allows for reporting of cybercrime in two categories, namely, 'cybercrimes against child/woman,' and 'other cybercrimes.' Additionally, the crime can be reported anonymously or with a "report and track" feature. In the latter category, the reporting individual/entity shall have to create an account on the server by sharing certain personal details.

Further, the portal is bilingual and supports the menus and command buttons in English and Hindi. But, in a huge country with about 74% literacy rate, only about 44% Hindi speaking population, and about 12% population speaking/understanding English, the only bilingual portal automatically excludes a majority of victim population out to fend on their own.

In addition, given the social stigma associated with reporting crimes against children and women in India, only a very small fraction of cyber-crimes is reported. In a news report titled "Cybercrime cases in India are under-reported" published on June 5, 2021, the Mint has quoted Pavan Duggal, a Supreme Court of India lawyer practicing in cyber laws as saying, "the reporting of cybercrimes may be as low as mere 1% of actual cases." In another post-demonetization[4] news story published in Times of India dated January 10, 2021, the newspaper has claimed that 80% of cybercrimes are unreported. Loopholes in the Indian cyber laws and long-drawn legal battles and social stigma are the ingredients of vastly underreported cybercrimes in India.

1.4.2   CHALLENGES RELATED TO FIRST RESPONSE TO CYBERCRIME

A 'first responder,' as the very nomenclature suggests, is an official or a private investigator, to whom a reported crime is assigned for investiga-tion. Typical jobs that a first responder is entrusted with are identifica-tion, gathering, preservation, and transportation of electronic evidence. *In criminal investigations, the first responders* have entrusted the job of securing evidence at the crime scene. Cybercrime is no exception in regard. Depending on the category of the cybercrime, the evidence might include the 'target' device(s) [for cyber-dependent crime] or the ICT infrastructure [cyber-enabled crime]. Depending on the cybersecurity policy and the victim's profile, the first responder can be a police officer [for a common citizen], a digital forensics expert working in the security division of an

---

[4]On November 8, 2016, the currency bills of ₹500 and ₹1000 denomination were withdrawn by a Government of India edict.

organization [for a work-related cybercrime being reported by an employee], a military police officer [for an army sepoy], or even a private investigator, so entitled under law of land. But notwithstanding who is the first responder, the procedure of collection of evidence needs to be valid and admissible in a courtroom (UN, 2000).

Hinduja (2007) has emphasized the profile of the officer who responds first to a cybercrime report. (S)he is entitled the critical importance, for, very often than not, the cybercrime evidence is intangible. It is paramount that the data stored on a mobile phone, a computer, a network device, or removable media are not deleted or changed in the process. Even the powering down of the gadget can alter the important attributes of the files such as the date and time of the last modification or last access. Any investigative experimentation with such intangible evidence must be done with a copy of the evidence so that the public prosecutor's charge(s) and the veracity of the evidence are not challenged during prosecution. Accordingly, to defend the prosecutor's stand, the first respondent officer must exercise due care while examining the crime scene. The following challenges have been assorted from the viewpoint of the first responder:

- Identification of electronic evidence relevant to the prosecution of the crime;
- Collection of electronic evidence without causing any change to the evidence;
- Preservation of electronic evidence to support the claim of prosecution;
- Examination of electronic evidence to conclude the findings of the investigation;
- Documentation of complete investigation of cybercrime.

1.4.3   CHALLENGES RELATED TO THE MULTIPLICITY OF AGENCIES

In several jurisdictions, multiple agencies may be entrusted with the investigation of the same cybercrime. The individual agencies involved in the investigation of cybercrime depend on the type of cybercrime being investigated. In India also, there are multiple central and state bodies that deal with cybercrimes, and each has a different reporting structure. Different central government ministries and departments thereunder deal with different aspects of cybersecurity. A description of the agencies involved in security against cybercrimes follows in Table 1.9.

TABLE 1.9   Agencies Looking into Cybercrime Breaches

SL. No.	Agency/Office	Description
1.	National Security Advisor (NSA) and National Security Council Secretariat	A cabinet minister rank apex officer in charge of national security threats posed by cybercrime [including].
2.	National Technical Research Organization (NTRO)	Primary technical intelligence agency headed by NSA.
3.	National Critical Information Infrastructure Protection Center (NCIIPC)	An arm of NTRO established u/s 70A of the Information Technology Act (2000).
4.	National Cyber Security Coordinator (NCSC)	An officer of PMO works as Nodal Officer for cybersecurity-related issues. Coordinates with other agencies like CERT-In at the national level.
5.	CERT-In	Indian arm of Cyber Emergency Response Team–an international effort to respond the cyberspace emergencies.

The Standing Committee on Information and Technology of Indian Parliament's lower house (Lok Sabha), in its '52nd Report on Cybercrime, Cyber Security and Right to Privacy' has noted that in India, multiple agencies are working towards securing its cyberspace. The report has recommended to minimize responsibility overlays between such agencies and adopting a public-private partnership mechanism for realizing a secure cyber ecosystem in India.

1.4.4   CHALLENGES RELATED TO TECHNICAL SKILLS

Since both the target and the means of the cybercrime involve technology and the criminals involved in the high impact cybercrimes are invariably cyber geeks, investigating officials, public prosecutors, judicial officers dealing with cybercrimes must possess specialized knowledge, skills, and abilities (KSA), beyond those required to deal with traditional offline crimes. Characteristics of cybercrime that warrant KSA are discussed in subsections.

1.4.4.1   ANONYMITY

The Internet offers very strong anonymizing services to internet users free of cost. And user *anonymity* presents a big obstacle in the investigation of cybercrimes. Anonymity empowers individuals to communicate and prowl on the Internet without revealing their identity to others. Proxy server connects a client machine to a server. An anonymous proxy server is one such user anonymizing technology that hides users' identity tokens by masking their IP addresses and substituting them with a different set of IP addresses (Maras, 2016).

The likes of TOR, I2P, and Freenet are also used by cybercriminals to conceal their internet identities by hiding the IP address of the sender. In TOR network, the machine at any level is in the possession of the IP address of the next-level recipient machine only and at no level the recipient machine knows the address of the sender machine, thereby making it impossible to trace back the communication (Maras, 2016).

1.4.4.2   ATTRIBUTION

Another impediment faced during cybercrime investigations is *attribution*, i.e., mapping a cybercrime to who and/or what is responsible for it. This process involves mapping the cybercrime to a user, a device, an organization, a state, or any other entity in cyberspace. Attribution can become a herculean task owing to rampant use of anonymity services and botnets and zombie machines–machines that are highjacked and used to commit cybercrimes, unbeknownst to the owner of the machine (Lin, 2016).

1.4.4.3   TRACEBACK

It is the process of tracing cyber activities back to their source after the occurrence, reporting, and detection of a cybercrime. Internet log files help to reveal important information (including IP addresses of the devices used) about the modus operandi of the cybercrime. But traceback may be quite time-consuming and depends on the KSA attributes of both–the perpetrators and the investigators. Having known the IP address of the device used in cybercrime, the relevant Regional Internet Registry (RIR)[5] can be requested to identify the internet service provider (ISP) associated with the IP address under question. Backed by proper legal documents, the cybercrime

---

[5]Regional Internet Registries (RIR) are mechanisms for registration of IP addresses in their respective regions. The world has been divided into five regions for this purpose, namely: (i) Africa; (ii) Asia Pacific; (iii) America; (iv) Latin American; and Caribbean; and (v) Europe.

investigators may approach the ISP associated with the IP address to supply the information about the person or the body who has subscribed to the IP address under question. Legal and constitutional frameworks governing the liabilities of ISPs in these respect vary from land to land (Lin, 2016).

1.4.4.4   THE VOLATILITY OF EVIDENCE

Digital evidence needs to be handled properly, for it is volatile and fragile. Digital evidence handling protocols are important to ensure its integrity and safety against any alteration or damage while accessing, packaging, transporting, or storing it. Such evidence is collected in order of its volatility. Accordingly, the most volatile evidence is collected first and the least volatile is collected last. Brezinski & Killalea (2002) have provided the following ordering of devices based on their volatility:

- Internal registers, cache memory;
- The routing table, process table, kernel statistics, memory;
- Temporary internet files;
- Disk media;
- Remote logs;
- Physical configuration, network topology;
- Archival media.

Digital evidence relevant to cybercrime may be resident on a diverse array of devices and may span several services and jurisdictions. Digital evidence can thus be stored in part or whole by several service providers at multiple. Collecting evidence from multiple devices, multiplayer, and possibly in multiple jurisdictions is quite a challenging task.

1.4.4.5   DELETED CONTENT

Digital evidence that has been deleted from the file system records poses another challenge for cyber forensics. Normally, files deleted from a computer are not deleted physically, but merely marked for deletion and are placed in what is called a Recycle Bin or Trash Bin. Their entries remain in the system's registry, i.e., the file allocation table. In case the Bin is emptied, the deleted files' entries from the file allocation table are deleted and the space occupied by the deleted files is marked 'free' for allocation to new files. Chances of recovery of files whose entries have been deleted from the Bin as well depend on whether the particular storage space freed by those files has been taken up by some newly created file thereafter or not. In case the space freed by deleted files has still not been taken up by any other file, the contents of the deleted file are recoverable, otherwise not (Maras, 2016).

1.4.4.6 MOBILE NUMBER PORTABILITY

Mobile number portability is a service enforced by telecom regulators and provided by telecom operators to the customers where any subscriber can switch between mobile services of different operators in a hassle-free manner, even without changing the subscriber information module (SIM) card. Although the service is very beneficial for the customers, it adds to the difficulties of cybercrime investigation and poses a challenge to track the cyberspace activities of criminals.

1.4.5 CHALLENGES RELATED TO JURISDICTION

The continuously evolving digital technologies coupled with the ambiguous and inefficient cyberlaws pose a challenge to all involved in the investigation, prosecution, and adjudication of cybercrimes cases. Issue of the jurisdiction in cybercrime cases has come to the fore in the discussions in courtrooms, techno-legal professionals and academicians, bureaucrats, and senior management of tech companies. A single internet operation may span the jurisdictions of several countries, e.g.: (a) laws of the land where the user is located; (b) laws of the land hosting the services; and (c) laws of the land where the merchant is registered. So, to investigate cybercrimes effectively, updated international laws must exist and the more importantly, the international community must cooperate to enforce these laws. However, under the garb of sovereignty, some nation-states provide heaven to cybercriminals.

In India, u/s 75 of Information Technology Act (2000), there exist provisions to try any person, irrespective of his nationality, for the cybercrimes committed outside India, provided that such offense involves a computer resource located in India. Despite such provisions in the Indian cyber laws, extradition of foreign citizens remains a far-fetched dream.

1.5   CONCLUSIONS

The thesis of this chapter can be concluded by stating that acknowledging and identification of the problem is the first step to solve the problem. Here, the endeavor was to identify and understand the challenges faced by the community of investigators, prosecutors, and adjudicators in solving, trying, and delivering justice in cybercrime cases. Essentials of cyber forensics, Indian cyber laws, various models of cybercrime investigation, and the challenges, in general, were discussed, followed by an India-specific discussion. Accordingly, the challenges were classified into five categories, namely, crime reporting, first response, multiple agencies, technical, and jurisdictional.

It may be held that, of late, most countries have enacted the laws to administer cyberspace but cybercriminals' technical profile and handsome dividends accruing from cybercrimes, criminals are encouraged to espouse the latest technology in innovative ways to misuse and abuse it to achieve their vicious goals. Most cybersecurity professionals approach cybercrime as a reaction to an incident of cyberattack instead of a proactive approach. The research body has suggested the use of machine learning (ML) techniques to train the security systems to identify and block malicious traffic by launching fake but harmless cyberattacks. This approach works the same way as vaccines works in the human body.

In addition, challenges of law enforcement also assume gravity in the absence of a globally accepted description of cybercrime, inadequate access and seizure authority to investigators, the difference in investigation procedure followed by agencies in different countries, and complex extradition process for cross-border criminals. Further, international events need to be organized to provide a platform for sharing the best practices in cybercrime investigations.

KEYWORDS

- **abstract digital forensics model**
- **integrated digital investigation**
- **knowledge, skills, and abilities**
- **National Institute of Justice**
- **onion router**
- **virtual private network**

REFERENCES

Aiyengar, S. R. R., (2010). *National Strategy for Cyberspace Security*. New Delhi: KW Publisher.

Ajayi, E. F. G., (2016). Challenges to enforcement of cybercrimes laws and policy. *Journal of Internet and Information Systems, 6*(1), 1–12.

Albert, A. B., & Venter, H., (2017). A model for digital evidence admissibility assessment. In: Peterson, G., & Shenoi, S., (eds.), *Advances in Digital Forensics XIII. Digital Forensics 2017: IFIP Advances in Information and Communication Technology* (Vol. 511). Springer, Cham. https://doi.org/10.1007/978-3-319-67208-3_2.

Brezinski, D., & Killalea, T., (2002). Guidelines for evidence collection and archiving. *RFC, 3227*, 1–10.

Carrier, B., & Spafford, E., (2004). *An Event-Based Digital Forensic Investigation Framework*. Presented at Digital Forensic Research Workshop 2004. Available online: http://www.digital-evidence.org/papers/dfrws_event.pdf (accessed on 10 January 2022).

Casey, E., Ferraro, M., & Nguyen, L., (2009). Investigation delayed is justice denied: Proposals for expediting forensic examinations of digital evidence. *Journal of Forensic Sciences, 54.*

Clancy, T. K., (2011). *Cybercrime and Digital Evidence: Materials and Cases*. New York: Matthew Bender & Company, Inc.

Cybercrime Investigation Handbook for Police Officers, (2014). Ministry of Home Affairs, Governement of India. In: Poonia, A. S., (ed.), Cybercrime: Challenges and its classification. *International Journal of Emerging Trends & Technology in Computer Science, 3*(6), 119–121.

Dhupdale, V. Y., (2010). Cybercrimes in India and the challenges ahead. *Indian Journal of Law and Justice, 1*(2), 102–114.

Govil, J., (2007). Ramifications of cybercrime and suggestive preventive measures. In: *International Conference on Electro/Information Technology* (pp. 610–615). IEEE Chicago, IL.

Grance, T., Chevalier, S., Kent, K., & Dang, H., (2006). *Guide to Integrating Forensic Techniques into Incident Response*. Special Publication (NIST SP), National Institute of Standards and Technology, Gaithersburg, MD, [online], https://tsapps.nist.gov/publication/get_pdf.cfm?pub_id=50875 (accessed on 10 January 2022).

Hinduja, S., (2007). Computer crime investigations in the United States: Leveraging knowledge from the past to address the future. *International Journal of Cyber Criminology, 1*(1), 1–26.

https://www.businessinsider.in/india/news/cybercrimes-are-underreported-half-of-the-adolescent- victims-never-report/articleshow/74645801.cms (accessed on 10 January 2022).

Iqbal, J., & Beigh, B. M., (2017). Cybercrime in India: Trends and challenges. *International Journal* of *Innovations & Advancement in Computer Science, 6*(12), 187–196.

Kumar, S., Koley, S., & Kumar, U., (2015). Present scenario of cybercrime in INDIA and its preventions. *International Journal of Scientific & Engineering Research, 6*(4), 1971–1976.

Lin, H., (2016). Attribution of malicious cyber incidents: From soup to nuts. *Columbia Journal of International Affairs.* Hoover Institution Aegis Paper Series on National Security, Technology, and Law.

Lok Sabha, (2014). *Standing Committee on Information and Technology, 52nd Report on Cybercrime, Cyber Security and Right to Privacy*. Available at: https://eparlib.nic.in/bitstream/123456789/64330/1/15_Information_Technology_52.pdf#search=Inf ormation%20 Technology% 20cyber%20crime (accessed on 10 January 2022).

Maras, M. H., (2016). *Cyber Criminology*, 1e. Oxford University Press.

McGuire, M., & Dowling, S., (2013). *Cybercrime: A Review of the Evidence Summary of Key Findings and Implications Home Office Research Report 75* (p. 30). Home Office, United Kingdom.

Meity, (2021). *Report of Committee–D on Cyber Security, Safety, Legal and Ethical Issues.* Ministry of Electronics and Information Technology, Govt. of India.

Micheli, M., Lutz, C., & Büchi, M., (2018). Digital footprints: An emerging dimension of digital inequality. *Journal of Information, Communication and Ethics in Society, 16*(3), 242–251. doi: https://doi.org/10.1108/JICES-02-2018-0014.

Palmer, G., (2001). *A Road Map for Digital Forensic Research.* Technical Report (DTR-T001-01) for Digital Forensic Research Workshop 2001 (DFRWS2001), New York.

Prerna Sindwani, (2020). Business Insider India, March 16, 2020. https://www.businessinsider.in/india/news/cybercrimes-are-underreported-half-of-the-adolescentvictims-never-report/articleshow/74645801.cms

Reith, M., Carr, C., & Gunsch, G., (2002). An examination of digital forensic models. *International Journal of Digital Evidence, 1.*

Rishi, M. S., (2018). *Cyber Forensics and Cybercrime: A Multidimensional Study of Techniques and Issues.* Available at https://www.academia.edu/39221650/Sub_Theme_Cyber_Forensics_and_Electronic_Evidence_and_Investigation_Title_Of_The_Paper_Cyber_Forensics_and_Cyber_Crime_A_multidimensional_Study_of_Techniques_and_issues (accessed on 10 January 2022).

Rogers, M. K., Goldman, J., Mislan, R., Wedge, T., & Debrota, S., (2006). Computer forensics field triage process model. *Journal of Digital Forensics, Security and Law, 1*(2). doi: https://doi.org/10.15394/jdfsl.2006.1004.

Roshan, N., (2008). *What is Cybercrime? Asian School of Cyber Law.* Access at http://www.http://www.asclonline.com/index.php?title=Rohas_Nagpal (accessed on 10 January 2022).

Swaathi, B., & Kannappan, M., (2018). Cybercrime-an Indian scenario. *International Journal of Pure and Applied Mathematics, 119*(17), 1053–1061.

UN, (2000). *Report of the Tenth United Nations Congress on the Prevention of Crime and the Treatment of Offenders.* Vienna.

Wall, D., (2007). *Cybercrime: The Transformation of Crime in the Information Age.* Wiley.

Yanbo, W., Dawei, X., JiangMing, G., & Yun, W., (2019). Research on investigation and evidence collection of cybercrime Cases. *IOP Conf. Series: Journal of Physics: Conf. Series 1176*, 042064. IOP Publishing. doi: 10.1088/1742-6596/1176/4/042064.

CHAPTER 2

A Comprehensive Review on Digital Forensics Intelligence

SHIJI ABRAHAM, K. ALAKANANDA, and NAMITA A. AMDALLI

Department of Computer Science and Engineering, Sahyadri College of Engineering and Management, Mangalore, Karnataka, India

ABSTRACT

Forensics science provides the information needed for the investigation and evaluation for identifying the origin and life history of human beings. Digital forensics is a branch of forensics science which investigates the material found in digital devices which is related to computer crime. There are different kinds of digital forensics that is mobile forensics (MF), social forensics, drone forensics and networks forensics. These different kinds of forensics are used to investigate and solve the different kinds of cases such as intellectual property theft, industrial espionage, fraud investigations, forgeries related matters and Bankruptcy investigations. The digital forensics helps to provide the integrity of the computer system, provides evidence in the court that leads to punishment of culprit, provides the efficiency to track cybercriminals and also helps in protecting the organization's money and valuable time. Here artificial intelligence (AI) and machine learning (ML) are combined with digital forensics, which helps in the investigation process at different stages of investigation and provides the significant potential for the digital investigators.

Advancements in Cybercrime Investigation and Digital Forensics. A. Harisha, Amarnath Mishra, & Chandra Singh (Eds.)
© 2024 Apple Academic Press, Inc. Co-published with CRC Press (Taylor & Francis)

2.1　INTRODUCTION

Forensic science provides the description about investigation, explanation, and evaluation for the events for the legal relevance which consists of identity, origin, and life history of humans, materials, substances, and artifacts. Some of the scientific methodologies are used for describing, interfering, and reconstructing events from fragmentary physical evidence and other relevant information. Based on certain analysis, the scientists will provide certain answers for the criminal investigation like who, what, where, when, how, and why. A Digital Forensics is a branch of forensic science that encompasses the recovery and investigations of the materials collected from the digital device which is related to computer crime. Digital forensics has different applications. The major application is in order to give a maximum hypothetical support in front of civil courts. The forensics helps for the investigation purpose like intrusion or internal corporate's investigation. Some of the problems in digital forensics are having magnetic residue data that deals with intrusion.

In Figure 2.1, the first step is the collection of evidence. By considering the logs, repairing the system, keeping track of hackers, keystroke loggers and finding the spy. The solution that can be provided for such kind of problems must be having completeness, authenticable, admissible, and must be accurate. Digital evidence must be admissible which must conform to the currently held legacy systems. This should be able to give a proper proof for some records. The digital evidence should have appropriate authentication, it must also be reliable. Authentication is one of the most important relate a data to the physical persons. It should have a strong accessible control and self-sustainable at place, logs, and audit in a good way. The term accurate specifies a processing of data reliability that helps to decide the contents of reliability and sometimes there will be a timings issues that leads to overhead (Vidya et al., 2014).

In digital Forensics the AI and ML techniques are also applicable. The AI ML helps digital forensics investigators in order to analyze a large volume of data that are generated by mobile devices and stored in a cloud. Similar to industries, while reaching modern criminal cases the demands are outstripping. With the support of artificial intelligence (AI), automation of some process and quick flag contents or some insight will be taken for the purpose of investigation longer in order to uncover. Some of the AI ML techniques are used for digital forensics: regression, classification, and clustering.

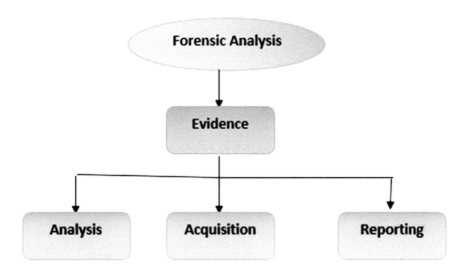

FIGURE 2.1 Flowchart of forensics analysis.

2.1.1 REGRESSION IN DIGITAL FORENSICS

Some of the value depends on the values of independent features with the help of existing data based on the past events. And this knowledge helps to handle new events. In digital forensics, fraud detection can be solved by using this method. Once the model is learnt based on the previous transaction database, by observing the features of current transactions the fraud transactions are identified. The AI ML provides support vector machine (SVM), linear regression, polynomial regression, decision tree and other regression methods for the analysis of regression.

2.1.2 CLASSIFICATION

Classification is done based on supervised machine learning (ML) techniques. In digital forensics the spam detection is done by using AI ML algorithms. In order to categorize whether the given mail is spam or not the AI and ML algorithms plays the main role. Some of the algorithms are used for the classification like logistic regression, K-nearest neighbor (K-NN), and SVM.

2.1.3   CLUSTERING

Clustering is an unsupervised ML model which takes general patterns from a dataset even though there is no labeling in datasets. Groups of similar events which constitute a cluster as they share common features and define a particular pattern. In cyber security clustering is used for forensics analysis, anomaly detection, for the analysis of malware, etc. Some of the ML clustering algorithms like K-means, K-Medoids, DBSCAN, Gaussian Mixture Model, Agglomerative clustering are used in cyber security (Sharma et al., 2018).

2.2   LITERATURE SURVEY

Shailendra Singh et al. (2009) has introduced a method for ensemble approach feature selection of cyber-attack dataset. They used a dataset called DARPA KDDCUP99. This consists of two phases. Phase one is filter phases that are used to select the information that has the highest gain and also has appropriate guidance for the sake of initialization for the search process for the phase of rapping output. These kinds of outputs are the final subset features. The feature subset of final is then passed by using some supporting algorithms such as K-NN classifier. This algorithm helps for the classification of different kinds of attacks.

Quamar Niyaz et al. (2016) has worked on network intrusion detection systems (IDS) with the help of deep learning approach for efficient and flexible NIDS. Here they used NSL-KDD datasets which is benchmark dataset for network intrusion. They utilized a validation method called 10-fold cross in order to train a data for the evaluation to classify accuracy. They used STL (self-taught learning) for the classes like two classes, five class and 23 class. They made a comparison on the performance with the regression called SMR (soft max regression) without any feature learning. By comparing STL gave a good classification accuracy rate of around 98% compared to all other types of classification.

Govindarajan et al. (2009) introduced a new K-NN classifier and applied it on IDS. They evaluated performance based on Run time and Error rate for malicious and normal datasets. The new K-NN classifier gave more accurate accuracy compared to existing K-NN classifiers.

Mohammadreza Ektela et al. (2010) used two techniques for intrusion detection in networks that are classification tree data mining (DM) and SVM techniques. They made a comparison with C4.5 and SVM. By comparing these two algorithms C4.5 algorithm gave the best performance for identifying the rate and false alarm rate compared to SVM. In case of U2R attack SVM gave good performance.

Deepthy et al. (2010) implemented various types of DM techniques for detecting the intrusions. They have given a brief description about the functioning of IDS classifications and how this kind of classification is working for the IDS. For the large amount of traffics which is caused by the networks by comparing classification the clustering will be more efficient for the identification of intrusions. The main aim is for classification purpose, a huge bulk of data is required by comparing with clustering.

Amudha et al. (2011) introduced DM techniques for intrusion detection. They used KDD CUP'99 as datasets. They evaluated the performance of DM classification algorithms such as 48, Naïve Bayes, NBTree, and Random Forest. In this work they mainly gave importance to correlation feature selection (CFS) measures. Based on the accuracy and detection rate NB tree and Random Forest is best outperformance by analyzing with other two algorithms.

China et al. (2017) implemented three techniques for DM, such as SVM, Ripper rule and C5.0 tree for the identifications of intrusions for the identifications of intrusions. By experimentation result, the C5.0 decision tree gave more efficiency compared to other algorithms. All three DM techniques are given higher accuracy around 96% detection rate.

Nahla et al. (2004) used algorithms for intrusion detection, i.e., Naive Bayesian. They considered the dataset called KDD'99 intrusion. Finally, they compared winning strategy, decision trees, and Naive Bayes. Naive Bayesian gave good accuracy compared to the other two algorithms.

Rahul et al. (2018) focused on predicting the attacks on network IDS. They used KDDCup-99 and epochs datasets for training and benchmarking the networks. In order to perform comparison training is done by using the same datasets with several classical algorithms and using deep neural networks (DNNs) of layers which range from 1 to 5. By considering the result the DNN of 3 layers gave the best performance compared to classical ML algorithms.

Praneet et al. (2021) introduced a method for edge node detection against multitudes of cyber-attacks on the internet of things (IoT). This process can be done with the support of deep learning methods. The evaluation of the model is done by deploying actual edge nodes with the representation of Raspberry Pi by using current cyber security datasets, i.e., UNSW2015. By comparing the convolution Deep learning method techniques from the previous result, this experimental result gave the best accuracy of 99% approximation.

Antonio et al. (2014) used supervised ML techniques for multi-class classification like decision tree and Bayes classifiers. Depending upon the rules ANN (artificial neural network) and nearest neighboring algorithms, the classifiers are evaluated based on performances like accuracy and Cohen's kappa. This experiment is designed based on 4-cross validation with 30 repetitions for non-deterministic algorithms. In order to get reliable results, the result averaging is performed by 120 runs. An analysis is done by comparing each pair of algorithms with the support of t-tests using both accuracy and Cohen's kappa metrices.

Xiaoyong et al. (2017) used a deep learning method for Denial-of-Service attacks. They compared performance with conventional ML. Here ML is another method for improving performance for static features. But this algorithm is limited by shallow representation models. They proposed deep learning based on a DDoS attack detection approach called deep defense. The main aim of using deep learning is to learn patterns from sequence of network traffic and trace network attack activities by using DNNs. The experiment result gave better performance compared to conventional ML models. They also reduced the error rate from 7.517% to 2.103% compared with conventional ML methods in larger datasets.

Sumaiya et al. (2013) used to evaluate different tree-based classification algorithms that are used to classify network events for the IDS. They used NSL-KDD 99 datasets for their experiment. The dataset dimensionality is reduced. The result shows that Random Tree models more accuracy and also reduced the false rate of alarm. To compare a better predictive accuracy the Random Tree model is evaluated by making a comparison with other intrusion detection models.

Meng Jianliang et al. (2009) used K-means algorithm for detecting intrusion. They used KDD-99 as a dataset. The main reason to use this dataset is it will give a stable accuracy and efficiency for the algorithms. By different setting rate, the detection rate gave more than 96% and false rate of alarm gave less than 4%. The time complexity is very less. By comparing with other algorithms, K-means methods are more effective algorithm for partitioning a huge set of data. For the intrusion detection K-means algorithms are most widely used compared to other algorithms.

2.3   TYPES OF DIGITAL FORENSICS

The use of drones is rapidly growing and expanding into criminal enterprises and terrorist organizations.

The new term 'computer forensics' was coined, and has recently been renamed 'digital forensics.' The term is further classified into; drone forensics, and social network forensics, network forensics, and mobile forensic. This chapter provides an overview of digital forensics, network forensics, and mobile forensic, drone forensic and social network forensics. Digital forensics is the application of electronic inquiry and analytical methods to identify potential evidence.

Identifying fingerprints is a classic example of a forensic task. With the rapid increase in criminal activity on computers, the word 'forensic has become more similar for the law enforcement and for IT community (Son et al., 2012).

Computer forensics in traditionally entails occurring a data arriving a devices called electronics device gathering data in order to use it in an investigation or for law enforcement purposes. Instead of taking each file, it creates an image like a digital forensic examiner (bit-by-bit copy).

These researchers should extract it from the computer and scan the data image for relevant information such as files, folders or e-mail conversations, as well as metadata about files stored by the computer. Digital forensic analysts are able to detect active files, deleted files, password-protected files, and files that have been overwritten.

Digital technology hardware, especially software, has developed at an exponential rate over the last decade. This increase in popularity has prompted not only regular users, but also malware developers and conventional computer hackers, resulting in cyber threats and malware threats.

The stolen phone rates are increased in the past few years. The main reason is the mobile phone has some of the confidential information of a person and it is very dangerous if the phone is stolen, or the phone is lost.

Digital forensics is the result of cybercrime committed by cybercriminals using IT infrastructure or by utilizing many technologies (Brown et al., 2009; Bassett et al., 2006). Digital Forensic is classified into different types such as computer forensic, drone forensic, social forensic, memory forensic, multimedia forensic, network forensic, small-scale device forensic/mobile forensic and android forensic (Brown et al., 2009; Son et al., 2012). In this chapter, it is classified as drone forensic, network forensic, social forensic, and mobile forensic (Figure 2.2).

2.3.1 NETWORK FORENSIC

The focus of traditional computer forensics is static data. The hard drive, CD, or flash drive, for example, requires for the recovery of data. There are various tools that are used to assist digital forensics investigators in order to retrieve and reconstruct the evidence of data which has arrived from any medium. Network forensics is mainly used for data transfer from one computer to another.

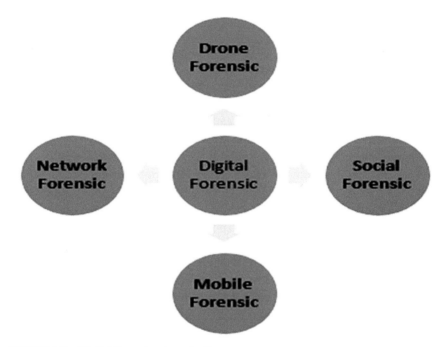

FIGURE 2.2   Digital forensic categorization.

Network forensics is used to analyze, capture, record, and interpret network events. Network Forensics is related to ephemeral data. Some create and distribute viruses, spyware, and malware over the network.

A computer forensic expert can capture and search, a network forensic expert can only search for firewalls, network packet filters, and wireless frames. Network investigators need the same skills as hackers to identify the motive (what happened and why) for a crime. For the understanding of potential attacks and gather evidence from these attacks, digital forensic investigators must effectively monitor network activities.

In a case involving network forensics, the situation is exacerbated by very large log files related to firewalls, IDS, or web servers.

In case of data logs are available from ISPs, further the availability of data that is available in social media. For example, a potential dataset would be further perspective of investigation related to crime: Facebook (indicating contacts, friendships, locations, etc.), Flickr (containing metadata name/location tags), YouTube Videos (Alastair Irons et al., 2001).

In the case of intelligence, some of the algorithms, such as ANN and SVM are used for identifying the intrusion and protecting the integrity and confidentiality of the information infrastructure.

2.3.2 SOCIAL MEDIA

In the case of social media, the websites are now available on mobile devices like smartphones and PDAs (Azhar et al., 2018). Due to the widespread use of social media, many people deliberately advertise their location, religion, medical status, friends, material status, personal e-mail ID, mobile numbers, photographs, status updates, and whereabouts to others. For many organizations and individuals, social media sites have become part of communication. There are currently several criminal cases involving crimes committed using social media or social networking sites. Several criminal cases are currently pending. After the social media sites revolution social network forensics has occurred. Recently, digital forensics has been widely used to investigate computer crime, which has been reviewed in court.

In digital forensics, the social network forensics are the portion of digital forensics. This social network forensics deals with collecting the digital evidence from the social media platform. In digital forensics case investigation, we need to ensure all the evidences are valid legally and this evidence must be presented in a forensics way (Table 2.1).

TABLE 2.1   Social Network Site Characteristics

Types	Details
User participation	Encourages users of social networking sites to participate and provide feedback.
User community	Individuals can form their own group and share common interests. This group will act as a forum for sharing and gathering knowledge.
Help public participation	Many social networking sites encourage users to give feedback and participate in ways that allow them to communicate with others in a public environment.
User communication	Traditional media only allow one-way communication, while SNSs allow two-way communication.
Connection with people	Sites can also help to reconnect with people who don't see in their daily life.

To conduct an effective digital forensic investigation, you need a broad understanding of software and hardware. By having appropriate knowledge about a various tools for the investigation that are not saving any time but the advantage preventing the actions and evidences modifications negligence. As a result, investigators do not need to go back and re-examine the facts. One of the suitable way is to present the evidence from SNSs. In order to present a evidence, a digital forensic investigators need a range of resources and techniques, and studying the capabilities of extraction tools can aid the investigative process more effectively.

2.3.3 MOBILE FORENSIC

Mobile technology hardware, especially software, has developed at an exponential rate over the last decade. The increase in popularity has promoted regular users, malware developers, computer hackers that leads to cyber threats and malware threats.

The stolen rate of phone has increased in past few years due to the phone has a confidential information's of a person that tends to dangerous if they are stolen or phone is lost.

Application developers and users that made forensic analysis of phones and devices more aware of how and what data should be stored on these devices or not. This sensitive information stored can be used by the attacker as a pretext to steal a person's true identity. Identifying fingerprints is a classic example of a forensic task. With the rapid increase in malicious activity, the word 'forensic' are more popular in law enforcement and IT community (Figure 2.3).

FIGURE 2.3   Mobile forensic recovery evidence.

Mobile forensics (MF) is used to retrieve digital evidence from mobile devices using specific and appropriate scientific forensic systems (Kavrestad et al., 2020). The world is changing as a result of mobile technology. Millions of mobile devices, including phones, watches, laptops, and GPS units also increased. People use cell phones even in places where computers are inaccessible, such as villages, hamlets, deserts, and mountains. Mobile devices carry everything. The different types of cases involving mobile devices include intelligence gathering, drug/narcotics, harassment, pursuit, homicide, counterterrorism, juvenile delinquency, financial crime, corporate compliance, and property ownership (Figure 2.4).

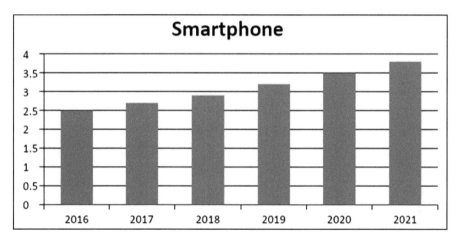

FIGURE 2.4   This graph shows a steady increase in smartphone usage.

The first type of forensic investigator can detect and detect mobile device evidence. These data can be physically stored in three different locations. Some types of data can be found in more than one place, e.g., contacts on SIM and handset, pictures on handset, and memory card. They can be the same set of images or completely different (Figure 2.5).

2.3.3.1   SIM CARDS (SUBSCRIBER IDENTITY MODULE)

A subscriber identity module (SIM) card is mandatory on GSM networks. The SIM card is a smart card: it contains the processor and storage. The SIM card is contained digitally: ICCID (Personal Card Identifier) and IMSI (Subscriber Identity), both of which are very important in the investigation. Text-based user data (e.g., contact and calls), also 3GPP.

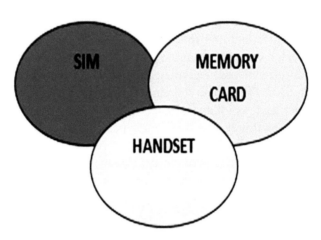

FIGURE 2.5 Three different locations of data.

From SIM card to memory card, different physical form components such as MMC, MicroSD, and memory stick M2 exist, remember, not all memory cards are SD cards, but 1 TB cards are currently available on each memory card. The most common types of information and media files that are expected to be recovered from a memory card. Depending on how they were used, including images, audio, video, documents, smartphone applications, databases, and facts; they can contain any type of file.

2.3.3.2 HANDSETS

It's important information for any type of device, categorized as mobile, calls, contacts, SMS, media messages, other messages (applications), images, audio and video recordings, web history, and location information. Information categories can have subcategories (Figure 2.6):

1. **Investigator Recovering the Data, Dealing with Live Devices:** When performing more basic logical extractions of handsets, the vast majority of the time the handset will be switched on.

   If the data from the devices is separated from the network, the investigator can get the data. There are three steps for connection interface they are:

   • Cable is the fast and secure way to get the data;
   • Bluetooth slower than cable, handset ID could be visible to others; and

- Wi-Fi slower than cable, handset has to be connected to a Wi-Fi network, results from reading the same handset with different connection interface may vary.

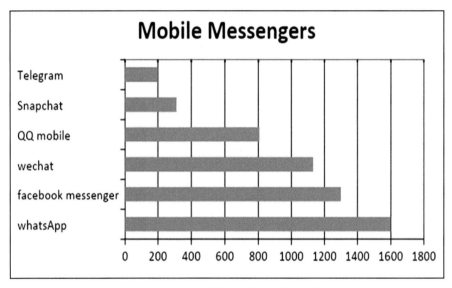

FIGURE 2.6   Mobile messenger apps – millions of monthly active users.

A tool called Simbrash has been created in order to extract the entire file system from the devices for the sake of mobile phones, Linux and windows platforms (Casadei et al., 2001; Husain et al., 2010). Additional steps with Smartphone OS: When both operating systems (OS) store data internally in similar fashions, different additional steps are required when connecting handsets that are ready to extract.

2. **iOS-iPhone Trust:** When your iPhone connects to your Investigator PC you will be asked to trust the computer on which the handset is plugged in. If it clicks as a trust, at that point, those two devices are communicating and get the data. These are the key points in a mobile forensic investigation researcher in Mokhonoana et al., (2007) suggested a simple and inexpensive framework for analyzing iPhone forensics. It has the capability to extract all the digital evidence from the iPhones. This model consists of three processes: obtaining data, analyzing data, and reporting data. On the other way, in Azhar (2018), a forensics tool called phone forensics tool was developed which has

ability to extract evidence from mobile active files. In another study (Kao et al., 2019), the researchers proposed a new model to isolate contacts of phones, recording system of calls in phones, SMS, documents which is present in phone and all the acquisition steps which are available in the window mobile.

There are two methods in order to extract a data of phone from a handset, either logical extraction (like handshake) or physical extraction, which is a brute force method for recovering the data (Figure 2.7).

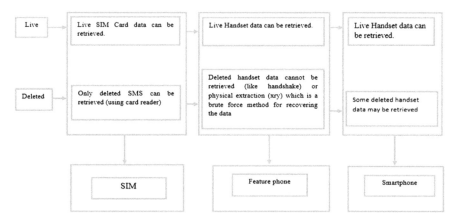

FIGURE 2.7   The logical extraction of phone data from a handset.

3. **Physical Extraction (HEX Dump):** This includes cable connection and appropriate software license. Data is recovered in a raw form, providing a lot of data, including deleted handset information. The decoding of raw data will vary based on the handset model (Figure 2.8).

4. **Present Challenges in Mobile Forensic:** SIM PIN protection, once it is protected, tries to extract data from the SIM card, the only thing it can recover is ICCID.

    i. **SIM PIN Protection:** SIM data can be protected with an optional 4–8-digit code, users can change the PIN, and entering the PIN incorrectly 3 times can cause the SIM card to be locked. PIN Unlock Key: Unlock requires 8 digits PUK, PUK code is usually with the service provider and cannot be changed by the user. After 101 incorrect PUK attempts, the SIM card is permanently disabled. These are the major challenges of MF.

    ii. **Bluetooth Extraction:** If the only option is to extract the Bluetooth.

    iii. **Handset Security:** If a device is locked, investigators need to find out if the lock on that particular handset can be recovered or bypassed in any form. This is usually done through different types of extraction depending on the software tool. If the device cannot be unlocked or bypassed by the software tools in the office, the phone may need to be sent away to be unlocked by vendors providing next level unlocking services.

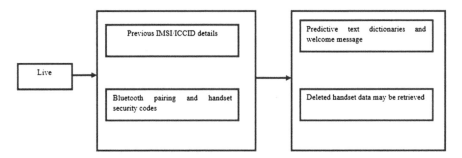

FIGURE 2.8   The physical extraction of phone data from a handset.

2.3.4   DRONE FORENSIC

Drone is another name for unmanned aerial vehicle (UAV). We call them as aircraft without human beings. The UAS (Unmanned aerial Systems) consists of a UAVs, ground-based controller, and a communication system between the two.

With the various degrees of autonomy UAVs are having ability to fly: either remotely commanded by a human operator or autonomously controlled by onboard computers. There are three components of drone forensics they are: UAV, Ground Control Station (GCS), and Communication Data-Link (Azhar et al., 2018; Shetty et al., 2017). (Figure 2.9).

A controller, mobile phones and drones are usually depending on Wi-Fi signals connectivity in order to communicate with each other at drone forensics. The ability to fly under the control of a drone is data transfer (Kao et al., 2019). When there is a connectivity of a drone with a mobile phone to the wires that carries the packets during the operation of the drone, the MAC address network interface card was found (Kavrestad et al., 2020).

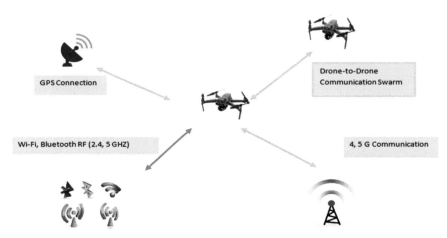

FIGURE 2.9   Drone communications network.

GPS spoofing is one of the current dangers and vulnerabilities in drone forensics. The information of unencryptions is easily collected, managed, and injected when there is a weak communication nature. These kinds of weak communication nature are like giving the attackers a complete authority for drone.

1. **Wi-Fi/GPS Jamming:** The drone hijacking is performed by transferring a de-authorized method between access point and device controlling drone or sometimes disruption of flight happens due to jamming GPS signal rendering the drone unless most of the commercial drones have availability of failsafe options.
2. **Data Interception and Interface:** In order to motor a vehicle a telemetry feeds are utilized. Further, the data is transferred through open non-secure wireless transmission gives the ability to perform man-in-middle attacks or upload malware on the drone.
3. **Malware Infection:** As drones become flying computers, malware can affect the operation of the drone or compromise the whole system.
4. **Drone Information can be Extracted like, GPS Position:** Extracting the GPS position (Lat, Long, Alt) can lead the examiner to visualize the flight path.
5. **Flight Path:** One of the most important findings since it can determine the places possibly affected by the drone flight.
6. **Take-Off Point:** Locating the take-off point can place the perpetrator in the Area of interest (AoI), and with the combination of local source (camera, witness) help identify the suspect.

7. **Firmware Version:** Information about drone, battery types, and verification about integrity.
8. **Error Recorded:** Commonly known as the crash-log can provide information of error leading to a drone crash.
9. **Media:** Pictures and videos can help understand the mission of the drone, and also the areas that the drone has been operating in.
10. **Smart Battery:** Contains information about the voltage consumption, battery life. Drone can log the serial of the battery so it can be tied to the owner of the drone.
11. **Fingerprints:** Although not relevant with cyber forensic, law enforcement agents can track down the pilot.
12. **Findings/Artefacts Common Location, Removable Storage:** Usually SD cards in the vehicle itself or supporting devices contain the majority of the logs.
13. **Internal Memory Chip:** Using chip-off technique to recover data stored on the drone itself. This may not be possible if the vehicle is significantly damaged.
14. **Supporting Devices:** Mobiles, tablets, laptops, and other portable devices can be used as GCS, and therefore contain data and logs from the drone. Different forensic acquisition and analysis techniques are used for each one.
15. **Cloud Storage:** It is possible that drone logs are stored on the cloud, and therefore should also be taken into consideration.
16. **Network PCAP Files:** Monitoring the network can provide findings about irregular activity within the interconnected components (Prashanth et al., 2015).

Drone's complete data extraction and analysis is drone data found in cloud, mobile app, and remote control, desktop app. Drone storages are found in external SD cards in a reader, internal memory via chip-off or physical (flight logs). Drone investigation proceeds if a drone is discovered to be a crime, based on preliminary analysis. Preliminary of the suspected drone involves three steps as follows (Figure 2.10):

➢ **Step 1:** Make a written record of everything so far. Disassemble the documented device.
➢ **Step 2:** Identify data storage locations.
➢ **Step 3:** Check each integrated circuit package.

Following the above techniques, forensic software is used to get a complete form of data acquisition of a media or a complete file system that is further extracted. The investigators of digital forensic starts from searching for drones, like SD cards, in order to capture media and analyze content for data storage areas. Further, the use of tools like forensic toolkit (FTK) for access data to recover data from drones also conduct investigations thoroughly. In the field of digital forensics, drone forensics is a growing field. As drone incidents around the world increase, it is crucial that law enforcement agencies around the world prepare counter-drone plans. VTO Labs' forensic images on NIST provide a platform for the creation and testing of drone forensic equipment (Airbail et al., 2021).

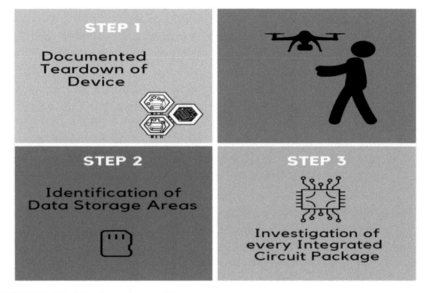

FIGURE 2.10   The three phases of drone.

Now, investigators can film the media from a drone, extract flight logs, photos, videos, and sensitive data, locate its user and owner, also locate suspects in related to the drone crimes. A digital forensics method for extracting data from drones are inadequate for investigation. Digital forensic software tools and technologies need to be updated with the introduction of advanced drone models.

2.4   AUTOMATION IN CYBER SECURITY

There is a very huge amount of data in the cloud, nearly about 1 exabyte. Cybercrimes are increasing everyday causing various security threats in day-to-day life. Organizations cannot keep pace with the growing threats. Manual detection has slowed down the pace of providing security. We need to fill this gap and take a leap or upgrade to automation with the help of technologies like AI. AI provides a means to identify anomalous behavior, take crucial action and help in alleviating the glitch (Singh et al., 2009).

Earlier traditional approach was used to prevent the systems from cyber-attacks. Humans' involvement was required for detection of threats, solving it using software. Over a period of time the AI and cybersecurity were allied. CAPTCHA (Completely Automated Public Turing Test to Tell Computers and Humans Apart) (Shaukat et al., 2020) is a good illustration for this. Automation reduces human interaction, involvement, struggles, and saves time instead of doing repetitive work, functioning in unpredictable environments, speed, and scalability. Humans can be involved in other business modeling.

Cybersecurity automation is a process of automating configuration, alerting and other responsibility that is achieved from the admirations of security and from the products which they manage. These kinds of automations we call it as orchestration. These kinds of offers are good metaphor of what the core advantage is: If we think that the products in your security architecture are as instruments, then the security administrator is the conductor of an orchestra.

Cybersecurity automation has the ability to take in different forms, but automation tasks can be managed by the devices without any interactions of human beings (Kirti Raj, 2019). We can accomplish the automation neither by scripts nor by using the automation tools.

2.5 TECHNIQUES USED IN AUTOMATIONS

Cyber criminals have pervaded the IDS by very powerful futuristic tools. Incline towards more advanced techniques like ML, AI (Ektefa et al., 2010), deep learning is increasing rapidly. Advanced automation technique ML is being used on both sides' cyber security experts and by attackers. To be ahead of the cybercriminals experts have adapted two methods. One is ANNs where this neural network works just like a human brain, there are nodes or neurons interconnected to each other. They co-ordinate and work together (Calderon et al., 2019; Wiafe et al., 2020).

There are many automation tools available SOAR products, robot process automation (RPA) tools and custom-developed software code (Robert Nagy et al., 2019) (Figure 2.11).

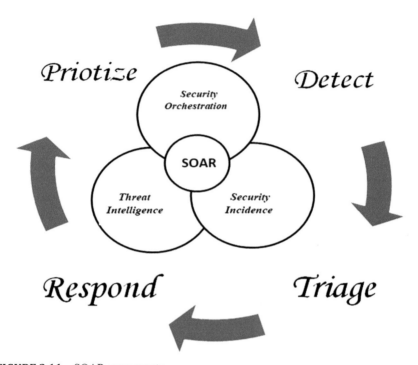

FIGURE 2.11   SOAR components.

2.5.1   RPA TOOLS

The BOTS of AI help in implementing general data protection regulation (GDPR). It prevents attacks or viruses in various sectors like defense, finance, business, and marketing. RPA reduces manual validation and reviewing. It provides encryption, allows only users who have genuine access credentials and secures data from all kinds of attacks.

RPA can be employed in situation where a shortage of professionals occurs. It can manage audits, create timesheets. It professionals can involve in other high value added procedure. Intimate high security alerts, notifications 24/7 tirelessly (Figure 2.12).

2.5.2   CUSTOM-DEVELOPED SOFTWARE CODE

It helps in planning, creating, deploying, and maintaining software for a dedicated set of organizations, people, or groups of community. Especially in E-commerce field, custom developed software helps in protecting the data; particularly, the security level is elevation (Robert Nagy et al., 2019).

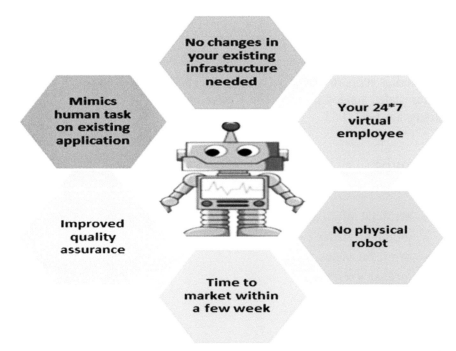

FIGURE 2.12 Robot process automation.

2.6 CONCLUSION

Digital forensics helps in investigation for the legal purpose involving the analysis of digital evidence. Digital forensics utilizes specialized tools and techniques for the investigation of various computer-oriented crimes such as fraud, illicit use, and many forms of computer intrusion. There are different kinds of digital forensics, such as drone forensics, network forensics, social forensics, and MF. Different types of forensics help in providing integrity of the computer systems and help in giving evidence to the court that leads to punishment of the culprit, which gives the appropriate efficiency to keep track of cyber criminals. Here AI and ML algorithms play a major role in digital forensics for intrusion detection, identity theft, cyberstalking, and slander spreading. Automated techniques such as RPA tools help in implementing data protection regulations with the support of AI and ML. It also provides encryption which helps the user to secure all kinds of data, and customer developed software helps in protecting the data, especially in e-commerce sites.

ACKNOWLEDGMENT

This work is supported by VGST, Department of ITBT and ST, Government of Karnataka, funded project, The Center of Excellence Digital Forensics Intelligence GRD 853.

KEYWORDS

- **artificial intelligence**
- **digital forensics**
- **drone forensics**
- **forensics science**
- **machine learning**
- **mobile forensics**
- **network forensics**
- **social forensics**

REFERENCES

Airbail, H., Mamatha, G., Hedge, R. V., Sushmika, P. R., Kumari, R., & Sandeep, K., (2021). Deep learning-based approach for malware classification. *International Journal of Intelligent Defense Support Systems, 6*(2), 61–80.

Amor, N. B., Benferhat, S., & Elouedi, Z., (2004). Naive bayes vs decision trees in intrusion detection systems. In: *Proceedings of the 2004 ACM Symposium on Applied Computing* (pp. 420–424).

Amudha, P., & Rauf, H. A., (2011). Performance analysis of data mining approaches in intrusion detection. In: *2011 International Conference on Process Automation, Control and Computing* (pp. 1–6). IEEE.

Azhar, M. A., Barton, T. E. A., & Islam, T., (2018). Drone forensic analysis using open-source tools. *Journal of Digital Forensics, Security and Law, 13*(1), 6.

Bassett, R., Bass, L., & O'Brien, P., (2006). Computer forensics: An essential ingredient for cyber security. *Journal of Information Science & Technology, 3*(1).

Bhatele, K. R., Shrivastava, H., & Kumari, N. (2019). The role of artificial intelligence in cyber security. In *Countering Cyber Attacks and Preserving the Integrity and Availability of Critical Systems* (pp. 170–192). IGI Global.

Boyd, D. M., & Ellison, N. B., (2007). Social network sites: Definition, history, and scholarship. *Journal of Computer-Mediated Communication, 13*(1), 210–230.

Brown, C. L., (2009). *Computer Evidence: Collection and Preservation.* Charles River Media, Inc.

Calderon, Ricardo, (2019). "The Benefits of Artificial Intelligence in Cybersecurity." *Economic Crime Forensics Capstones.* 36. https://digitalcommons.lasalle.edu/ecf_capstones/36 (accessed on 14 February 2023).

Casadei, F., Savoldi, A., & Gubian, P., (2006). Forensics and SIM cards: An overview. *International Journal of Digital Evidence, 5*(1), 1–21.

Casey, E., (2001). *Handbook of Computer Crime Investigation: Forensic Tools and Technology.* Elsevier.

Chaudhari, R. R., & Patil, S. P., (2017). Intrusion detection system: Classification, techniques and datasets to implement. *International Research Journal of Engineering and Technology (IRJET), 4*(2), 1860–1866.

Ektefa, M., Memar, S., Sidi, F., & Affendey, L. S., (2010). Intrusion detection using data mining techniques. In: *2010 International Conference on Information Retrieval & Knowledge Management (CAMP)* (pp. 200–203). IEEE.

Govindarajan, M., & Chandrasekaran, R. M., (2009). Intrusion detection using k-nearest neighbor. In: *2009 First International Conference on Advanced Computing* (pp. 13–20). IEEE.

Husain, M. I., Baggili, I., & Sridhar, R., (2010). A simple cost-effective framework for iPhone forensic analysis. In: *International Conference on Digital Forensics and Cybercrime* (pp. 27–37). Springer, Berlin, Heidelberg.

Javaid, A., Niyaz, Q., Sun, W., & Alam, M., (2016). A deep learning approach for network intrusion detection system. *EAI Endorsed Transactions on Security and Safety, 3*(9), e2.

Jianliang, M., Haikun, S., & Ling, B., (2009). The application on intrusion detection based on k-means cluster algorithm. In: *2009 International Forum on Information Technology and Applications* (Vol. 1, pp. 150–152). IEEE.

Kao, D. Y., Chen, M. C., Wu, W. Y., Lin, J. S., Chen, C. H., & Tsai, F., (2019). Drone forensic investigation: DJI spark drone as a case study. *Procedia Computer Science, 159*, 1890–1899.

Kävrestad, J., (2020). *Fundamentals of Digital Forensics.* Springer International Publishing.

Mohammad, S. M., & Lakshmisri, S., (2018). Security automation in information technology. *International Journal of Creative Research Thoughts (IJCRT), 6.*

Mokhonoana, P. M., & Olivier, M. S., (2007). Acquisition of a Symbian smart phone's content with an on-phone forensic tool. In: *Proceedings of the Southern African Telecommunication Networks and Applications Conference* (Vol. 8).

Naiping, S., & Genyuan, Z., (2010). A study on intrusion detection based on data mining. In: *2010 International Conference of Information Science and Management Engineering* (Vol. 1, pp. 135–138). IEEE.

Prashanth, D. S., Agrawal, A., & Raj, M., (2015). Reduction of sample impoverishment problem in particle filter for object tracking. In: *2015 Annual IEEE India Conference (INDICON)* (pp. 1–5). IEEE.

Robert, N., Todd, C., & Geoff, H., (2019). *"Cybersecurity Automation for Dummies.* ISBN 978-1-119-57580-1 (pbk).

Sharma, S. (2021). *Overview of Machine Learning in Cybersecurity Comparative Analysis of Classifiers Using Weka, 23*(8), p.334.

Shaukat, K., Luo, S., Varadharajan, V., Hameed, I. A., & Xu, M., (2020). A survey on machine learning techniques for cyber security in the last decade. *IEEE Access, 8,* 222310–222354.

Shetty, S., Vishwakarma, V., Harisha, & Agrawal, A. (2017). "Design and implementation of video synopsis using online video inpainting, *2017 2nd IEEE International Conference on Recent Trends in Electronics, Information & Communication Technology (RTEICT),* Bangalore, India, pp. 1208–1212, doi: 10.1109/RTEICT.2017.8256790.

Singh, P., Pankaj, A., & Mitra, R., (2021). Edge-detect: Edge-centric network intrusion detection using deep neural network. In: *2021 IEEE 18th Annual Consumer Communications & Networking Conference (CCNC)* (pp. 1–6). IEEE.

Singh, S., & Silakari, S., (2009). *An Ensemble Approach for Feature Selection of Cyber Attack Dataset.* arXiv preprint arXiv:0912.1014.

Son, J., (2012). *Social Network Forensics: Evidence Extraction Tool Capabilities.* Doctoral dissertation, Auckland University of Technology.

Tallón-Ballesteros, A. J., & Riquelme, J. C., (2014). Data mining methods applied to a digital forensics task for supervised machine learning. In: *Computational Intelligence in Digital Forensics: Forensic Investigation and Applications* (pp. 413–428). Springer, Cham.

Thaseen, S., & Kumar, C. A., (2013). An analysis of supervised tree based classifiers for intrusion detection system. In: *2013 International Conference on Pattern Recognition, Informatics and Mobile Engineering* (pp. 294–299). IEEE.

Vidhya, B., & Vaijayanthi, R. P., (2014). Enhancing digital forensic analysis through document clustering. *International Journal of Innovative Research in Computer and Communication Engineering, 2*(1).

Vigneswaran, R. K., Vinayakumar, R., Soman, K. P., & Poornachandran, P., (2018). Evaluating shallow and deep neural networks for network intrusion detection systems in cyber security. In: *2018 9th International Conference on Computing, communication and Networking Technologies (ICCCNT)* (pp. 1–6). IEEE.

Wiafe, I., Koranteng, F. N., Obeng, E. N., Assyne, N., Wiafe, A., & Gulliver, S. R., (2020). Artificial intelligence for cybersecurity: A systematic mapping of literature. *IEEE Access, 8,* 146598–146612.

Yuan, X., Li, C., & Li, X., (2017). Deep Defense: Identifying DDoS attack via deep learning. In: *2017 IEEE International Conference on Smart Computing (SMARTCOMP)* (pp. 1–8). IEEE.

CHAPTER 3

Specific Security Mechanisms for Information Protection and Transmission Over Open Networks

PRIYA R. SANKPAL and P. A. VIJAYA

Department of Electronics and Communication Engineering, BNM Institute of Technology, Bangalore, Karnataka, India

ABSTRACT

Healthcare essential's delivery has changed drastically with the use of information technology. The adoption of digital health information is transforming healthcare from a paper-pushed device to a new era where fitness records are instantly transmitted and shared. The overall requirements for healthcare delivery are that it must be kept personal, possess integrity, critically accurate, accessible round the clock and in every location it's needed.

The digital fitness document system contains huge quantities of facts. These sensitive records must tour to the multiple sites of care through laptops, mobile phones and tablets which are largely used by healthcare experts/assistants. This raises substantial protection issues due to the human intervention and data risks over the mobile devices. This makes information security vital on mobile and storage devices.

Information security intends to provide protection to the information in terms of "confidentiality, integrity, and availability" (CIA triad). In case of a security breach, it is certain that the confidentiality, integrity, and availability of information needs to be protected via information security mechanism tools.

Advancements in Cybercrime Investigation and Digital Forensics. A. Harisha, Amarnath Mishra, & Chandra Singh (Eds.)
© 2024 Apple Academic Press, Inc. Co-published with CRC Press (Taylor & Francis)

In this chapter, specific security mechanisms such as encryption and watermarking are explored and investigated for information protection and transmission over open networks. In the proposed method, the information protection techniques are analyzed for "digital imaging and communications in medicine (DICOM) images." The information of the patient is made secured by using fusion techniques of image encryption and watermarking. Two sets of images related to the patient are transmitted securely by hiding one image into the other. One of the images is addressed as cover image (CI) and the second is addressed as a watermark image (WM). WM, which contains sensitive patient information is made unreadable by encryption followed by embedding or insertion into the CI which enhances security. Chaotic map image encryption technique is used to encrypt WM. "Discrete wavelet transform (DWT) along with singular value decomposition (SVD)" are employed for embedding the encrypted WM into the CI.

Choice of chaotic maps for encryption is mainly due to their sensitive dependency on initial conditions, which is an essential aspect of encryption. The patient information is embedded into the DICOM image using DWT along with SVD. DWT converts image pixels into wavelet coefficients. These wavelets coefficients are used for image compression and coding. DWT captures input frequency content along with temporal content. The patient data is encrypted using a combination of logistic and skew tent maps. This encrypted information is embedded into the DICOM image using DWT and SVD.

The performance evaluation of implemented encryption and embedded algorithms are investigated using objective evaluation metrics such as the "number of changing pixel rate (NPCR), unified averaged changed intensity (UACI) for encryption techniques, mean square error (MSE), peak signal to noise ratio (PSNR), correlation coefficient (CC), normalized correlation coefficient (NCC) and structural similarity index (SSIM)" for watermark embedding process (Nguyen, Marie, & Azeddine, 2010; Silpa & Aruna, 2012). Performance of the proposed chapter is accessed by tracking the minute changes of the objective metrics when subjected to watermarking artifacts. With the implementation of the proposed scheme, it can be inferred that PSNR can be considered as a good metric for watermarking embedding strength variation. However, PSNR fails to measure image quality for different watermarking artifacts. It is observed that SSIM serves as a better metric for measuring image quality.

3.1 PREAMBLE

Internet usage has increased the multimedia exchange many folds, which makes it necessary to safeguard the confidentiality of the details being shared, from any forbidden access. Security contravention have invaded the solitude and reputation of user. Often the data being exchanged are textual details, image, audio and video information. Different category of multimedia statistics has different characteristics, and different methods are adopted to maintain confidentiality and avoid unauthorized access. Hence means of hiding information are essential for securing it when transmitting over open networks like the web, wherein the applications of multimedia are rapidly increasing. Cryptography comprises studying techniques related to encryption, authentication, and key distribution algorithms provides secured communication even in the presence of an adversary. Image encryption algorithms provides security by transforming the images into an image of unreadable form. Applications of image encryption vary from multimedia entertainment, military communication, internet communication, telemedicine to life and health sciences.

3.2 IMAGE ENCRYPTION

The raw method of encrypting images involves transforming an image of two dimension into a one-dimensional data stream and this stream was used to encrypt any text-based information (Ali, Zulkarnain, & Jan, 2012). This method of image encryption is applicable for texts, audio with smaller bit rates, images, and videos that may be transmitted over dedicated channels. Such algorithms are not suitable for various image data formats like BMP, GIF, JPEG, PNG, TIFF, etc. Use of basic cryptographic techniques for image encryption may not be suitable as image are more voluminous in contrast to textual data. Also, when the textual data is decrypted, complete recovery is mandatory, but for image data, this is not critical as the decrypted image contains a small amount of perturbations which is typically acceptable due to the persistence of vision.

Generally, encryption techniques for images may be categorized on the methods adopted to set up encryption scheme as: Chaotic maps oriented and Non-chaotic maps. Alternatively, there are also grouped as partial and full encryption, in consistent with the share of the information being encrypted. Literature survey provides many selective methods for image and video encryption, which describe the techniques that have been developed so far (Sankpal & Vijaya, 2014).

3.2.1   CHAOS THEORY FOR CRYPTOGRAPHY

A chaotic dynamic system is deterministic in nature which possess random behavior. Their sensitive dependence on initial conditions makes them random in nature. The changes created in output for small changes in input cannot be predicted with finite precision. This makes the behavior of chaotic systems completely unpredictable that mimics noise and is a desirable feature for cryptography. This striking relationship between chaotic systems and cryptography makes chaos-based cryptographic algorithms as preferred choice for secured communication over open networks. Chaotic system and cryptographic algorithms have identical properties of long periods of periodic orbits that are unstable, behavior that is pseudo-random and sensitive dependence on initial and control parameters. Chaos-based image encryption involves the basic principle of generating random sequence numbers that are used for encrypting the messages (Alireza & Abdolrasoul, n.d.). This makes the system output to be pseudo-random in nature, thereby making the system output to appear as random in the opponent's eye, though it is revealed as well-defined to the recipient and correct decryption is achieved (Somaya, Afnan, & Turki, n.d.). Traditional encryption transformations deal with finite sets, whereas chaotic systems deal with real numbers (Alireza & Abdolrasoul, n.d.). Chaotic systems have parameters that resemble the secret/encryption key used in cryptography. Comparison between cryptographic algorithms and chaotic systems with various parameters as indicated in Figure 3.1.

There are two ways, a chaotic system may be used for implementing a cryptosystem system:

- Pseudo random key generation using random sequences of chaotic maps; and
- Control parameters and initial conditions are replaced as secret keys.

3.2.2   ARCHITECTURE FOR IMAGE ENCRYPTION

The architecture of image encryption gleaned from chaotic system mainly comprises of two levels: the Confusion and the Diffusion levels. The graphical representation of the architecture adopted for encryption purpose, is as shown in Figure 3.2 (Sankpal & Vijaya, 2014).

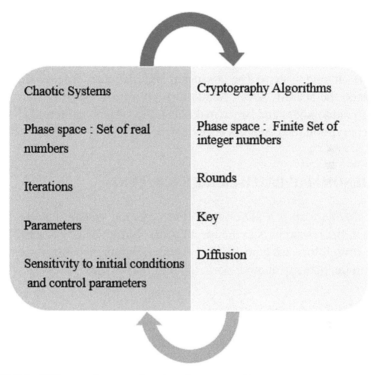

FIGURE 3.1   Difference between chaotic and cryptographic systems.

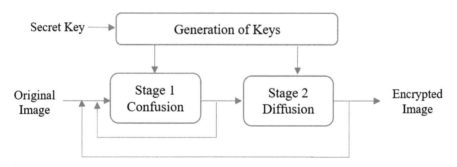

FIGURE 3.2   The confusion and diffusion levels of image encryption architecture.

The confusion level comprises of permutating pixels, where the pixel positions are distributed over the entire image. Only pixel positions are modified while pixel values. This renders the image unrecognizable. For confusion stage, control parameters and initial conditions of the chaotic sequence serve up as the secret key. Permutation stage alone cannot provide security. It can be easily compromised by any attacks.

To augment safety, the encryption process second stage changes the worth of every pixel within the entire image. Sequentially modify the values of the pixels in the diffusion level, with the aid of random sequence created using the chaotic system. The confusion and diffusion process are iterated to enhance the security level. Iteration of confusion and diffusion process results in randomness and hence makes chaotic systems preferred choice for image encryption (Sakthidasan & Santhosh, 2011).

3.3 HENON MAP-BASED IMAGE ENCRYPTION

Henon chaotic map is a 2D discrete time dynamic system that has kindled interest in the researchers in the recent years. Introduced by Michel Henon, it is the straightforward type of Lorenz model. Henon map transforms a point (x_n, y_n) in the plane to another point (x_{n+1}, y_{n+1}) as defined below:

$$x_{i+1} = y_{i+1} + 1 - \alpha x_i^2 \tag{3.1}$$

$$y_{i+1} = \beta x_I \tag{3.2}$$

Henon map has two dependent parameters α, β are called the initial parameters. The canonical values are = 1.4 and = 0.3, which make the Henon map chaotic, and iterations appear to be a boomerang-shaped chaotic attractor. With these values, the plane's beginning point can either map to the Henon attractor or diverge to infinity. The Henon map may appear chaotic or intermittent or converge to a periodic orbit for any values other than the canonical (https://en.wikipedia.org/wiki/H%C3%A9non_map).

Figure 3.3 shows a Henon map for a 2D plane formed from a well-defined number of rounds starting at the initial point (0.1, 0.1). Slightest changes in the initial point leads to drastic variations and hence in random behavior.

3.4 PERFORMANCE PARAMETERS FOR IMAGE ENCRYPTION

Performance analysis of encryption algorithm involves finding ways to attack the cryptosystem to extract partial or complete information from the encrypted image or to determine the secret key. These attacks are performed without the prior information about the decryption key. Parameters used for evaluating the performance of encryption algorithms mainly depends on what the attacker or the opponent wants to access. An attacker or opponent may want to acquire information about either the encrypted image or original image or keys used in the algorithm. Most widely used attacks on encrypted output images are weighed up as in subsections (Ali, Zulkarnain, & Jan, 2012; Chattopadhyay, Mandal, & Nandi, 2011).

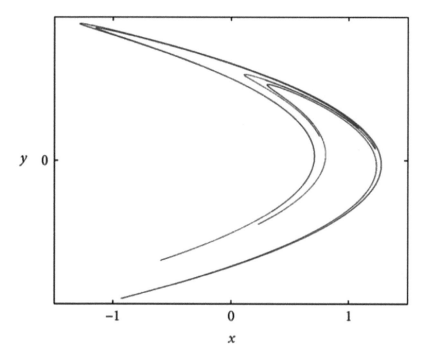

FIGURE 3.3   2D Henon map representation.

Source: https://en.wikipedia.org/wiki/H%C3%A9non_map.

3.4.1   KEY SPACE

It is the total attempts required in determining the decryption key by doing an exhaustive search of all combinations of keys. Key space and key size are exponentially related. Increasing the key size, increases the key space of the algorithm many folds. As an example, for a key size of 128 bits, the key space would be 2128 even with the use of high performance computers. Therefore, for a cryptosystem can sustain a brute force attack with an exceptionally large key size.

3.4.2 KEY SENSITIVITY

The secret or encryption key used in an efficient image encryption approach must be sensitive to changes in the secret. For every minor modification in the encryption key, the encrypted and/or decrypted image is completely different.

3.4.3 STATISTICAL ANALYSIS

Investigation of image statistics exhibit connections linking the actual and the encrypted image. As a result, an efficient encryption technique requires the image being encrypted to differ from the actual image. Various methods are used to see if the encrypted image reveals any details about the actual image. Histogram analysis and Correlation co-efficient analysis make use of pixels within the actual and encrypted image as statistical analysis tools for judging the encryption algorithm (Sankpal & Vijaya, 2014). The distribution of pixels at different levels of intensity is represented by the histogram of an image. The statistical nature of the image may be understood from the image histogram. The encrypted and original image histograms are identical, indicating that the encryption algorithm is resistant to statistical attacks. Histogram of encrypted image needs to have steady distribution of pixels at all the intensity levels.

3.4.4 CORRELATION CO-EFFICIENT ANALYSIS

Another tool adopted for performing statistical analysis is the correlation coefficient (CC). It is an indicative measure of neighboring pixel correlation in all directions, i.e., horizontal, vertical, and diagonal of the actual image as well as encrypted image. For actual image, the neighboring pixels will have very high correlation in all directions with correlation value close to unity. Unlike to this, in encrypted images neighboring pixels are highly uncorrelated with correlation value nearly equal to zero indicating the good balance of confusion and diffusion process which enable in thwarting the statistical attacks. Correlation value close to unity of encrypted image implies failure of the algorithm. Correlation value close to zero implies the similarity between the actual and decrypted image.

$$\text{Cov}(x, y) = E(x - E(x))\, E(y - E(y))$$

$$r = \frac{cov(x, y)}{\sqrt{D(x)}\,\sqrt{D(y)}} \tag{3.3}$$

3.4.5   *INFORMATION ENTROPY ANALYSIS*

For access the ruggedness of the algorithm, additionally entropy analysis is adopted. Ruggedness of encryption algorithm refers to the sustaining ability of the algorithm for various attacks. This may be measured by performing entropy analysis of the encrypted image.

In general, a system that is truly random generates all symbols with equal probability. Original and encrypted image entropies are compared, and it should be close to 8 indicating stubbornness of algorithm to entropy attack.

$$H(m) = -\sum_{i=0}^{2^N-1} P(m_i) log_2[P(m_i)] \tag{3.4}$$

where; $P(m_i)$ is the probability of a pixel, and N is the bit-depth of each pixel.

3.4.6   *DIFFERENTIAL ANALYSIS*

It is an indicative measure of sensitiveness of the encryption algorithm for very minor changes in the input to the algorithm. If an attacker is able to observe drastic changes in the encrypted image by making a minuscule change to the algorithm input, then he will able to determine a valid link between the encrypted and original images. If the opponent is successful, then the algorithm is not sustainable to differential attacks. Differential attacks may be thwarted with tools such as: "number of pixels change rate (NPCR) and unified average changing intensity (UACI)."

The sensitiveness of the algorithm is tested using NPCR and UACI for a one-bit modification in the actual image. NPCR is a metric for difference in the pixels in total of two encrypted images when their plain images differ by one pixel. UCAI gives the measure of average pixel intensity of the number of pixel differences between two images of the same category, i.e., two same actual image or the two same encrypted images. With a high UACI value, even the tiniest modification in the actual image causes the encrypted image to exhibit significant modification.

In all, if the encryption algorithm exhibits high values of NPCR and UACI, then it can sustain any differential attacks.

3.5   IMAGE WATERMARKING

In the digital era, quick access to the internet and image processing, solitude, and security of knowledge are paramount concern. In maintaining the upcoming threats to digital information, Digital image watermarking provide aide in tamper resistance, ownership, content authentication and protection of information. When implementing a watermarking scheme, different requirements got to be considered to support the applications involved. The essential requirements of watermarking schemes are imperceptibility, payload, and robustness. These requirements play a pivotal function in evaluating watermarking system performance. Often these requirements are conflicting, and trade-offs between them are considered based on the watermarking system applications. A watermarking scheme is claimed to be effective when there's a balance between these requirements. Research literature during this area has discussion of various transformation techniques that satisfy requirements of watermarking system. One of the most preferred combinations for watermarking process is the "discrete wavelet transforms (DWT) and singular values decomposition (SVD)" (Begum & Uddin, 2020).

3.5.1   DISCRETE WAVELET TRANSFORMS (DWTS)

Discrete wavelet transform (DWT) is a wavelet transform that breaks down a signal into wavelets rather than frequencies. In DWT, wavelets are sampled discretely. In terms of temporal resolution, DWT outperforms another Fourier transforms such as "discrete cosine transform (DCT) and discrete Fourier transform (DFT)." DWT captures multiple information, such as time and frequency location. DWT-based image watermarking techniques break down the image into different frequency subband levels. The sub-bands frequencies cover the entire image spectrum (Begum & Uddin, 2020).

3.5.2   SINGULAR VALUE DECOMPOSITION (SVD)

Singular values are the useful tools of linear algebra that finds application in various signal processing operations including image watermarking. Singular values of any image (of the dimension, i.e., $m \times n$), is given as:

$$\text{Image} = U * S * V^T \qquad\qquad (3.5)$$

where; the orthogonal elements are represented by "U and V" and diagonal elements by "S."

Choice of SVD for image watermarking applications is especially because, they represent an outsized portion of the signal energy, are fairly resistant to noise, i.e., when a small uncertainty is added to the intensity of the image, there is no significant changes in the image singular values and theses values are often used with square and rectangular images. SVD based watermarking techniques robustness is found to be fairly good when utilized in combination with transform domain techniques (Akshya & Mehul, 2012).

3.6   PERFORMANCE PARAMETERS FOR IMAGE WATERMARKING

To measure the performance of watermarking algorithms, characteristics of the watermarked cover image (CI) is compared with the actual CI in terms of image degradation or distortions. This is can be done with various quality assessment metrics which broadly fall under two categories: Subjective and Objective evaluation metrics. Subjective evaluation involves measuring the variations in statistics between the original and watermarked CIs, in terms of visual impairments. Subjective evaluation does not have a standardized process. Pooja & Biju (2013) described the watermarking process as comprising of distortions and evaluating the same by physical observance. Observers rated the impairments on a five scale as: imperceptible (value 5), perceptible though not annoying (value 4), slightly annoying (scale value 3), annoying (scale value 2) and very annoying (scale value 1). Watermarking algorithms with scale value higher than 4, perform fairly with regard to robustness and imperceptibility. However, Subjective evaluation process is expensive, tedious, time taking and as a result, they are unsuitable for real-time applications. Objective evaluation process assesses the standard of the watermarking algorithms by using automatic tools such as: "peak signal to noise ratio (PSNR), mean square error (MSE), normalized correlation efficient (NCC), structural similarity (SSIM), universal quality index (UQI), komparator, reduced reference image quality assessment (RRIQA) and C4." The suggested nested watermarking system is evaluated using objective picture quality metrics, namely: PSNR, NCC, and SSIM in this section (Nguyen, Marie, & Azeddine, 2010).

3.6.1   MEAN SQUARED ERROR (MSE)

Error linking the CI and the watermarked image is accessed using MSE. It is defined as cumulative of the squared error of the images under consideration, as below. An efficient watermarking algorithm exhibits a very small value of MSE:

$$MSE = \frac{1}{NxM} \sum_{i=0}^{N-1} \sum_{j=0}^{M-1} [X(i,j) - Y(i,j)]^2 \tag{3.6}$$

where; "M and N" represent the rows and columns of the image.

3.6.2   PEAK SIGNAL-TO-NOISE RATIO (PSNR)

The PSNR along with MSE are collectively adopted for comparing the image compression quality after the embedding process. The PSNR is a measure of the image's PSNR. The greater the PSNR value for a particular image, the better the watermarked image and/or restored image quality.

$$PSNR = 10*\log(255*255/MSE) \qquad (3.7)$$

3.6.3   STRUCTURAL SIMILARITY INDEX (SSIM)

The virtue of watermarking algorithm may be measured using SSIM. It calculates the degree of resemblance between the cover and watermarked images. The similarity is calculated in terms of image characteristics such as: structure, contrast, and luminance.

$$SSIM(x,y) = \frac{(2\mu_x\mu_y + c_1)(2\sigma_{xy} + c_2)}{(\mu_x^2 + \mu_y^2 + c_1)(\sigma_x^2 + \sigma_y^2 + c_2)} \qquad (3.8)$$

where; μ_x is the average of x; μ_y is the average of y; σ_x^2 is the variance of x; σ_y^2 is the variance of y; σ_{xy}^2 is the covariance of x and y; $c_1 = (\kappa_1 L)^2$, $c_2 = (\kappa_2 L)^2$ two variables used for stabilizing the division for weak denominator; 'L' is the dynamic range of the pixel values; $\kappa_1 = 0.01$ and $\kappa_2 = 0.03$ by default (Priya & Vijaya, 2021).

3.7   IMPLEMENTATION OF IMAGE ENCRYPTION

The image encryption algorithm encrypts the actual image by means of chaotic sequence created using the Henon map. Figure 3.4 shows a 256-by-256-pixel DICOM grayscale picture, that is inputted to the algorithm. Henon chaotic map generated using the canonical values of $\alpha = 1.4$ and $\beta = 0$. The Chaotic sequences generated for two different iterations are as shown in Figure 3.5. The encryption function is applied to the actual image in conjunction with the Henon chaotic map. The final encrypted image is as shown in Figure 3.6. Performance parameters discussed above are evaluated. Histogram analysis of the actual image before encryption and after encryption give a measure of security against statistical attacks. Histogram of actual image give the total pixels distribution at each intensity level whereas for the histogram of encrypted image, the total pixels distribution is uniform, which is required to thwart the statistical attacks. Histogram of actual image as in Figure 3.7, gives the total pixels distribution at each intensity level where as for the histogram of encrypted image as in Figure 3.8, the total pixels distribution is uniform, which is required to thwart the statistical attacks. The histogram of decrypted image as in Figure 3.8, is similar to actual image indicating complete decryption of the actual image as shown in the Figure 3.7.

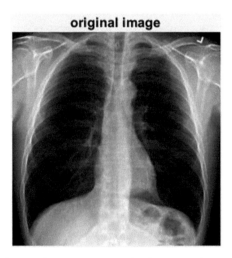

FIGURE 3.4   Actual image of chest X-ray used for demonstrating encryption.

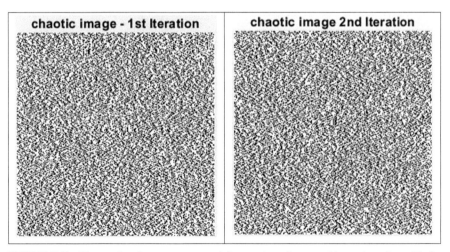

FIGURE 3.5   Henon chaotic maps generated for two iterations used in conjunction with the actual image.

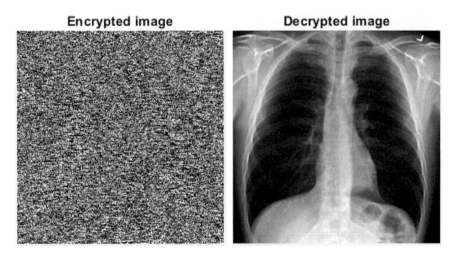

FIGURE 3.6   Encrypted and decrypted actual image using chaotic maps.

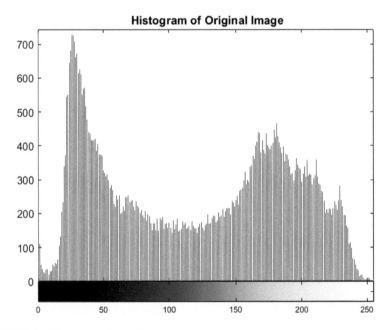

FIGURE 3.7   Histogram of actual image.

CC analysis of actual image and encrypted image is done to thwart the statistical attacks. Correlation values are considered in all three directions, i.e., horizontal, vertical, and diagonal.

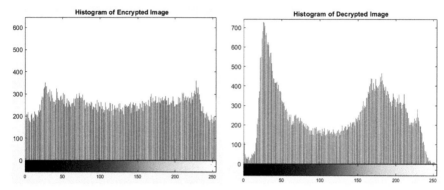

FIGURE 3.8 Histogram of actual image after encryption and decryption process.

Actual image correlation value close to unity indicates highly connected pixel values, whereas the correlation values of encrypted images indicate no correlation between any pixel values. For the implemented encryption algorithm, NPCR and UACI values that are higher are obtained, as indicated in Table 3.1. This infers that the implemented algorithm is able to sustain differential attacks. Experimental correlation values of both actual and encrypted image are shown in Table 3.2.

TABLE 3.1 Differential Attack Values and Information Entropy

Performance Parameter	Obtained Value
NPCR (%)	85.58807
UACI (%)	31.14382
Elapsed time (sec)	18.40824

TABLE 3.2 Values of Correlation Coefficient for Actual and Encrypted Image in All Directions

Pixel Position	Actual Image	Encrypted Image
Horizontal	0.99129	0.20878
Vertical	0.99310	−0.00323
Diagonal	0.98549	−0.00037
Entropy analysis	7.73614	7.98644

3.8 IMPLEMENTATION OF IMAGE WATERMARKING

Literature work in the area of watermarking describes many transformation techniques for image watermarking. Recent developments in this field show that a combination of "DWT and singular value decomposition (SVD)" have been the most preferred one (Ganic & Ahmet, 2004). Image watermarking was accomplished with the use of "Discrete Haar wavelet transforms and SVD." The process of making communication secure over open networks is achieved by adopting fusion of image encryption and image watermarking techniques. Image encryption is done prior to embedding it into the CI and then performing image watermarking.

Input to fusion algorithm are the CI and WM. Image database used for evaluating the performance of the fusion algorithm comprised of 20 different "digital imaging and communications in medicine (DICOM)" images with dimension of 256 by 256 in JPEG format. Both the CI and the WM are of the same patient with different viewing angles. Figure 3.9 shows the DICOM images used as CI and the WMs.

Cover Image for watermarking

Watermark

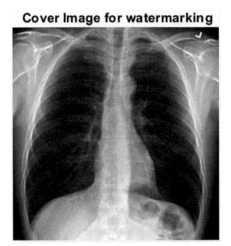
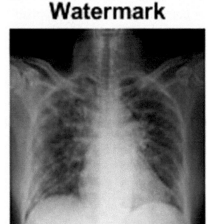

FIGURE 3.9	DICOM cover image and watermark image as input to watermarking algorithms.

3.8.1 ALGORITHM: WATERMARK EMBEDDING

1. Read the cover image cover image.
2. Apply DWT to cover image. Decomposition results in LL, HL, LH, HH bands.
3. SVD is applied to the LL band of the cover image.

$$LL = Uc* Sc* Vc^T$$

4. Read the watermark image for embedding into the cover image.

5. Encrypt the watermark using Henon chaotic map.
6. Compute performance parameters of image encryption
7. Compute the SVD of watermark image:

$$Wml = Uwl * Swl * Vwl^T$$

8. Singular values of cover image LL band are superseded by the singular values of the watermark image.
9. Chose the appropriate embedding strength.
10. Compute watermarking bits for cover image as:

$$Smarkl = SVD\ (LL) + Embedding\ strength\ *SVD\ (Wml)$$

11. Rebuild LL band of cover image by recomputing singular values as

$$LL\_wl = Uc* Smarkl * Vc^T$$

12. Inverse DWT is applied to obtain the watermarked image of cover image
13. Compute performance parameters of image watermarking.

3.8.2 ALGORITHM: WATERMARK EXTRACTION

1. Read the watermarked cover image.
2. Apply inverse DWT to it. Decomposition results in LL, HL, LH, HH bands.
3. Compute Singular values of watermarked image.
4. Compute watermarked recovery bits, using singular values of cover image and WM2 as:

$$Swrec = (SVD\ (Watermarked\ image) - SVD\ (Cover\ image))/\ Embedding\ strength$$

5. Extract Watermarked image from the watermarked cover image, by computing singular values as:

$$WM = Uw * Swrec * Vw^T$$

3.9 FUSION IMPLEMENTATION OF IMAGE WATERMARKING AND ENCRYPTION

By applying Haar DWT to the CI, the WM is embedded or watermarked into the CI. Both the cover and WMs are broken down into their four constituent parts of various frequencies as: LL, HL, LH, and HH. The different wave-band s is termed as low frequency band (LL) which provides approximate details of the image, mid frequency bands (HL and LH) give the horizontal and vertical attributes of an image. High frequency band (HH) provides the image diagonal details. Singular values of both the CI and WM are computed for all the four bands. Singular values of the WM are used for modifying the singular values of the CI during the embedding process. The WM is encrypted before it is embedded into the CI. Figure 3.10 shows a flow diagram of how the fusion technique is implemented. The process of encryption enhances the confidentiality and watermark enhances integrity and authenticity. The encrypted WM as shown in Figure 3.11, is embedded into the LL frequency band using SVD.

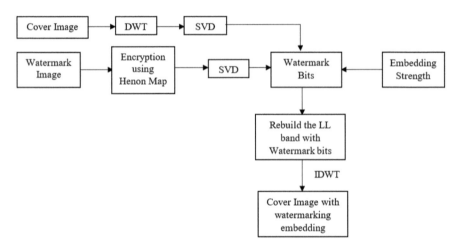

FIGURE 3.10 Flow diagram of fusion algorithm implementation.

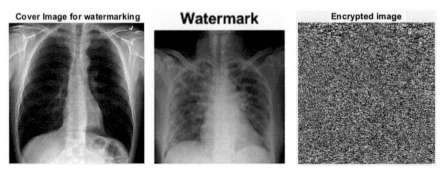

FIGURE 3.11 Cover image for embedding encrypted watermark image.

CI and encrypted WM computed singular values were used alongside a multiplicative factor called the embedding strength in determining the watermarking bits to be embedded into the CI. Embedded strength plays a vital part in deciding the watermarked image's imperceptibility. Embedding strength is chosen in such a way to maintain a balance between the imperceptibility and PSNR. The LL band of the CI is rebuilt with the modified SVD bits consisting of CI and WM SVD bits. When inverse DWT was applied, it results in watermarking the encrypted image in the main CI. WM is preprocessed to possess singular values that match with those of the cover or the target image. For the removal or extraction of the watermark from the CI, again DWT and SVD is employed, as discussed within the algorithms above (Priya & Vijaya, 2021) (Figures 3.12 and 3.13).

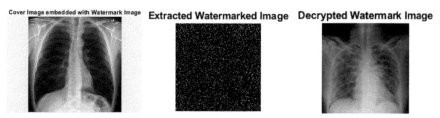

FIGURE 3.12 Embedded cover image with encrypted watermark image; extracted watermark image that is encrypted; and final watermark image after decryption.

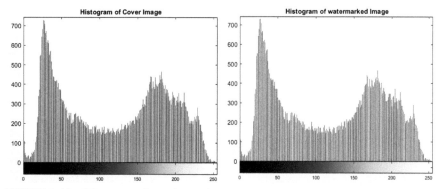

FIGURE 3.13 Histogram of actual cover image and embedded cover image with encrypted watermark image.

Variation of embedding strength gives the amount of imperceptibility associated with the watermarked CI as depicted in Table 3.3. Smaller the embedding strength, higher the PSNR value. As the embedding strength increases, watermarked CI quality is degraded, and the embedded watermark becomes visible.

TABLE 3.3   Performance of Watermarking Scheme for Different Values of Embedding Strength

Embedding Strength	PSNR
0.1	32.4198
0.01	40.9832
0.001	45.3561

3.10   CONCLUSION

Security contravention has been of paramount aspect in designing security measures for transmitting vital information over open networks. When information is being shared over open networks, integrity, and confidentiality have to be maintained. Fusion of security mechanisms such as encryption and watermarking were explored for protection of vital information. In the implemented algorithm, one image is embedded into another image. Embedded image is encrypted to provide additional security. Fusion algorithm implementation was discussed with performance parameters used for evaluating the same. Implemented fusion algorithm was able to sustain statistical and differential attacks with 86% NPCR and 31% UACI. The watermarking also exhibited good imperceptibility. In the future, this algorithm may be extended for multiple watermarks being embedded into a single CI.

ACKNOWLEDGMENT

The authors would like to thank all the people involved in the earlier works on image encryption and watermarking, which kindled enthusiasm to carry out the work done in this chapter. All the work done and images used in this chapter are for educational purpose and not for commercial purpose. We express our heartfelt gratitude to the college management of BNM Institute of Technology, Bengaluru for enabling us to carry out the research work with all the resources and subsequent publication. Also, we express our sincere gratitude to Visvesvaraya Technological Institute (VTU), Belagavi for providing us the research platform.

KEYWORDS

- **confidentiality, integrity, and availability**
- **digital imaging**
- **discrete wavelet transform**
- **image encryption**
- **noise ratio**
- **open networks**
- **singular value decomposition**
- **watermarks**

REFERENCES

Abhinav, S., (2012). A survey report on different techniques of image encryption. *International Journal of Emerging Technology and Advanced Engineering* (Vol. 2, No. 6). ISSN 2250-2459.

Akshya, K. G., & Mehul, S. R., (2012). A robust and secure watermarking scheme based on singular values replacement. *Sadhana, 37*(Part 4), 425–440. Indian Academy of Sciences.

Ali, S., Zulkarnain, Md. A., & Md Jan, N., (2012). *A Survey on Principal Aspects of Secure Image Transmission* (Vol. 66, pp. 247–254). World academy of science, Engineering and Technology.

Alireza, J., & Abdolrasoul, M. (2010). An image encryption approach using chaos and stream cipher. *Journal of Theoretical and Applied Information Technology*, 117–123.

Anto, S., & Dipesh, S., (2013). Modified algorithm of encryption and decryption of images using chaotic mapping. *International Journal of Science and Research (IJSR)* (Vol. 2, No. 2). India Online ISSN: 2319-7064.

Begum, M., & Uddin, M. S., (2020). Digital image watermarking techniques: A review. *Information, 11*(2), 110. doi.org/10.3390/info11020110.

Chattopadhyay, D., Mandal, M. K., & Nandi, D., (2011). Symmetric key chaotic image encryption using circle map. *Indian Journal of Science and Technology, 4*(5), 593–599. ISSN: 0974-6846.

Chong, F., Jun-Jie, C., Hao, Z., Wei-Hong, M., Yong-Feng, Z., & Ya-Wen, (2012). *A Chaos-Based Digital Image Encryption Scheme with an Improved Diffusion Strategy* (Vol. 20, No. 3, pp. 2363–2378). Optical Society of America.

Enrico, M., Florent, A., Patrick Le, C., & Patrizio, C., (2007). Evaluation of standard watermarking techniques. *Electronic Imaging, Security, Steganography, and Watermarking of Multimedia Contents* (pp. 6505–6524). San Jose, United States. hal-00250682.

Ganic, E., & Ahmet, E. M., (2004). Robust DWT-SVD domain image watermarking: Embedding data in all frequencies. *Proceedings of the Workshop on Multimedia and Security*, 166–174.

https://en.wikipedia.org/wiki/H%C3%A9non_map (accessed on 10 January 2022).

Kamlesh, G., & Sanjay, S., (2011). New approach for fast color image encryption using chaotic map. *Journal of Information Security, 2*, 139–150.

Monisha, S., & Manoj, K. K., (2010). Image encryption techniques using chaotic schemes: A review. *International Journal of Engineering Science and Technology, 2*(6), 2359–2363.

Nguyen, P. B., Marie, L., & Azeddine, B., (2010). Statistical analysis of image quality metrics for watermark transparency assessment. *Lecture Notes in Computer Science*, 685–696. doi: 10.1007/978-3-642-15702-8_63.

Pooja, M., & Biju, T., (2013). A survey on various encryption and key selection techniques. *International Journal of Engineering and Innovative Technology (IJEIT)* (Vol. 2, No. 7). ISSN: 2277-3754.

Priya, R. S., & Vijaya, P. A., (2021). Performance evaluation of nested watermarked scheme using objective image quality metrics. *Journal of University of Shanghai for Science and Technology, 23*(6), 306–314. ISSN:1007-6735.

Sakthidasan, S. K., & Santhosh, K. B. V., (2011). A new chaotic algorithm for image encryption and decryption of digital color images. *International Journal of Information and Education Technology, 1*(2).

Sankpal, P. R., & Vijaya, P. A., (2014). Image encryption using chaotic maps: A survey. In: *2014 Fifth International Conference on Signal and Image Processing* (pp. 102–107). doi: 10.1109/ICSIP.2014.80.

Shima, R. M., & Supriya, M., (2011). An uncompressed image encryption algorithm based on DNA Sequences. *Computer Science & Information Technology (CS & IT)* (pp. 258–270). CCSEA, CS & IT 02.

Silpa, K., & Aruna, M. S., (2012). Comparison of image quality metrics. *International Journal of Engineering Research & Technology (IJERT), 1*(4). ISSN: 2278-0181.

Somaya, A. M., Afnan, A. A. A., & Turki, A. (2012). A new chaos-based image-encryption and compression algorithm. Hindawi Publishing Corporation. *Journal of Electrical and Computer Engineering* (Vol. 2012). Article ID 179693.

Wang, Z., Bovik, A., & Sheikh, H., (2005). *Structural Similarity Based Image Quality Assessment.* Digital video image quality and perceptual coding, Ser. Series in Signal Processing and Communications. 10.1201/9781420027822.ch7.

CHAPTER 4

Wearable Devices Forensic Investigation

DEEPIKA DUBEY[1] and RICHA ROHATGI[2]

*[1]Research Scholar, Department of Forensic Science,
Amity School of Applied Sciences, Amity University, Haryana, India*

*[2]Assistant Professor, Forensic Science, LNJN NICFS, National Forensic
Science University, Delhi Campus, Rohini, New Delhi, India*

ABSTRACT

Wearable devices have become an integral part of our daily lives, and their adoption is increasing rapidly. Considering the importance and potential of these types of devices as evidence, the authors attempted to present a comprehensive sketch of wearable device forensics and its future directions. The chapter provides a comprehensive examination of the ways in which wearable technology is changing our daily lives and its potential implications for forensic investigations. It delves deep into the advantages of wearable devices for various user groups, including busy professionals, fitness enthusiasts, and the elderly and infirm. The chapter also provides an in-depth analysis of the types of data collected by these devices and their significance in forensic investigations. The various types of wearable devices available, including those worn as accessories, embedded in clothing, and even implanted in the human body are also covered in the chapter. The design principles, challenges, and opportunities of wearable device forensics, the admissibility of wearable devices as digital evidence, the digital forensics investigation process, and data extraction methods are also discussed in the chapter. The chapter concludes with a summary of key findings and recommendations for future research in wearable device forensics, providing a comprehensive and

Advancements in Cybercrime Investigation and Digital Forensics. A. Harisha, Amarnath Mishra, & Chandra Singh (Eds.)
© 2024 Apple Academic Press, Inc. Co-published with CRC Press (Taylor & Francis)

meaningful understanding of the technical, legal, and operational challenges and opportunities presented by this rapidly evolving technology.

4.1 INTRODUCTION

The adoption of wearable devices is increasing widely, and it is undoubtedly becoming an integral part of our lives. Wearable devices are revolutionizing the way we track our daily life activities. Such type of technology is very advantageous for busy professionals to stay informed with notifications while on the go, fitness enthusiasts track their workouts, and even the elderly and infirm by alerting emergency personnel if the device senses an unpleasant medical event. However, in order for these gadgets to function properly, they must collect information about the wearer, such as their location, biometric data, social media, calendar, and other information.

The geographical information, user's physical and health statistics, different activity logs (e.g., sleeping, walking, calories burnt, etc.), and other sensitive information such as account details of social media communication and notifications of instant messaging applications, media files, etc., that are collected by these devices can play a huge role in forensic investigations (Kasukurti & Patil, 2019). The devices are worn as accessories, embedded in clothing, and might even be implanted in human bodies and are a new emerging class of technology. These devices incorporate sensors, operating systems (OS), communications interfaces and numerous software applications that provide rich features like human-biometric monitoring, location tracking, video recording, navigation, calling, messaging, secretarial tasks, cardless payments and music streaming, among others. Such devices need to be connected to either a Wi-Fi or a cellular data network so that the recorded information can be synchronized with the cloud. The types of sensitive user information that may possibly be extracted from wearable devices are expected to play an important role in the future of digital forensics. The users have a close connection to the wearable devices which is continuously keeping track of what the user is doing, with large amounts of personally identifiable data being stored to a device to which it is relayed after being generated.

4.1.1 WEARABLE DEVICE TECHNOLOGY

Wearable device technology is a class of electronic device, which is used to track, analyze, and transmit personal data. These devices are physically carried as accessories, fixed somewhere_in the body wear, implanted in the users' body or sometimes may even be tattooed on the skin in order. Such smart technology can track heart rates, sleep patterns, calls, e-mails, messages, etc. Some of the most commonly used wearable devices are– Apple Watches, Fitbit, Samsung Galaxy Watch, Garmin Activity Tracker, Huawei Watch, etc. (Hayes, 2020).

These wearable devices usually rely on a short-range wireless technology such as Bluetooth, Wi-Fi or a login account to pair with a companion device. During setup, a wearable device is registered with the manufacturer and may be linked to a cloud server account, which is used to back up the user's configuration, on board sensor data and logged activities. A smartphone application may also be used to manipulate the local configuration settings of the wearable device, update its firmware, load additional features, synchronize data with the cloud and visualize the collected data.

There is a wide range of wearable devices. However, the devices characteristically have similar engineering design constraints due to their small form factors. Without the assistance of paired devices, they may be unable to operate for long hours, store large quantities of data or provide rich applications and features. To mitigate these downsides, manufacturers design wearables with custom chips that consume low amounts of energy, and their sensors and applications are usually put in a standby mode when they are not needed. Storage capacity limitations are usually resolved by compressing and offloading data to cloud servers or paired smartphones. Strong transport protocols are engaged to increase the synchronization speed of the data. Some wearable devices even slot in cellular chipsets for increased mobility and performance (Loomis, 2019).

Smart glasses, fitness trackers and smartwatches collect large amounts of user data and store the data across the wearable devices, companion smart-phones and cloud storage. Since these devices collect data continuously, they contain considerable amounts of digital evidence that are valuable in forensic investigations. A wearable OS is designed to work seamlessly across various wearable devices. It varies based on the type of wearable device and its specifications. These wearable devices are typically operated using two predominant operational methods. Some wearable devices operate unaided in an autonomous mode, whereas the others typically require internet-connected smartphones.

For instance, most smartwatches have built-in features that allow them to connect to the user's smartphone to create a web interface. In such cases, the wearable OS acts similar to a mobile OS. This OS is designed to enable the smartwatch to provide a user interface. The watch is essentially a wireless Bluetooth adaptor that allows the smartphone's capabilities to be transferred to the smartwatch (Rowland, Goodman, Charlier, Light, & Lui, 2015). It facilitates the user in performing various daily tasks with ease and comfort such as managing calls, check, and dictate e-mail and text messages, get real-time weather and sports reports, listen to a song or speak to a digital assistant (Hayes, 2020; Tech Target Contributor, 2015) (Figure 4.1).

FIGURE 4.1   Some examples of wearable devices.

4.1.2   DESIGN PRINCIPLE OF WEARABLE DEVICES

The design of wearable devices is a complicated task. It is because of the various types of sensors working within an essentially interconnected system. Adebayo (2020) classified the sensors into three categories:

1. **Biosensors:** These comprise glucose, blood pressure, ECG, EMG, temperature, and brain wave sensors.
2. **Motion Sensors:** Examples include acceleration, gyroscope, geomagnetic, and atmospheric pressure sensors.
3. **Environmental Sensors:** For example, temperature and humidity, gas, PH, ultraviolet, ambient light, dust particles, and pressure sensors, as well as microphones (Adebayo, 2020).

The data to the CPU or display processor is sent by sensor. These sensors are coupled with a microcontroller, which is the most important component that facilitates the operation of the wearable technology. Microcontroller is a common component in small computers that enables the integration of the internet of things (IoT) with desired devices and applications. Most importantly, they help in performing different functions on a single component. Due to the ease in programming, low cost, small size, great connectivity with other sensors and the competency to handle complex functions, including graphic displays, microcontrollers are almost a must-use in wearable technology. A typical architecture of a device includes hardware to hold all the components, an allied applications running in the users' phone and a cloud server, apart from the wearable device itself (Aroganam, Manivannan, & Harrison, 2019).

Figure 4.2 depicts how fitness wearable technology could be used to track various lifestyle activities such as weight, calories burned, heart rate, GPS location, and so on (Chan et al., 2012). Sensors monitor the lifestyle activities of the users. The users input some of the data themselves, such as their food intake. These datasets are then conveyed to a smart phone and the cloud service. The data is then processed, making it suitable so that the user can understand it. Depending on the type of display, this processed data is then sent back to the paired smart mobile device or the wearable itself (Aroganam, Manivannan, & Harrison, 2019). Apart from the architectural design, the other design challenges that are important in the context of a wearable device are:

1. **Network Connectivity:** The architectural design of systems described above clearly shows that connectivity is the dominant feature in terms of connection of system elements with each other via different networks (Ngu et al., 2008; Stelvaga & Fortin, 2017).

2. **Limited Resources:** To make the device compact and to be carried it comfortably, several restrictions are obligatory such as, and a wearable device is created lightweight so as the user can easily use it. Secondly, the overall size is relatively small. The functions including computing capability, powering sensors and network connectivity suffer from the lack of energy resources and the design is highly constrained by this requirement (Nirjon, Gummeson, Gelb, & Kyu-Han, 2015; Stelvaga & Fortin, 2017).

3. **Asynchrony:** As stated by Rowland, for smartphones and desktops, it is attainable to handle network outages graciously and assume that the flow of interactions will be moderately smooth. However, with wearables, the entire system becomes primarily asynchronous. A wearable device cannot maintain continuous connectivity due to the limited amount of power available. It can, however, only be used for

intermittent connectivity. For example, the wearable device can be linked to a phone using a low-power protocol, such as Bluetooth low energy (Rowland, Goodman, Charlier, Light, & Lui, 2015; Stelvaga & Fortin, 2017).

4. **Sustainability:** According to Lee et al., the conception of wearable device is that an individual ought to carry it for extended periods of time. The device should even be resistant to the jolting, possible physical damage, and environmental changes. These factors, for example, have a significant impact on sensor readings, as do factors such as humidity, which can result in an increased skin conductivity signal on the output of a galvanic skin response sensor (Lee, Kim, Ryoo, & Shin, 2016; Stelvaga & Fortin, 2017).

5. **Convenience and Ergonomics:** As the wearable devices are worn directly onto the body of the user, it should be comfortable to the wearer. This requirement limits the types of materials, size, etc., that can be used in the product design (Ferraroa, 2011).

6. **Service Focus:** According to Chan et al., users frequently tend to discern the wearable as a distinct object. However, the device behavior is often influenced by a software algorithm that is deployed somewhere elsewhere on the network. This implies that the service designed around the wearable is simply often more critical than the device itself in delivering the proper user experience (Chan et al., 2012; Stelvaga & Fortin, 2017).

FIGURE 4.2   Image depicts the utility and usability of wearable devices in day-to-day life.

4.1.3   *WEARABLE DEVICES FORENSICS*

Wearable device forensics focuses on the data recovery and its analysis from various types of wearable devices. The goal of the investigation process is to extract and analyze the data that is stored on a wearable device without altering/ modifying the data present on the device in a manner that is acceptable by the law (Loomis, 2019). Although the concept of wearable device forensics has been around for a while, but it is still in its infancy. Due to the wide variety of devices and their varying characteristics, forensic investigations are still challenging.

Potential information of evidentiary value that can be recovered from wearable devices include:

- Global positioning system (GPS);
- Heart rate;
- Biological data;
- Near field communication (NFC) devices;
- Bluetooth information;
- e-mail;
- CDR (call detail records);
- Messages;
- Application data, etc.

4.2 BACKGROUND AND RELATED WORK

Digital forensics is a branch of forensic science that deals with the type of investigation that is conducted when evidence of an event must to be found. It can be performed in various ways, such as by analyzing a particular time of day or determining if a particular individual has committed a crime. Because of the importance of the investigation, digital forensics has grown into a thriving academic field. It entails gathering and analyzing digital evidence in order to answer questions about a case. In conventional digital forensics, data is acquired in the order of volatility, i.e., from the most volatile to the least volatile. Whereas data stored on the secondary storage devices have less possibilities to be corrupted or lost and may be acquired and recovered later. Digital forensics tools such as write blockers are used to prevent any alterations or modifications being made to the original evidence connected to the forensic workstation (Yoon & Karabiyik, 2020).

Digital forensics research also scrutinizes the development of new models and procedures for investigation, which raises new challenges and gaps in existing ones. Various studies in digital forensics have attempted to refine the way that forensic evidence is stored. Currently, most software platforms such as EnCase, Cellebrite UFED or AccessData's FTK are able to store digital copies of devices in various formats. The way these formats are structured can affect the speed at which they are processed and the information they contain. Various analytical and investigative models have been developed time and again to provide the investigators and law enforcement agencies with a clear and concise workflow in order to conduct digital forensics investigations with the least amount of error.

The wearable devices are inextricably conjugated with mobile device forensics because these mobile devices typically use an application that manages and reports data collected by the wearable devices. There are several studies already been conducted that focuses on formulating models for mobile forensic investigations depending on how and where data is stored by various applications. Numerous studies have also been conducted on the security of various application stores for smartphones. These studies were conducted in response to the growing number of security issues that are affecting the operation of these apps. Due to the nature of these applications, forensic examiners have to deal with the storage and analysis of data in order to extract usable data. For instance, social media applications like Facebook have to store user data in a way that is difficult to extract. Due to the pervasiveness of iOS devices, forensic investigations had been performed on them. Previous research found out where and how important portions of records are saved on Apple mobile devices. Data from Facebook and Skype have additionally been analyzed to discover wherein crucial information inclusive of contacts and login occasions are saved.

Various studies on wearable tracking devices including Fitbit, Smart-watches, etc., have provided us with a framework for safely extracting data from such devices. Although the available tools are still robust and have been approved by authorities, they are not yet suitable for extracting data directly from the wearable devices as the data extraction from the memory of such devices is a complex procedure and requires necessary equipment's and skills (Yoon & Karabiyik, 2020).

4.3 ADMISSIBILITY OF WEARABLE DEVICES AS DIGITAL EVIDENCE

Digital evidence is evidence that is collected and transmitted in binary form. This includes data such as the timestamp, connection type, IP addresses, user profiles, etc. It includes all such information that has evidential value and can help the investigators in pinning down the culprit. Every technology innovation has some risks associated with it. These wearable devices are no exception. There are several challenges associated with wearable devices, particularly when the court of law comes into the picture. Although the value of evidence obtained from these devices has been identified as potential evidence in several cases previously. Most jurisdictions have specific legal requirements for the admissibility of digital evidence in legal proceedings. For example, Sections 65A and 65B of the Indian Evidence Act (IEA) establish the criterion for the admissibility of electronic records as digital evidence in courts. These requirements equally apply to all sources of digital evidence including evidence extracted from wearable devices (Khairallah, 2018). The requirements included are:

1. **Admissible:** For the evidence to be admissible, it must be related to the fact being proved.
2. **Authentic:** For the evidence to be admissible, it should be proven with none doubt that the evidence comes from the claimed source. Moreover, it must be real and concerning the incident directly.
3. **Complete:** The evidence must be complete, which means it must either prove or disprove the consensual fact in the litigation.
4. **Reliable:** To prove the evidence as admissible, there must be no doubt about the authenticity or veracity of the evidence. The evidence should be handled carefully while maintaining the chain of custody record.
5. **Believable:** The submitted evidence must be clear, comprehensible, and easily understandable by the judges (Khairallah, 2018; EC-Council's Computer Hacking Forensic Investigator v9, Module 1).

4.4 METHODOLOGY

4.4.1 DIGITAL FORENSICS INVESTIGATION PROCESS AND ITS IMPORTANCE

The digital forensics investigation process is a set of methodological approaches for preparing for digital evidence investigation including identification, collection, and analysis (Karie & Venter, 2015), and managing the case right from the time of reporting to the conclusion. Forensic investigations are conducted in various circumstances, such as, but not limited to, establishing proof for legitimate procedures, and investigating security arrangement infringement. There are a variety of strategies and devices utilized for this type of investigation. However, due to day-to-day advancements in the digital world, acquiring, analyzing, and preserving information in a forensic manner becomes very challenging. This is especially so when it comes to extracting data for a trial. There are various proposed models for digital forensics investigations but in most of the cases, the model including three investigative phases namely, pre-investigation phase, investigation phase and post-investigation phase is utilized (EC-Council's Computer Hacking Forensic Investigator v9, Module 4).

4.4.2   *PROCESS OF WEARABLE DEVICE FORENSIC INVESTIGATIONS*

Generally, the wearable device forensic examination goes through the following three investigative phases. This investigative process ensures that the steps are repeatable, defensible, reliable, and forensically sound. The step-by-step description of investigative phases are detailed in subsections (Figure 4.3).

FIGURE 4.3   Image depicts phases of investigation involved in wearable device forensic investigation.

4.4.2.1   PRE-INVESTIGATION PHASE

The pre-investigation phase deals with the preparations before the commencement of the actual investigation process. The investigators cannot jump into action immediately after receiving a complaint or report of a security incident, but they have to follow a specific protocol that includes gathering of essential information about the incident, e.g., type of incident and obtaining permissions and warrants for taking further action. It consists of building a forensic workstation, the wearable devices investigation toolkit, the investigation team, getting necessary approvals and consent from the relevant authorities and stakeholders. It includes the following steps:

1. **Preparation:** Completion of the proper intake paperwork as well as the chain of custody log.
2. **Reception:** Receiving the wearable device for the forensic examination; identifying the device (e.g., make/model and serial number); recording the presence of paired and companion devices.
3. **Evaluation:** Setting the goals of the examination; document the physical and operating state of the wearable device.

4.4.2.2 INVESTIGATION PHASE

The investigation phase involves analyzing and securing the evidence collected by various means, such as by acquiring and preserving the data collected from wearable digital devices. Each step in this process is crucial for acceptance of the evidence in a court of law and prosecution of the perpetrators. It includes the implementation of the technical skills and expertise to identify and inspect the evidence, examine, document, and preserve the findings as well as the evidence by professionals trained to analyze wearable digital devices upholding the quality and integrity of the findings to the highest level.

Some noteworthy points to consider before conducting the investigation are as follows:

1. **Examination/Investigation Goals:**
    *   Before an investigator embarks on an examination, they should have a clear idea of the goals of the investigation.
    *   They should have an in-depth technical understanding about the inner workings of what is being examined.
    *   The investigators should have the capability to take a systematic approach to examine evidence based on the request made.
2. **Hypothesis Formulation:** In order to successfully carry out a case, the investigator must formulate a hypothesis that will prove or disprove the main ideas of the case.
3. **Experimental Design:**
    *   The investigators should first draw up their hypothesis before carrying out the experiment. They should then test the methodology as well as the test results.
    *   The test system should have an environment like that of the suspect machine to yield accurate results.

4. Tool Selection:
   - Each case is different, and the tools used for it will vary depending on the platform and OS used.
   - Digital forensics tools can be: hardware or software, commercial or open source and designed for specific purposes or with broader functionality.
   - No single tool is all-inclusive thus; it is suggested to have multiple tools at hand (EC-Council's Computer Hacking Forensic Investigator v9, Module 4).

The investigation phase includes the following steps:

1. **Identification:** Identifying the potential data that is of probative value and the acquisition method(s) supported by the wearable device as well as the paired and the companion devices; determining the forensic equipment and tools that are required/supported.
2. **Documentation:** Documenting each detail about the digital crime scene and evidence is necessary to maintain a record of all the forensic investigation process applied to identify, acquire, analyze, and preserve the evidence. The details should include the identification information including make, model, serial numbers of the digital evidence, status of the evidence, type of cable/interface needed to connect it with forensic workstation, etc. It should also include the different steps involved in the forensic investigation. The commonly used methods for documentation include photography, videography, and notetaking.
3. **Acquisition:** The investigation and analysis process can have both positive and negative impact over the evidential data, and sometimes these processes can alter this data in such a way that it is no longer acceptable in the court. Therefore, the investigators should make copies of the evidence and work on it to prevent any damage to the original data in cases of accidents or mishaps. Data acquisition includes bit-by-bit copying of the original evidence using a software or hardware tool. The acquired data should be an exact copy of the original evidence. The common digital forensic used for this purpose include Cellebrite UFED, EnCase, FTK, Magnet Axiom, etc.
4. **Verification:** Or validation of digital evidence is one of the most important aspects of digital forensics. It is essential to ensure that the evidence is not altered or modified. Validating the digital evidence requires hashing algorithm utility, which was developed to create

a binary or hexadecimal number, called digital fingerprints, which represents the uniqueness of a file or acquired image (Panhalkar, n.d.). Examples of hash values include MD5, SHA-1, SHA-256, CRC-32, etc.

5. **Preservation:** It entails securely storing the forensic data as read-only, as well as maintaining duplicate copies across multiple systems and storage media.

6. **Data Filtering and Examination:** Data filtering is a technique that enables the investigator to sort through all of the information extracted. It is utilized to pick a portion of it that investigator would like to review or examine. Moreover, the examination involves the assessment of the reliability and significance of the data, ascertaining if the data meets the goals and the requirements of the examination.

7. **Analysis:** It includes evaluation of the examined data for relevance as evidence; understanding, structuring, relating, and reasoning about the data from a holistic perspective. The evidence is evaluated based on the type of incident, the objectives required to be performed and fulfilled, the loopholes discrepancies that exist for incident occurrence, etc. (Panhalkar, n.d.).

8. **Presentation:** Includes preparing and formatting the analyzed evidence.

4.4.2.3 POST-INVESTIGATION PHASE

The post-investigation phase includes the preparation of reports that details all the actions undertaken and findings during the course of the investigation. It is important that the report is legally sound, easy to understand and meets the requirements of the applicable jurisdiction. It includes the following steps:

1. **Reporting:** Documenting the examiner, date/time, processes/tools, analysis, and the results and findings.

2. **Documentation:** Compiling all the examination deliverables (e.g., reports, videos, presentations, and exhibits) in legally acceptable formats.

3. **Dissemination:** Distribution of evidence, case reports and deliverables to the authorities through forensically sound and secure channels (Loomis, 2019; EC-Council's Computer Hacking Forensic Investigator v9, Module 4).

4.5   DATA EXTRACTION METHODOLOGY

We can extract the data from wearable devices using one of the methods which are discussed in succeeding subsections.

4.5.1   MANUAL DATA EXTRACTION

Manual data collection involves navigating a wearable device via its LCD display to inspect the active user content visually, generally capturing screenshots of the) display. The content may include pictures, call logs, messages, fitness activities and other user-accessible data. Manual data collection methods are limited and time-consuming in the amount of data they can provide. Unintentional modifications of the system and user data may occur during manual data collection, e.g., inadvertently deleting files, modifying timestamps, update metadata, etc. Manual data collection should be performed only when more sophisticated collection methods are not available.

4.5.2   LOGICAL EXTRACTION

Logical data collection involves taking out data using interfaces such as USB or wireless in an automated manner in the original format. However, if the device encodes the data, encoded data is extracted. The collected data is often stored in a readable format in the forensic examination machine, allowing for immediate analysis. Additionally, a wearable device may have an associated SDK (software development kit) which can be installed on a forensic examination workstation. The SDK provides manufacturer-level access to the hardware and software of the device via specialized commands.

4.5.3   PHYSICAL EXTRACTION

Physical data extraction method involves the acquisition of a bit-by-bit copy of wearable devices storage using software and hardware techniques. The extracted data is stored in a bit stream image file, which may comprise of more than one nested system of files which can be mounted on a forensic examination tool for subsequent analysis. Other bit stream images must be processed using advanced data reconstruction techniques to obtain individual data files, deleted data and fragments. A data reconstruction technique searches for different file signatures. Once the file signatures are recognized, the associated data is extracted from the bit stream image to produce separate files with their corresponding file extensions. After processing the entire binary image for active, deleted, and fragmented data, the results may be analyzed by a forensic examiner (Loomis, 2019).

4.5.4   *INVASIVE DATA EXTRACTION*

Invasive data collection involves exposing the internal hardware components of a wearable device to enable access to the data and is only used when other methods of data collection are ineffective. Such techniques may also be applied to severely damaged or inoperable devices or to correlate the data retrieved using other data collection techniques. Invasive data collection techniques require specialized equipment and training, which may involve soldering wires, removing memory chips, repairing damaged hardware and applying cutting-edge computer engineering methods. Invasive data collection enables the subsequent use of logical and/or physical data collection techniques. An invasive data collection technique is time-consuming, but it provides access to data that is otherwise unobtainable (Loomis, 2019; Rainwater, 2014).

4.6   WEARABLE DEVICE FORENSICS VS. MOBILE FORENSICS (MF)

A mobile device is typically a device that is created to be carried on the user's body or is reasonably portable and can provide computing and communications functionality. It is normally capable of providing access to commercial mobile services. A wearable device, for instance, is designed to be carried on the user's body; it would not fall within the mobile device category if it cannot provide access to commercial mobile data services (Chauriye, 2016). The data extracted from mobile devices are admissible in the court of law as an evidence under certain laws and regulations but the admissibility of wearable device's data as a crucial evidence in court still requires an extensive research, which can prove the legitimacy and accuracy of the data directly retrieved from such devices.

4.7　CHALLENGES AND OPPORTUNITIES

The forensics of Wearable devices is challenging because it requires specialized techniques. The challenges include device diverseness, operational, and legal functionalities, artifact characteristics, device security, anti-forensic features, inadvertent data modifications, and the need of specialized tools and resources for forensic analysis (Loomis, 2019).

4.7.1　LEGAL CHALLENGES

Digital evidences are highly fragile in nature and require utmost care while handling and processing, and not every police or law enforcement agency has the necessary skills and expertise to properly collect, secure, and store digital data collected from Wearable devices, which often leads to data loss or contamination.

4.7.2　OPERATIONAL CHALLENGES

The time-stamp on the wearable devices can be altered to an incorrect time, which will prevent the events from being recorded accurately, and causing an inability to correlate with other digital devices. These devices rely upon GPS for data transmission and tracking, and the GPS requires a direct communication with the satellite, which it is using for data transmission. Therefore, if the GPS signal is lost, blocked or jammed due to any of the reasons such as the user is out of range, underground or in bad weather conditions, then the data collected will no longer be accurate. Moreover, not all devices provide the same functionality and interpretation of users' movement. For example, some device models will not distinguish arm movement from actual steps, and hence their data records will be unreliable. In addition, this can have an effect on the data collected and the reliability of the reports.

4.7.3　MISUSE CHALLENGE

It is possible that multiple people will be wearing the same wearable device. That means the data records from these devices are accurate only if the user is wearing it properly. For example, wearing the device on the ankle instead of the wrist will produce inconsistent data (Ferraroa, 2011).

4.7.4   DEVICE DIVERSITY

Wearable devices are ubiquitous and heterogeneous. They have rapid evolution cycles, which lead to new models being released by more than a dozen major manufacturers every year. The explosion of new devices and models is problematic to forensic examiners because each device requires extensive research. Wearable devices also lack standardization. Manufacturers leverage diverse hardware components, sensors, OS, software applications and file systems. Devices often incorporate specialized debugging interfaces, proprietary data formats, support for specific brands of companion mobile devices, cloud interoperability and communications mechanisms. In order to develop the most accurate and comprehensive forensic capabilities, forensic examiners need to conduct extensive research on the various features and software versions of devices (Loomis, 2019).

4.7.5   DATA CHARACTERISTICS

Data stored in wearable devices is in various formats (e.g., databases, raw binaries and pictures). Even when data is readable, its meaning may not be apparent, or the data might not be accurate. In a criminal trial, sleep data from a Fitbit device was determined to be inadmissible due to its inherent inaccuracy. The lifetime of data stored on wearable devices is a challenge. Since wearable devices generate data continuously and the devices have limited storage capacities, data is overwritten or deleted frequently. Data synchronized to mobile devices or cloud servers, on the other hand, may last longer. However, additional search warrants would be required to obtain the data residing in these platforms. Additionally, wearable devices are often designed for single users. However, some wearable devices support multiple users, making it important to identify whether or not a particular device supports multiple users.

4.7.6   SECURITY FEATURES

Wearable devices implement security features by design, typically in hardware and in the OS. Hardware components may provide protection features such as restricting memory readout and bypassing such protection requires advanced techniques involving expensive micro-probing stations or focused ion beams to modify fuse bits (Helfmeier et al., 2013). Other wearable device security features include secure boot chains, mandatory code signing, sandbox environments, dedicated security processors, kernel patch protection and data protection (Loomis, 2019). Bypassing such techniques to collect evidence requires significant resources and advanced techniques. OS used in wearable devices are updated frequently along with updates to core system libraries, utilities, and software applications including security patches that address newly discovered vulnerabilities which often limit data collection.

4.7.7 ANTI-FORENSICS (AF)

Wearable devices often incorporate anti-forensic measures to hinder data collection and analysis either by obfuscating data using proprietary formats or by obfuscating data by XOR-ing it with a key or by using a data encoding such as Base64. Wearable devices may support robust encryption algorithms. Data in flash memory or read-only memory (ROM) at rest, as well as data in motion between a wearable device and a companion device or cloud service may also be encrypted. Wearable devices that incorporate open-source data formats and OS may be more susceptible than devices without publicly disclosed designs. Open-source data formats enable malicious users to create fabricated datasets that might appear to be legitimate. Wearable devices with sophisticated anti-forensic features may support secure wiping which may be triggered by initiating a particular button sequence on the device during the boot process, by failing to enter the correct lock screen pin code or upon installing specialized forensic tools. Note that some wearable devices may delete user data, but this data can still be recovered via a physical data collection technique.

4.7.8 ACCIDENTAL DATA MANIPULATION

Wearable devices are compact by design. Their touchscreen displays are small, and their buttons are close together. A forensic examiner can accidentally interact with such a device and contaminate the evidence by introducing new data. Therefore, wearable devices should be handled and maintained carefully by forensic examiners throughout the investigative process and beyond. These devices should be packaged safely during their transportation to a forensic laboratory. At the laboratory, the devices should be safely secured using clamps (holders) to mitigate accidental user interactions. A best practice is to remove lithium-ion batteries, when possible, to ensure that no accidental modifications are made.

4.7.9 FORENSIC RESOURCES AND TOOLS

Forensic examiners are expected to achieve high case turnover rates. Since wearable device forensics is a new practice, forensic examiners typically acquire wearable data only from paired mobile devices and cloud service providers. It is relatively rare to leverage physical access to wearable devices. Regardless of the source of wearable data, the analysis of the data and its significance in forensic investigations are not well-established. Other challenges include inadequate personnel training, equipment, and tools. Forensic examiners may not be able to replicate techniques described in the literature because of their lack of experience or training. Furthermore, forensic techniques frequently necessitate the use of specialized equipment and software tools that are not readily available. A few forensic tools have been developed to automate wearable device data collection and analysis. These tools target specific wearable devices (e.g., Shattered developed for Google Glasses) (Loomis, 2019; Khairallah, 2018).

4.8 DISCUSSION

Wearable devices are a new form of technology that could be worn as accessories, embedded in clothing, or implanted in humans. Due to their increasing popularity, forensic investigators have been working on developing tools and techniques to analyze the data stored by these devices. However, the advancement of new devices that store such sensitive data has been strenuous to forensic analysts as well as the law enforcement agencies. Computers, mobile devices and different types of digital data storage devices (e.g., USB drives, thumb drives, etc.), are few common sources of digital evidences; however, it has been proven from a many legal cases that wearable devices are becoming more prevalent digital evidence source. Wearable devices store immense amounts of user data including the geo-location, calories burnt, eating patterns, and sleeping and moving habits; thereby creating a detailed record of the users' day-to-day life. Due to varying reliability and accuracy concerns, wearable devices have not been widely used in court proceedings. Furthermore, their reliability, verifiability, and authenticity are improving with new age technology and data retrieval through various software tools hence, increasing the admissibility of such evidence in legal proceedings. Digital evidence such as smart wearable devices hold the potential to be used in investigations. Once ascertained that the digital evidence is valid to be used as an evidence in the court of law, it may become a crucial breakthrough in cases involving wearable devices (Khairallah, 2018).

4.9   CASE STUDIES

There is no doubt that the wearable devices are becoming popular and the trend will only keep on increasing in the coming future. As already discussed, these type of devices contain huge repository of sensitive user information about the owner that can help the police and forensic investigators track the activity of the person related to crime a bit more easily. There are several criminal cases where the use of personal user data tracked using these devices has been used to solve the crime. A few examples of the noteworthy case investigations which involved Fitbit fitness trackers that provided law enforcement agencies with evidence vital to prosecution includes the Mollie Tibbetts missing person case, the Richard Dabate murder case and a few others.

In the mid of 2015, a murder case was reported where the accused claimed that his wife was killed by an assailant. He gave a detailed account of the events leading up to his wife's death. According to the statement given, the accused heard his wife scream from the garage before he saw an intoxicated man inside his home. He tried to counter attack and save his wife, but the assailant was eventually able to kill her. The assailant then left the house and walked away. Even though accused was partially tied up, he was still able to activate their alarm system. After obtaining the accused's statement, the police searched for evidence related to the events that led to his wife's murder, which included his and his wife's cell phone records, computer records, and Facebook records, text messages sent to and from his wife and his girlfriend and Fitbit records of his wife, which included data from home's alarm system, geo-location, activities, and her last movements. They also gathered various electronic records from all of his devices. The final analysis of all the digital evidence contradicted with the statement given by the deceased's husband. The evidence proved that the statement given by the husband differed from the detailed records and geographical locations identified from their personal digital devices (Black, 2017).

In December 2015, Connecticut resident Richard Dabate was investigated for the death of his wife, Connie Dabate. Richard Dabate, who proclaimed his innocence, said that his wife was shot and killed by a home intruder. However, Mrs. Dabate's Fitbit provided a different timeline of events. Data extracted from the device was correlated with surveillance footage, Facebook posts and Internet trace to place Richard Dabate at the crime scene at the time of the killing (Lartey, 2017).

In the 2016 Wisconsin case, fitness trackers data was used to annihilate the possibility of a woman being killed by her live-in partner. The information extracted from the accused's wearable device was presented as an evidence. However, the judge ordained that an affidavit should be submitted to establish the authentication of the Fitbit and allowed the lawyers to present its step-counting data (Smiley, 2019), at the trial; the police department's analyst confirmed the reliability of a man's particular wearable device (Gill, n.d.).

In March 2017, Kelly Herron, a Seattle runner who was viciously attacked in a park's restroom while wearing a Garmin Vivo smart GPS device, provides graphic evidence of how detailed and accurate the data from these devices can be. Herron, who had recently taken a self-defense class, was able to fight off her attacker and escape in the midst of a desperate fight for her life. When she got outside, a bystander assisted her in locking the assailant in the park's restroom until police arrived. Herron's Instagram page depicts her violent movements, which have become a battle cry for victims to fight back (Mccoy, 2017).

In July 2018, University of Iowa student, Mollie Tibbetts, was reported missing after going on a run (Boyette & Simon, 2018). Police correlated Tibbetts' Fitbit cloud data (GPS locations, heart rates, steps, and activity spikes) with social media and surveillance footage. The evidence led law enforcement to Cristhian Rivera of Brooklyn, Iowa, who confessed to the murder (Boyette & Simon, 2018).

In October 2018, San Jose's case, women's Fitness tracker helped imparting a clue to police investigating her death, which resulted in her 90-year-old stepfather's arrest. The accused denied the charges of murder but according to police the deceased's fitness tracker recorded rapid rise in her heart rate followed by a significant slowdown at the time accused was with her, helping investigators compile clues of her death. The accused's presence was established by investigation of the video footage that police collected from a neighbor's camera that pointed towards victim's house (CBS News, 2018).

Aside from the use of wearable data in criminal investigations, these devices can also assist users in preventing drug abuse and overdoses. According to a Cellebrite's article, these devices can detect the use of narcotics by using electrodermal activity (EDA), skin temperature, and tri-axis acceleration data generated by a wrist-worn biosensor that detects whether a person has used opioids. Doctors and hospital staff who need to identify incoming patients quickly in order to administer overdose antidotes stand to benefit significantly. The same technology could help law enforcement determine whether someone they encounter is under the influence of opioids (Watson, 2019).

4.10 CONCLUSION

There is a lot yet to be done in the field of Wearable devices forensics as Wearable devices have rapid evolutionary cycles. Development of databases that maintain detailed device information, including makes, models, hardware upgrades, software versions, communications interfaces, security, features, data types, and supported data collection and analytical techniques would be a good starting point. These databases would greatly assist practitioners, researchers, and vendors in ensuring that digital forensic tactics, techniques, and procedures keep abreast of advancements in wearable devices. In addition, like smartphones, wearable devices are beginning to incorporate security mechanisms in their hardware and OS that hinder data recovery. We have to focus on techniques such as fuse disabling, side-channel analysis, and hardware and software reverse engineering to recover protected data during forensic investigations. However, these devices are being upgraded continuously, their accuracy, reliability, and thereby admissibility is enhancing mainly when they are synced to other devices like mobile phones, clouds, and computers which are already used as admissible evidence in court of law. Provided that if the evidence meets the defined requirements of admissibility, wearable devices should be considered as one of the sources of digital evidence that can be useful in many legal cases. Wearable devices collect and process copious amounts of data. There are leaps yet to be made in the data analytics and data mining (DM) algorithms that can perform automated trend analysis, user behavior analysis and anomaly detection.

4.11 FUTURE SCOPE

Wearable devices are a rising category of technology that are worn as accessories, embedded in clothing, even engrafted in human bodies. Advanced digital forensic analysis tools and techniques are still being developed to be able to process and analyze data from such type of devices. Wearable devices, however, have not yet been potentially used on a wide scale due to arguments regarding the reliability, validity, and accuracy of these devices. This chapter focuses on the comprehensive study of wearable devices with respect to its forensic evidentiary value in crime cases and possesses tremendous scope (Khairallah, 2018).

KEYWORDS

- **forensic evidentiary value**
- **global positioning system**
- **Internet of things**
- **near field communication**
- **read-only memory**
- **software development kit**

REFERENCES

Adebayo, K. O., (2020). *Digital Forensic Analysis of Smartwatches*. Bachelor's Thesis, School of Information Technologies, Tallinn University of Technology.

Aroganam, G., Manivannan, N., & Harrison, D., (2019). Review on wearable technology sensors used in consumer sport applications. *NCBI, Sensors (Basel), 19*(9), 1983.

Ballard, B., (2015). *A Periodic Table of Wearable Technology*. |Online| Available on: https://techcrunch.com/2015/06/10/a-periodic-table-of-wearable-technology/(accessed on 10 January 2022).

Black, N., (2017). *Fitbit Data, other Digital Evidence Used by Prosecution in Murder Case*. |Online| Available on: http://www.legalnews.com/detroit/1442488 (accessed on 10 January 2022).

Boyette, C., & Simon, D., (2018). *Her Fitbit May be the Key to Finding a Missing University of Iowa Woman*. |Online| Available on: https://edition.cnn.com/2018/07/26/us/missing-university-of-iowa-student-mollie-tibbetts/index.html (accessed on 10 January 2022).

CBS News, (2018). *Fitbit Helps Lead Police to San Jose Woman's Alleged Killer–Her Stepfather*. |Online| Available on: https://www.cbsnews.com/news/fitbit-helps-lead-police-san-jose-womans-alleged-killer-her-stepfather/?intcid=CNM-00-10abd1h (accessed on 10 January 2022).

Chan, M., Estève, D., Fourniols, J. Y., Escriba, C., & Campo, E., (2012). Smart wearable systems: Current status and future challenges. *Artificial Intelligence in Medicine*, 56.

Chauriye, N., (2016). *Wearable Devices as Admissible Evidence: Technology is Killing Our Opportunity to Lie* (Vol. 24, No. 2). Catholic University of America, Columbus School of Law, Article 29.

EC-Council's Computer Hacking Forensic Investigator v9, Module 1, "Computer Forensics in Today's World", EC-Council.

EC-Council's Computer Hacking Forensic Investigator v9, Module 4, "Computer Forensic Investigation Process. EC-Council.

Ferraroa, C., (2011). A new approach to wearable systems: Biodesign beyond the boundaries. *ICoRD2011 Conference Proceedings*, 283–291.

Gill, J. (2019). *Fitbit Data Provides Clues in Murder Case: eDiscovery & Criminal Investigation.* |Online| Available on: https://ipro.com/resources/articles/fitbit-data-ediscovery-criminal-investigation/ (accessed on 10 January 2022).

Hayes, A., (2020). *Wearable Technology.* |Online| Available on: https://www.investopedia.com/terms/w/wearable-technology.asp (accessed on September 3, 2021).

Helfmeier, C., Nedospasov, D., Tarnovsky, C., Krissler, J., Boit, C., & Seifert, J., (2013). Breaking and entering through the silicon. *Proceedings of the ACM SIGSAC Conference on Computer and Communications Security*, 733–744.

Karie, N. M., & Venter, H. S., (2015). Taxonomy of challenges for digital forensics. *Journal of Forensic Sciences, 60*(4), 885–893.

Kasukurti, D. H., & Patil, S., (2019). Wearable device forensic: Probable case studies and proposed methodology. *Security in Computing and Communications.* Springer Singapore.

Khairallah, T., (2018). *Wearables as Digital Evidence.* Preprints. doi: 10.20944/preprints 201812.0313.v1.

Lartey, J., (2017). *Man Suspected in Wife's Murder After her Fitbit Data Doesn't Match his Alibi.* |Online| Available on: https://www.theguardian.com/technology/2017/apr/25/fitbit-data-murder-suspect-richard-dabate (accessed on 10 January 2022).

Lee, J., Kim, D., Ryoo, H. Y., & Shin, B. S., (2016). Sustainable Wearables: Wearable Technology for Enhancing the Quality of Human Life, *MDPI Sustainability, 8*, 466; doi:10.3390/su 8050466.

Loomis, M. E., (2019). *Wearable Device Forensics.* The University of Tulsa, ProQuest LLC, a dissertation report |Online| Available on: https://www.proquest.com/openview/bb6f13b1 5f875493bd2003302ef2429e/1?pq-origsite=gscholar&cbl=18750&diss=y (accessed on 10 January 2022).

Mccoy, J., (2017). *Seattle Runner Attacked Midrun Fought Like Hell to Defeat Her Offender.* |Online| Available on: https://www.runnersworld.com/news/a20850315/seattle-runner-attacked-midrun-fought-like-hell-to-defeat-her-offender/ (accessed on 10 January 2022).

Ngu, A., Gutierrez, M., Metsis, V., Nepal, S., & Sheng, Q., (2008). IoT middleware: A survey on issues and enabling technologies. *IEEE Internet of Things Journal, X*(X), 2016.

Nirjon, S., Gummeson, J., Gelb, D., & Kyu-Han, K., (2015). TypingRing: A wearable ring platform for text input. *MobiSys'15 Proceedings of the 13ᵗʰ Annual International Conference on Mobile Systems, Applications, and Services.* Florence, Italy.

Panhalkar, T. (2020). *Data Analysis & Evidence Assessment.* |Online| Available on: https://info-savvy.com/data-analysis-evidence-assessment/ (accessed on 10 January 2022).

Panhalkar, T. (2020). *Understand Acquiring RAID Disks.* |Online| Available on: https://info-savvy.com/understand-acquiring-raid-disks/ (accessed on 10 January 2022).

Rainwater, S., (2014). *Physically Invasive Forensic Data Recovery Techniques.* Ph.D. Dissertation, Tandy School of Computer Science and Department of Electrical and Computer Engineering, University of Tulsa, Tulsa, Oklahoma.

Rowland, Goodman, Charlier, Light, & Lui, (2015). *Designing Connected Products: UX for the Consumer Internet of Things* (p. 726). O'Reilley Media, ISBN: 978-1-449-37256-9.

Smiley, L., (2019). *A Brutal Murder, a Wearable Witness, and an Unlikely Suspect.* |Online| Available on: https://www.wired.com/story/telltale-heart-fitbit-murder/ (accessed on 10 January 2022).

Stelvaga, A., & Fortin, C., (2017). Design principles of wearables systems: An IoT approach. *21st International Conference on Engineering Design, ICED17: Product, Services and Systems Design* (Vol. 3). Vancouver, Canada, 21.-25.08.2017.

Tech Target Contributor, (2015). *Wearables OS (Wearables Operating System).* [Online] Available on: https://whatis.techtarget.com/definition/wearables-OS-wearables-operating-system (accessed on 10 January 2022).

Watson, A., (2019). *How Wearables Are Being Used to Solve Homicides, Missing Person and Illicit Drug Cases.* |Online| Available on: https://www.cellebrite.com/en/how-wearables-are-being-used-to-solve-homicides-missing-person-and-illicit-drug-cases/ (accessed on 10 January 2022).

Yoon, Y. H., & Karabiyik, U., (2020). Forensic analysis of fitbit versa 2 data on android. *MDPI Journals, Electronics, 9*(9). 10.3390/electronics9091431.

CHAPTER 5

Combating Cybercrimes with Digital Forensics

SHIPRA ROHATGI[1] and SAKSHI SHRIVASTAVA[2]

[1]Assistant Professor, Amity Institute of Forensic Sciences, Amity University, Noida, India

[2]MSc Student (Information and Cyber Security), NSHM College of Management and Technology, Kolkata, West Bengal, India

ABSTRACT

The advent of technological changes in communication and data exchange has committed completely modern forms of crime, such as cybercrime and computer crime. Therefore, the advent of the Internet and cyberspace as highly effective communication media, on the other hand, has brought many advantages. This clearly realizes that everything is also constructive with some dark side. Data and information, at least in the form of currencies that can be traded to criminals, legally target companies that have access to this data and information. Whenever any offense related to cyberspace happens, every time it comes with the same questions:

- How did the incident happen?
- How can we prevent the future?

These questions' answer is difficult to judge, which is based on the complexity of the particular case. The forensic part is important in relation to the main question. Evidence collected at the crime scene is "carefully investigated to understand who, what, where, and why." Cyber forensic investigators reveal all questions and use the information gathered. Leverage and report incidents accurately and easily to predict future comparative

Advancements in Cybercrime Investigation and Digital Forensics. A. Harisha, Amarnath Mishra, & Chandra Singh (Eds.)
© 2024 Apple Academic Press, Inc. Co-published with CRC Press (Taylor & Francis)

attacks. As cybercrime grows, law enforcement agencies that rely on computer forensics and digital forensics and expertise in all areas are more importantly needed to find cybercriminals (http://www.ummedcyber.com/Digital-Forensics-Combating-Cybercrime-ummed-meel-kolkata-police-THE-PROTECTOR-magazine-hurdles-in-Digital-Forensics.pdf).

Digital forensics or cyber forensics may "collect and investigate information from computer framework systems, telecommunication streams (remote) and capacity media in a court-approved manner." These include the main hypothetical and methodological perspectives of computer hacking and advanced robbery, monetary cheating and online extortion, obscenity, and online sex crimes, cyberbullying and cyber stalking, cyber terrorism and extremism, advanced legal investigations and the scope of cybercrime in its legitimate setting is included arrangement (Harbawi & Varol, 2016).

5.1 INTRODUCTION

Computer crime and computer-aided criminal activity is growing rapidly. Criminals, scammers, and psychological two lions appear to attack whenever they have the opportunity. Electronic information and data collection has been a central issue of increasing conflict and crime due to injuries in the modern sector of science, science (e.g., cyber forensics and computer forensics).

It is almost impossible to prove with a computer that is reliable enough to stand in court and persuade. Computer forensics is one of the greatest development missions of the 21st century. The term cyber forensics has many equivalent words. Also known as computer forensics. It is the youngest department in the legal world. Cyberlaw specifically addresses web torts or legal issues. We discovered its roots in the late 1980s to suggest that experts would take advantage of this to scan stand-alone computers for advanced proof of crime (https://www.unodc.org/e4j/en/cybercrime/module-4/key-issues/standards-and-best-practices-for-digital-forensics.html).

Advanced forensic medicine is a discipline that applies law that can provide acceptable evidence in the process of criminal investigation. Forensics integrates advanced evidence collection, preservation, and investigation. Computerized certificates are at the heart of cybercrime investigations. Cybercrime can be committed using electronic devices, hacking or damaging electronic devices. Prior to 2000, cybercrime was under the category of financial crime investigated under the framework of common law. In 2000, the IT Act was enacted continuously, recognizing the increased use of computers and computer violations. IT Law, Computers, and Computer Assets Crimes committed by retrieving information from or using computer systems or such assets fall under the category of cybercrime (https://www.unodc.org/e4j/en/cybercrime/module-4/key-issues/standards-and-best-practices-for-digital-forensics.html).

5.2 STEPS

Digital forensics is used to solve private sector internal problems such as abuse of non-default corporate policies that are not stored in court and are sometimes drawn into the "crime" category. In our daily lives, advanced forensics is additionally utilized to recover accidentally deleted information on mobile phones, computers, or other electronic devices.

This fusion of computer science and law is largely in cyber forensics. Editing information and exploring basal embryo data can be associated with this definition (Majed, Noura, & Chehab, 2020). Traditionally cyber forensics includes:

- Collection;
- Identification;
- Validation;
- Interpretation;
- Preservation;
- Analysis;
- Documentation;
- Presentation.

Cyber forensics is the method of collecting, protecting, analyzing, and interpreting computer capacity data and data, ensuring accuracy and quality (Table 5.1).

In 2002, another computer forensics demonstration was launched, based on the 2001 Computer Science Research Conference and the United States Government's Mistakes Scene Appearance Convention (for actual crime). This proves has nine stages:

1. **Identification:** Identifies the occurrence from the index and predicts type.
2. **Prepare:** Set up tools, procedures, review orders, and comply with authorizations and return operations.

TABLE 5.1 A Road Map for Digital Forensic – Protocol for Physical Crime Scene Searches

Collection	Preservation	Examination	Identification	Analysis	Presentation
Preservation	Case management	Preservation	Event/crime detection	Preservation	Documentation
Approved method	Imaging technologies	Traceability	Resolve signature	Traceability	Expert testimony
Approved software	Chain of custody	Validation techniques	Profile detection	Statistical	Clarification
Approved hardware	Time synchronization	Filtering techniques	Anomalous detection	Protocols	Mission impact statement
Legal authority	–	Pattern matching	Complain	Data mining	Recommended countermeasure
Lossless compression	–	Hidden data discovery	System monitoring	Timeline	Statistical interpretation
Sampling	–	Hidden data extraction	Audit analysis	Link	–
Data reduction	–	–	–.	Spatial	–
Recovery techniques	–	–	–	–	–

3. **Approach Strategy:** Create a strategy to use that maximizes the collection of uncontaminated evidence with minimal impact on victims.
4. **Preservation:** Separation, preservation of the state of matter and of advanced evidence.
5. **Collection:** of improved field recording and trial recording to use recognized and standardized strategies.
6. **See:** An insightful overview of evidence of suspected misconduct.
7. **Analysis:** Make sure of the importance, piece together parts of the information.
8. **Presentation:** Conclusion and interpretation.
9. **Return of Proof:** Material and advanced property returned to good owner.

Cycles measures implied the application of science for a legitimate process, so Cyber legally fast, the recent digital evidence recently provided the court. Electric debris is the death of evidence can be easily broken and can be easily set. Disconnect of these, hooded, hoodlums, immoral, and a little real time is genuine wiping, covering, near, clothes, masks, turbo charging and destroying the increasing presence of the web as the main assets and the key must ensure the resources of the company to safely and safely to protect. Whenever these resources are under the attack or if they are abused, security professionals must collect electronic proof for the attack or misuse, they use the proven debris to bring equity to those who abuse this innovation (https://www.unodc.org/e4j/en/cybercrime/module-4/key-issues/standards-and-best-practices-for-digital-forensics.html).

Cyber forensics is a science and a craftsman. With the evolution and rapid change of innovation, the rules for overseeing the application of cyber forensics in the areas of inspection, security, defense, and legal requirements are also changing. Modern strategies and technologies are summarized to donate information security professionals much better. The mystery and social trends that the web gives are suggestions for budget and social debate to achieve data innovation, the criminal component secures basic communications assets.

Recognizing cyber risks quickly and responding to those who, in recent times, may be a real risk, are central to the success of cyber fraud and advanced forensic countermeasures. Since not all businesses can do this on their own, it can be a basic association of production to get professionals and equipment that can do the job without compromising the legal judgment of the proof that can be erased from behind (https://www.unodc.org/e4j/en/cybercrime/module-4/key-issues/standards-and-best-practices-for-digital-forensics.html).

5.3   DIGITAL FORENSIC INVESTIGATOR

Imagine a security breach in your company and your data stolen. In these instances, a computer forensic analyst would break out and find out what the hacker was doing on the network, regardless of the access method, places through the network, and whether or not it was running malicious code. In these circumstances, the digital forensic investigator's role is to restore data, such as documents, photos, and e-mails, by deleting, destroying, or otherwise tampering with the computer's hard drive or other data storage devices, such as zip drives or flash drives. A Digital Forensics Investigator is someone who has a desire to effectively solve crimes based on evidence (https://www.eccouncil.org/what-is-digital-forensics/).

The application of digital forensics includes the collection, investigation; analysis and reporting of incidents involving computers, networks, and mobile devices. The work of digital forensics professionals spans both the public and private sectors, and their roles typically include investigation of data breach, recovering, and checking data from computers and electronic storage devices, dismantling, and rebuilding corrupted systems to get lost data, determine additional systems to attack an attack on an attack and make a decision by creating evidence of legitimacy.

The ultimate objective of a digital forensic investigator is to identify the perpetrators of cybercrime, secure solid evidence against the perpetrators, and bring the evidence to court (https://www.eccouncil.org/what-is-digital-forensics/).

5.3.1   HOW DIGITAL FORENSICS IS USED IN INVESTIGATION?

The digital foot printing is information about the users of the system, such as the web page accessed and the device used in the active state. Tracking digital footprints, investigators get the data they need to resolve criminal cases (https://www.eccouncil.org/what-is-digital-forensics/).

Digital forensic investigators are experts in investigating encrypted data using a variety of software and tools. Depending on the type of cybercrime the investigator deals with, the methods you can use in the future will vary. Cyber investigators' duties include recovering deleted files, decryption, security, and understanding the cause of breaches. The evidence collected can be preserved and translated and submitted to court or used for additional police investigations. The role of cyber forensics in crime can be understood in case studies (https://www.eccouncil.org/what-is-digital-forensics/).

5.3.2 CASE STUDY: DIGITAL FORENSICS MAKES A DIFFERENCE TO UNRAVEL CYBER SURVEILLANCE

In 2008, the most serious cyber-attack in US military history spilled exceptional military secrets into remote areas. Advanced forensic personnel were dispatched to determine the source of the Pentagon attack that was not prepared for the internally launched attack they claimed and whether a breach had occurred. The investigation accurately pointed out the infringement of a US military base in Center East. The cause was a continuous USB drive built in by one of the professors who claimed to be inside a military computer organization. Therefore, it bypassed all security measures (firewalls) built by the cyber security group. Encouraging the inspection to determine that the person is not a specialized hairdresser working in the US military, a trusted employee who thinks they have found a free drive. They did not doubt it to choose it in a car stop outside the military base, where hundreds of drives containing malware were scattered. Individual people have grown them because it does not have to know that they do not need to know anything to choose one, and use them on their computers. Experts for cyber forensics– working with cyber security specialists – have played a significant part of the decision of the infringement of the violation and, in turn, input measures to ensure that such violations do not happen again. Experts for cyber forensics– working with cyber security professionals – have played an important part of the decision source for violations and, in turn, entering measures to ensure that such a violation does not happen again (Van & Jean-Paul, 2015).

5.3.3 STANDARDS USED IN DIGITAL FORENSIC INVESTIGATION

For comprehensive procedures for digital forensic research and electronic evidence, there are existing standards that provide collaborative support, ideal models and procedures. The convention on electronic evidence (CEE) 2016, a contract, is a general guideline for the recognition and containment of electronic evidence for foreign lawyers and encourages judges and lawyers to infer the concept of digital evidence containment (Table 5.2).

Standards and processes in digital forensics include identification, collection, retrieval, and storage to facilitate investigations. However, digital forensic investigation methods can be safeguarded by maintaining the storage chain and ensuring that digital evidence is accepted, and convincing accurate and complete judges based on criteria regardless of the process employed (Garfinkel, 2007).

TABLE 5.2   Digital Forensic Investigation Standards

1.	**ISO/IEC27043**	It provides pre-prepared advice covering idealized patterns in different situations and ensures that surveys are repeated in all situations to achieve the same results.
2.	**ISO/IEC 27037**	Designed for accident response. It maintains the integrity and reliability of digital certificates and provides guidelines for specific activities in the storage and processing of potential digital certificates.
3.	**ISO/IEC 27041**	This ensures that the methods used in incident management, evidence handling, archiving, and investigation are appropriate to the case being investigated and the resulting requirements.
4.	**ISO/IEC 27042**	This is a comprehensive guide of tools, techniques, and methods used to perform analyzes and interpretations to identify and evaluate digital certificates to aid in the understanding of ISO 27037-related events. A guide is provided.

5.4   CYBER FORENSICS TOOLS USED IN INVESTIGATION

As serious problems such as cyber terrorism, cyber stalking, and spam increase, cyber forensic tools can help you investigate these criminal cases, draft evidence, and create them. Some of these tools are:

1. **First on Scene (FOS):** It is a basic visual script code. It also works with other tools like PS Tools, Login Sessions, FPort, NTLast, PromiscDetect, and File Hasher to generate evidence log reports. This can be additionally analyzed by forensic professionals to extract important information.

2. **X-Ways WinHex:** Used for low level data processing, file scanning, digital camera card recovery, corrupted file system, original file recovery, etc. This is a powerful tool used to collect digital evidence.

3. **Rifiuti:** It is a tool that helps you find the last details of your system's Recycle Bin. Helps collect all files that have not been deleted or deleted.

4. **Galleta:** This means "cookie." Galleta helps you check the contents of the cookie file on your computer. Cookie files are temporary internet files used to keep their own logs for purposes such as tracking from websites.

5. **Pasco:** This means "ball" in Latin. Pasco helps analyze all searches run on your computer. In other words, it is useful for collecting a record of internet activity executed on the target computer.

6. **Network Mapper (NMap):** A port scanner tool that helps you find open ports on your remote computer. NMap has the ability to bypass the source system ID and operate without the intrusion detection system (IDS) alarm sounding. Mainly related to network security systems.

7. **Forensic Acquisition Utilities (FAU):** This is a set of forensic tools such as a checker. A file wiper used for various investigations and research, etc.

8. **Bin Text:** It is useful for moving collected evidence files, such as log files generated by other forensic tools. Used for pattern matching and filtering of these log files.

9. **Ethereal:** It is another network security tool that is a network packet sniffer. It provides interrogators with incoming data sent over the network by sniffing data packets across the network. However, strong encryption algorithms cannot be used if they are deployed on the source and destination computers.

10. **Encrypted Disk Detector (EDD):** It is a command line tool that quickly scans your system's local physical drives for encrypted volumes. Then you can investigate further and make a decision to determine whether you need to do a live search to protect and preserve the evidence that would be lost if unplugged.

11. **PyFlag Tools:** These are some of the tools used for log analysis and can be a very effective tool for investigators.

12. **MemGator:** A memory file inquiry tool that automates the extraction of data from memory files and edits investigators' reports. It can extract data about memory information, processes, network connections, malware detection, passwords, and encryption keys.

Another method used in cyber forensics that is not specifically included in the scope of Miscellaneous Steganography Tools. This is a way to convert facts or text files or insert image files to decrypt other files. However, there are several useful tools for detecting these injections. There are hackers or malicious users who came up with the idea of inserting data files as image files as well music and video files too (https://www.blackhawkintelligence. com/forensic-services/digital-forensics/cybercrime-digital-forensics-1/).

Individuals may try to encrypt criminal information by changing the extension of certain types of files to rename other types. This makes it difficult to determine the correct type of file. Encase is used to flag suspicious files like this one. Running a hash (#) applied to the hard drive interprets the file header and marks it as containing incorrect header information (https://

www.blackhawkintelligence.com/forensic-services/digital-forensics/cybercrime-digital-forensics-1/).

Creating an accurate image of the information is very important to make this information/evidence admissible in court. And because of this, our experts work very hard with every patient and every commitment to accuracy, all the confidentiality of the has no one knows about whether they are working, and to gather important information that can produce concrete evidence before the courts (https://www.blackhawkintelligence.com/forensic-services/digital-forensics/cybercrime-digital-forensics-1/).

Once the information and all evidence have been collected, the expert will produce a comprehensive report that can be submitted in court. These people have expertise and special education in the use of complex tools and techniques, so they can also testify in court on issues at work (https://www.blackhawkintelligence.com/forensic-services/digital-forensics/cybercrime-digital-forensics-1/).

Today, malicious angry employees are attacking many e-commerce websites, including viruses, wiretapping, and financial fraud, from independent businesses and businesses of various governments. This e-commerce attachment creates various economic difficulties for businesses. This is being observed as a common feature of individuals fired or insulted from headquarters, regardless of cybercriminals such as hackers (https://www.blackhawkintelligence.com/forensic-services/digital-forensics/cybercrime-digital-forensics-1/).

5.5 THE CHALLENGES OF DIGITAL FORENSICS

No matter how effective the technology is. As always, there were drawbacks. Similarly, storing data or information for use as evidence is in the court's favor, but on the other hand, there are certain technical and human barriers to the collection of such information. There are so many limitations which are as follows: The method used to store a particular file may not be an individual label for when and where the file was acquired. These files can be easily forged or altered. Some functions in browsers for saving WWW pages to disk are incomplete because they can save text but not related images. In most cases it is difficult for the system to find the last searched page. Looking at the whole series, it gets harder to see which is slow and which is fast. There is a difference between what is stored on disk as you can see on the screen. Many ISPs use proxy servers to speed up the transfer of popular pages on the web. Therefore, users may not be able to see what their ISP received from a particular website. Common mistakes such as changing dates and timestamps, terminating loose processes, and pre-investigation system patches can lead to data loss on disk. As a result, the e-file and evidence portion stored on your computer crashes (https://intesecurity.com/data-forensics-faqs/). New technologies will help engineers develop and create more powerful hardware and software for investigating computer-related crime. The advancement of cryptography is one of these challenges. As encryption standards rise, the complexity of algorithms increases, making it difficult and time consuming for experts to crack the code. Another challenge is maintaining trusted certifications and industry standards (https://intesecurity.com/data-forensics-faqs/).

5.6 ANTI-FORENSICS (AF): CAN IT BE HELPFUL FOR DIGITAL FORENSICS IN COMBATING CYBERCRIMES?

Anti-forensic tools, techniques are a major obstacle to the digital forensics' community. This is exactly a valid forensic investigation of digital evidence aims to clearly ascertain its meaning when the evidence must be reliable, accurate, and complete. However, little attention is paid to whether it is considered 'anti-digital forensics,' 'anti-forensics (AF)' or 'counter forensics,' especially in the form of academic research. However, some studies, such as cryptography, argue that it can be considered as AF (Anjani, n.d.).

AF is also receiving more attention from cybercriminal investigators and academia. As cybercrime increases and the amount of software that can be used to thwart forensic investigations increases, doctors can identify the same forensic prevention activities that have occurred in the past experience of others is needed. The formal definition of AF and the shared terminology of anti-digital forensics facilitate knowledge sharing and enable better mitigation strategies. So, first of all, past studies are worth emphasizing how to define anti-digital forensics (Harbawi & Varol, 2016; https://www.unodc.org/e4j/en/cybercrime/module-4/key-issues/standards-and-best-practices-for-digital-forensics.html).

AF in the digital world is the process of removal or obfuscation of digital forensic artifacts aimed at disabling digital forensic investigations. There are more of the strategies but some of them are commonly used:

- Data concealment;
- Data destruction;
- Trace obfuscation;

- Data contraception;
- Data generation;
- File system attacks.

AF aims to remove all finding of digital events, invalidate data, increase the complexity of an investigation, or delete evidence of their use, or question investigations in general. The various AF methods are detailed in subsections.

5.6.1   DATA HIDING

Concealment of data means adopting security measures by storing or hiding data in places where data is likely to be lost. Simple methods exist, such as renaming extensions or editing signatures, but these are usually readily available in modern forensic software. In data communication, data concealment refers to the technique of adding a host signal ambiguous message signal without recognizing the distortion of the host signal. This composite signal generally adopts a communication method different from data communication.

One of the easiest and most effective ways to hide data is Steganography. Hidden writing practices have been around for thousands of years, but the ability to hide digital data in any format in other carrier files presents a task for digital forensic investigators. In addition to hide data in a versatile way, Steganography technique is very difficult to detect. There is an only one open-source tool that effectively retrieves data that is hidden by latest Steganography Tools "StegDetect" by www.outguess.org. Some Steganography algorithms do not know the key to the algorithm and hide the information in such a way that it cannot be recovered (https://www.unodc.org/e4j/en/cybercrime/module-4/key-issues/standards-and-best-practices-for-digital-forensics.html).

5.6.2   ENCRYPTION

Unless you use a cryptographic algorithm to scramble your data and use a key to decrypt it, you're just protecting your data by making it unintelligible or undetectable. Encryption has long been used in some form to protect against message eavesdropping. Encryption is used in many aspects of digital data storage and transmission. In context with anti-forensic data concealment, cryptographic tools provide a very powerful tool for users looking to thwart the efforts of DF investigators. There are so many standard protocols such as

SSL (secure socket layer), SSH (secure shell) or TLS (transport layer security) which can encrypt Network traffic very easily so that it could help packets to get transferred safely (https://www.unodc.org/e4j/en/cybercrime/module-4/key-issues/standards-and-best-practices-for-digital-forensics.html).

5.6.3   DATA DESTRUCTION

Destruction of data by file wipe is an anti-forensic method that has been commonly used since ancient times. The wisest course of action for cyber-criminals is to remove all traces that effects adversely. A simple purging leaves the data intact by default. Although not all data is visible to the every computer user, such data can be easily retrieved using forensic tools. Using one of the free data deletion tools (Eraser, PGP, etc.), users can safely remove files by overwriting the clusters they occupy with random data.

These tools can perform safe deletion of the artifacts that are mentioned above to ensure, such remnants can't be backed up or recoverable, after erasing. These tools are easily configured (but only by experts who have knowledge about) to wipe all hard drive's free space securely, including slack space, whether it can be done manually or automatically at intervals. In addition to the destruction of data or file by the help of wipe or overwrite there is also Cybercriminals sometimes take drastic measures to return. These could help in such way when there's an injection of malware's we can wipe all the data or at time of attack we can save data (Van & Jean-Paul, 2015).

5.6.4   CASE

Anti-forensic involvement is the latest in unusual or high-profile case from the Federal Bureau of Investigation (FBI) vs. Apple. The FBI had to work out an anti-forensic technique to get the iPhone 5C owned by San Bernardino County, California government issued to its employee, Syed Rizwan Farook, one of the shooters involved in the December 2015 San Bernardino attack that killed 14 people and injured 22.

The hackers were killed, but the iPhone 5C was recovered. Built-in anti-forensic technology forcing encryption and automatic device erasure after multiple failed password attempts was locked with a four-digit password interfering with the forensic acquisition process. The legal case of ordering the FBI to help gain access to data on Apple devices is complex and beyond the scope of this document. In the end, zero-day abuse secured evidence for the iPhone 5C. The scale of this case underscores the need for both research and practice to gain a more comprehensive multidisciplinary understanding of the anti-forensic impact of the entire digital forensics' community.

As you can see from the FBI's Apple case, AF will make digital media research more difficult, more time consuming and more expensive. Users can use anti-forensic tools and technologies to remove, modify, or suspend criminal activity in systems, much as crime removes pieces of evidence from the physical realm of crime scenes increases (https://www.unodc.org/e4j/en/cybercrime/module-4/key-issues/standards-and-best-practices-for-digital-forensics.html).

5.6.5   *LIMITATIONS*

Everything has its coined side heads and tails, it depends how we utilize the given purposes. AF can be utilized in terms with the digital forensics to minimize the crime in cyberspace. One limitation of this task is that the number of software tools that can be considered "against the law" is intrinsically large and continues to grow; making it difficult to define the full scope of a domain becomes extremely difficult. It means, however, there is no "limitation" as this is an opportunity for future research efforts. A further limitation of taxonomic datasets and extensive classification methods is that semi-digital forensic fields are difficult to interpret clearly, and new domains cannot be fully explained. For example, not just the "cloud service forensic" domain. Finally, a way to automate the process of classifying anti-forensic tools can be done using machine learning (ML) by analyzing the tool's metadata online, which is of interest for scientist's works. This work will help alleviate ongoing research and increasing anti-digital forensics problems in the general area (Harbawi & Varol, 2016).

5.6.6   *NECESSARY SUGGESTION*

The section is allocated for figuring out the optimal solution to minimize the total cost of the system. The software allows us to investigate the anti-forensic resistance traces inside the system using a checklist to discover:

- From BIOS, virtualization should be disabled;
- Systems must be always up-to-date;
- Logs should be securely saved on servers;

- Encrypted backup should be stored properly;
- The BIOS should be secured by a strong password to avoid unnecessary modification;
- Block VPN access whenever or wherever needed.

To that end, the development of software is very useful and there's guideline to consider and follow, for protection of your data from any anti-forensic actions that might target your system (https://itexamanswers.net/question/what-procedure-should-be-avoided-in-a-digital-forensics-investigation).

5.7 CONCLUSION

Advancement with Technology, Digital forensics has illustrated a very important role. In addition, as cybercrime such as hacking increases, the need for cyber forensics has risen, and various tools and techniques that can track crime have been developed, and this is accurately reported and submitted to the court. I am doing it. Today, various industries, businesses, and government agencies are enthusiastic about appointing experts in this area to identify cyber malfunctions in their employees. The experts get appointed to investigate cyber related crimes. After the investigation is completed, the expert will accurately extract and prepare the evidence collected through various media and submit it to the investigative agency (Holt, Bossler, & Seigfried-Spellar, 2015; https://itexamanswers.net/question/what-procedure-should-be-avoided-in-a-digital-forensics-investigation).

Traditional forensic tools play an important role in retrieving data. Every tool has its own flaws. These tools and technologies need to be further developed and improved for computer forensics to be fully successful and legally effective.

The insight of the digital forensic is infinite. With the expansion of technology, this sector continues to grow with advantages and barriers. Use only tested and evaluated tests and methods to ensure accuracy and reliability. Evidence of experts who have been collected by experts must be adequately processed and stored so that it can be produced correctly in court. Each process or method for analyzing the Cyber Forensic implementation method finally leads to the risk of the case.

Cybercrime and Digital Forensics promptly identifying and responding to cyber threats because serious damage can effective the processes. Access to professional tools that can go without any compromise with the forensic integrity of evidence those criminals may have left behind that not all businesses can perform their tasks is an important partnership (Ofori & Akoto, n.d.).

The cyber forensic scenario of the law can be described as:

- Law enforcement agencies have insufficient education regarding collection and use of evidence using digital forensics.
- Laws need to react slowly to technological developments; it needs to be synchronized and updated in technology in the field of digital forensics to allow offenders to be punished.
- Digital evidence collected using digital forensic tools is accepted as evidence under current law to aid in criminal trials.
- Cyber forensic tools are widespread in criminal investigations and achieve better conviction rates.

Therefore, you need to:

- Identify tools used in criminal investigations and study their level of effectiveness;
- The proposed system is scientifically tested to test its effectiveness;
- Understand the use of judicial administration cyber forensics;
- We propose a new system (not already approved) used for criminal investigations.

Therefore, it is necessary to analyze the existing legal system regarding the utilization and legality of cyber forensics in criminal investigations and trials. For this reason, the various tools and techniques used for disk and device forensics need to be analyzed. These tools and techniques can be used more conveniently in criminal investigations and trials. These cyber forensic tools also need to analyze the legal provisions that can be used by law enforcement agencies, law enforcement agencies, and courts (Urvashi, n.d.).

KEYWORDS

- **anti-forensics**
- **cybercrime**
- **electronic evidence**
- **encrypted disk detector**
- **forensic acquisition utilities**
- **intrusion detection system**
- **network mapper**

REFERENCES

Anjani, S. T., (2014). *Cyber Forensics in Combating Cybercrimes.* www.worldwidejournals. com/paripex/recent_issues_pdf/2014/September/September_2014_1410775556_23.pdf (accessed on 10 January 2022).

Conlan, K., Ibrahim, B., & Frank, B., (2016). Anti-forensics: Furthering digital forensic science through a new extended, granular taxonomy. *Digital Investigation, 18*, S66–S75.

Cybercrime and Digital Forensics (2021). Blackhawk Intelligence London. Retrieved from: https://www.blackhawkintelligence.com/forensic-services/digital-forensics/cybercrime-digital-forensics-1/ (accessed on 10 January 2022).

Cybercrime Module 4 Key Issues: Standards and Best Practices for Digital Forensics. (2021). Retrieved from: https://www.unodc.org/e4j/en/cybercrime/module-4/key-issues/standards-and-best-practices-for-digital-forensics.html (accessed on 10 January 2022).

Fighting Cybercrime: Cybersecurity and Digital Forensics Are the New A-Team (2019). Tech-NewsWorld. Retrieved from: https://www.technewsworld.com/story/fighting-cybercrime-cybersecurity-and-digital-forensics-are-the-new-a-team-86198.html (accessed on 10 January 2022).

Frequently Asked Questions about Data Forensics, (2021). Retrieved from: https://intesecurity. com/data-forensics-faqs/ (accessed on 10 January 2022).

Garfinkel, S., (2007). Anti-forensics: Techniques, detection and countermeasures. In: 2nd *International Conference On I-Warfare and Security.*

Harbawi, M., & Varol, A., (2016). The role of digital forensics in combating cybercrimes. In: *2016 4th International Symposium on Digital Forensic and Security (ISDFS)* (pp. 138–142). doi: 10.1109/ISDFS.2016.7473532.

Holt, T., Bossler, A., & Seigfried-Spellar, K., (2015). *Cybercrime and Digital Forensics: An Introduction.* 10.4324/9781315296975.

Majed, H., & Noura, H., & Chehab, A., (2020). *Overview of Digital Forensics and Anti-Forensics Techniques.* 10.1109/ISDFS49300.2020.9116399.

Ofori, A. Y., & Akoto, D., (2020). *Digital Forensics Investigation Jurisprudence: Issues of Admissibility of Digital Evidence.* doi: 10.24966/FLIS-733X/100045.

Ofori, Y. A., Boateng, Y., & Yankson, H. G., (2019). *Relativism Digital Forensic Investigations Model.* IEEE Xplore.

Ummedcyber.com (2021). Retrieved from: http://www.ummedcyber.com/Digital-Forensics-Combating-Cybercrime-ummed-meel-kolkata-police-THE-PROTECTOR-magazine-hurdles-in-Digital-Forensics.pdf (accessed on 10 January 2022).

Urvashi, S. M., (2018). *Application of Cyber Forensics in Crime Investigation.* http://ijrar. com/upload_issue/ijrar_issue_1227.pdf (accessed on 10 January 2022).

Van, B., & Jean-Paul, (2015). Anti-forensics: A practitioner perspective. *International Journal of Cyber-Security and Digital Forensics, 4*, 390–403. 10.17781/P001593.

What is Digital Forensics? (2021). *Phases of Digital Forensics* | EC-Council. Retrieved from: https://www.eccouncil.org/what-is-digital-forensics/ (accessed on 10 January 2022).

What Procedure Should be Avoided in a Digital Forensics Investigation? (2021). Retrieved from: https://itexamanswers.net/question/what-procedure-should-be-avoided-in-a-digital-forensics-investigation (accessed on 10 January 2022).

CHAPTER 6

Clustering and Classification of Digital Forensic Data Using Machine Learning and Data Mining Approaches

E. FANTIN IRUDAYA RAJ

Assistant Professor, Department of Electrical and Electronics Engineering, Dr. Sivanthi Aditanar College of Engineering, Tamil Nadu, India

ABSTRACT

Nowadays, crime investigators collect a more significant amount of potential digital evidence from suspects, necessitating the use of digital forensic techniques. This evidence is commonly in the form of mostly unlabeled and unstructured data and seemingly unrelated information. Manually sorting and comprehending this type of data is a significant challenge, sometimes even a psychological burden, or at the very least, a prohibitively time-consuming activity. As a result, forensic research should investigate and leverage the development of robust and autonomous analysis tools for investigators confronted with this situation. With the advent of machine learning (ML) algorithms and data mining (DM) Tools, numerous autonomous systems are realized in the present modern world. The method of extracting usable information, patterns, and trends from a given data set is known as DM. The aim of DM is to make data-driven judgments from extensive data collection. ML is the process by which computers learn how to perform tasks without being explicitly programmed to do so. ML is trained using a trained data set, which teaches the computer to make decisions and find new patterns in the new data set. By adopting these two techniques in digital forensic research,

Advancements in Cybercrime Investigation and Digital Forensics. A. Harisha, Amarnath Mishra, & Chandra Singh (Eds.)
© 2024 Apple Academic Press, Inc. Co-published with CRC Press (Taylor & Francis)

the investigators can easily classify and cluster the digital data and make some quick decisions. It is also very helpful in creating autonomous forensic analysis tools for more accurate and quick analysis. The present work discusses various ML techniques and DM tools employed in digital forensic research to more effectively cluster and classify digital forensic data.

6.1   INTRODUCTION

Computers and networks have become so established in our culture, so much a part of our daily lives, that digital evidence will almost certainly be used in any investigation or legal dispute. Electronic discovery has become increasingly common in property litigation, and countries are changing their judicial systems to handle electronic evidence. In addition, computers are increasingly used in crimes such as homicide, terrorism, drug trafficking, fraud, and child abuse. In investigations of breaches into government and corporate information technology systems, digital evidence is becoming more crucial, yet obtaining it is becoming more complex as criminals get more adept at obscuring their tracks. Lawbreakers are growing more familiar with digital investigative and forensic skills, and they are increasingly employing networks and computers to commit crimes. A few are working on "anti-forensic" tactics and technology to hide their actions, erase forensic evidence, and undermine digital investigators in general.

Practitioners and researchers in the field of digital forensics have made great progress in recent years. Our technology expertise has grown, and now we've earned the essential experience to improve our techniques. We overcome key technical obstacles, facilitating practitioners' access to digital evidence easier. New forensic methodologies and techniques are being developed to aid network traffic analysis, remote system inspection, and forensic capture of volatile data. These breakthroughs hold great promise, but they also lay new digital investigations and forensics requirements, shifting the field's topography and allowing new approaches to flourish.

Today's crime investigators collect an increasing amount of potential digital evidence from suspects, necessitating the use of digital forensics procedures. Unfortunately, Digital evidence is frequently in the form of mostly unlabeled and unstructured data and information that appears to be unrelated. Manually organizing and comprehending this type of data is a significant challenge, sometimes even a psychological strain, or at the very least an activity that takes an excessive amount of time. As a result, forensic research should investigate and use cluster algorithms and unsupervised machine learning (ML) capabilities in order to develop independent and robust analysis tools for criminal investigators confronted with this predicament.

Data mining (DM) tools and ML (ML) methodologies have grown significantly and become more common in recent years. DM is a technique for extracting useful information, patterns, and trends from a set of data. The primary purpose of DM is to make data-driven decisions from large data collections. The process by which computers learn to do tasks without being explicitly programmed is known as ML. A trained data set is used to train ML, which teaches the computer how to draw conclusions and discover hidden patterns in new data. As a result, investigators can efficiently classify and cluster digital data and make speedy findings using these two digital forensic research strategies. The present work explores the various ML and DM strategies used in digital forensic research to better cluster and classifies digital forensic data.

6.2 REVIEW OF LITERATURE

Forensic information stored and created on computers is a double-edged sword, giving solid evidence in a wide range of investigations while also presenting complexity that can trip up even the most experienced investigators (Casey et al., 2009). Basic questions regarding a crime can be answered using digital evidence, such as who interacted with whom (linkage), who was accountable (attribution), when something occurred (sequencing), and where something came from (origins and source evaluation) (Kim et al., 2021). Simultaneously, the complexity of computer systems needs knowledge that individual components of digital evidence may be subject to alternative interpretations and that further evidence may be required to get the desired outcome (Razaque et al., 2021). Therefore, forensic investigators must regularly understand and use the scientific process for making the most of digital evidence. In combination with digital forensics technologies and methods, the scientific technique enables us to adapt to a wide range of scenarios and requirements while also ensuring that the results obtained are factually sound (Soltani et al., 2021).

The practitioner's thoroughness, expertise, experience, knowledge, and, in some cases, curiosity are all crucial factors in digital evidence forensic analysis (Rafique et al., 2013). Although each forensic investigation will change in certain ways depending on the resources, objectives, data set, and other circumstances, but the basic procedure remains the same (Chopade et al., 2021). Compared with traditional crimes, modern digital crimes have electronic exchange tools, money laundering, and leftovers. All these are put together to make a crime investigation more difficult for the investigators. In addition, digital forensic investigators have access to a wide variety of tools, both open-source and commercial. The main goal of the investigation is to collect the evidence using acceptable methods to make it accepted and admitted from a legal point of view (Sindhu et al., 2012).

Artificial intelligence (AI) is a well-established field of current computer science that may help solve complex and computationally huge issues in a reasonable amount of time (Mitchell et al., 2010). Digital forensics is a rapidly expanding area of computing that frequently requires the cognitive assessment of large amounts of complicated data (Costantini et al., 2019). As a result, AI looks to be a viable technique for addressing many of the current digital forensics challenges (Sikos et al., 2021). DM is a method for extracting meaningful facts, patterns, and trends from data collection. The primary goal of DM is to make data-driven judgments based on big data sets (Ageed et al., 2021; Tan et al., 2016). DM is also called knowledge discovery in database (KDD). It is one of the most important approaches for assisting entrepreneurs, researchers, investigators, and individuals in extracting vital data and making quick decisions (Nirkhi et al., 2012; Tallón-Ballesteros et al., 2014). Thus, DM techniques can be applied in digital forensics and help crime investigators investigate digital evidence.

ML employs a set of algorithms to analyze and interpret data, learn from it, and make the best judgments possible based on those learning (Usman et al., 2021; Raj et al., 2013; Qadir et al., 2021). Deep learning (DL), on the other hand, divides algorithms into numerous layers to build an "artificial neural network (ANN)." This neural network is capable of self-learning and making intelligent decisions (Ulloa et al., 2021; Chouhan et al., 2021; Shen et al., 2021). Moreover, ML and DL are both subsets of AI. Using these two methodologies in digital forensic investigation, investigators can efficiently classify and cluster digital data and arrive at quick conclusions.

We may infer from the literature that digital forensics deals with a substantial amount of data on a daily basis. Therefore, investigators desperately need automatic forensic analysis technologies for more accurate and speedy analysis. They can readily identify and cluster the massive amount of digital data and draw speedy conclusions using DM and ML algorithms. The present study gives insight into the important DM and ML approaches utilized in forensic analysis.

6.3 DIGITAL FORENSIC TOOLS

Digital forensic tools play a vital role in data collection for forensic analysis. These tools are divided into several categories, so the tool you choose will determine where and how you intend to utilize it. Here are some broad categories to give you a sense of the wide range of digital forensics tools available (Richard et al., 2006): (a) digital image forensics; (b) computer forensics; (c) disk and data capture; (d) file analysis; (e) memory forensics; (f) network forensics; (g) internet browsing analysis; (h) video/audio forensics; (i) E-mail analysis; and (j) database forensics. While this is not a full list, it provides an overview of what digital forensics tools are and what they can be used for. In addition, many tools are occasionally packed together into a single toolkit to help you tap into the possibilities of related tools. It's also worth noting that the lines between these groups can blur depending on the skill set of the employees, contractual requirements, current regulations, equipment availability, and lab circumstances. Regardless of these differences, digital forensics technologies provide a large array of options for gathering data throughout an investigation. It's also worth noting that the digital forensics arena is constantly changing, with new tools and capabilities being introduced on a regular basis to keep up with device updates. Figure 6.1 depicts the Digital Forensic process in detail.

The digital forensic process (Kohn et al., 2013) involves: (i) identification; (ii) preservation; (iii) collection; (iv) examination; (v) analysis; (vi) interpretation; (vii) documentation; and (viii) evidence presentation. The basic behind all these steps is a proper data forensic tool to collect digital evidence and manage the huge volume of collected data. It isn't easy to choose the best tool for our needs among so many possibilities. When making a decision, there are a few things to consider: (a) skill level; (b) output; (c) cost; and (d) focus are the factors to take into account. When it comes to choosing a digital forensics tool, the skill level is crucial. Some tools require rudimentary expertise, while others may necessitate extensive knowledge. A good rule of thumb is to compare your talents to what the tool requires, so you may select the most powerful tool you can use. Because tools are not all treated fairly, outputs will vary even within the same category. Some tools will only return raw data, while others will generate a complete report that can be shared with non-technical personnel right away. In some circumstances, raw data is sufficient because your information will be processed further anyhow; however, in others, having a structured report would make your task easier. Cost is, of course, a major consideration, as most departments are on a tight budget. One thing to bear in mind is that the cheapest tools may not have all of the functionality you desire because that is how developers keep costs down. Instead of choosing a tool only based on price, consider finding a balance between price and features. Another important consideration is the tool's focus area, as different jobs normally necessitate distinct tools. Tools for evaluating a database, for example, are significantly different from those required to study a network. Before purchasing, it is best to make a comprehensive list of feature requirements. As previously stated, some tools may perform many functions in a single kit, which may be a more cost-effective option than purchasing separate equipment for each operation. These tools will acquire a vast amount of digital data, which will be stored in a cloud database. From the investigator's standpoint, there is critical to evaluate such huge amounts of data and uncover meaningful patterns and information. DM and ML techniques are being used to accomplish this goal.

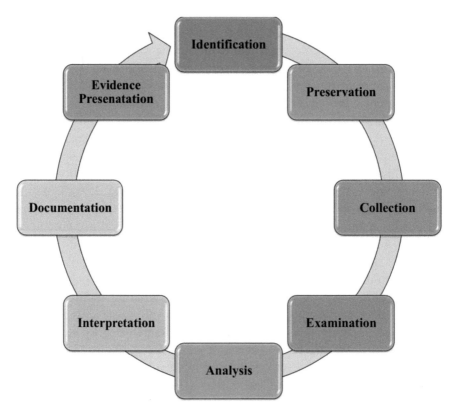

FIGURE 6.1   Digital forensic process.

6.4 DATA MINING (DM) TECHNIQUES FOR DIGITAL FORENSIC

Data mining (DM) is the technique of obtaining information from vast amounts of data in the database in order to identify significant data, trends, and patterns that will aid the investigator in making a data-driven decision. The goal of DM is to make data-driven judgments based on massive data sets. DM is associated with Data Science, which is performed by a person in a specific situation, on a specific data set, and with a specific goal in mind. DM technique comprises video and audio mining, social media mining, graphical DM, text, and web mining, among other services. It's done with either simple or extremely specific software (Quick et al., 2014). Although there is a huge amount of information available on multiple platforms, there is a scarcity of expertise. Therefore, the most difficult task in digital forensics is to evaluate data in order to extract essential information that may be used to solve cases.

DM can also be defined as a tool for examining data patterns that pertain to specific perspectives. It aids us in classifying the data and turning it into meaningful knowledge and information. This vital information is then captured, either for storage in database systems or for DM and to assist in making a decision. There are various procedures involved in DM implementation in digital forensics. The various DM procedures involved in Digital forensic is depicted in Figure 6.2.

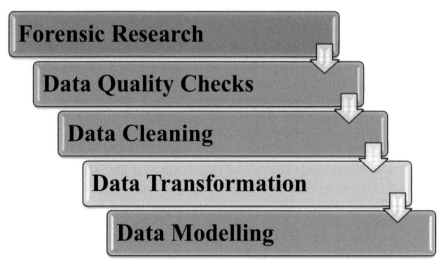

FIGURE 6.2   Data mining procedures involved in digital forensic.

Before we start, we must have in-depth knowledge about the current scenarios, objectives, and resources available that meet the criteria of forensics. Then, it would help you develop a precise data extraction plan to meet the digital forensic goals efficiently. In order to ensure that no data integration bottleneck exists, data must be revised and compared as it is acquired from numerous sources. Before mining, quality assurance examines any underlying data anomalies, such as misplaced data interpolation, to ensure the data is in good shape. Prior to mining, the selection, cleaning, structuring, and encryption of data are estimated at 80% of the effort. The steps involved in data transformation, which consists of six sub-stages, prepare data for final data sets (Suaib et al., 2020): (a) smoothing; (b) retrieving; (c) summary; (d) generalization; (e) normalization; and (f) data attribute construction is all part of it. Various mathematical models are used in the data set to help find data patterns based on a range of conditions.

6.4.1 TYPES OF DATA

DM can be applied to the following types of data (Ch et al., 2021). They are:

1. **Transactional Database:** It is a database that stores record that has been captured as transactions. Flight bookings, consumer purchases, website clicks, and other transactions are examples of these types of transactions. A unique ID is assigned to each transaction record. It also includes a list of all the elements that made the transaction possible.

2. **Data Warehouse:** It is a single data storage site that gathers data from various sources and organizes it into a coherent strategy. Data is cleaned, integrated, loaded, and refreshed before being stored in a data warehouse. A data warehouse's data is divided into numerous sections. We can obtain the summary of the information stored in the data warehouse even after a very long time.

3. **Data Stored in the Database:** Every database management system (DBMS) holds data connected in some way. It also features a suite of software packages for managing data and facilitating access to it. These programs are used for various tasks, including designing database structure, ensuring that stored data is consistent and secure, and managing various types of data access, such as concurrent, distributed, and shared. A relational database contains tables with names and properties that hold rows or records from massive data collections. A unique key is assigned to each record in a table.

4. **Additional Data Types:** Other kinds of data are known for their adaptability, semantical significance, and structure. They're employed in a variety of situations. Spatial data, multimedia data, graph data, sequence data, data streams, engineering design data, and more are just a few of the data types available.

6.4.2 VARIOUS DATA MINING (DM) TECHNIQUES

DM is the process of using advanced data analysis technologies to find previously unknown, valid patterns and linkages in big data sets. This technology can make use of mathematical methods like neural networks, statistical models, and ML approaches like decision trees altogether. As a result, DM is used to make predictions as well as data analysis. Different fundamental DM techniques have been created and used for contemporary DM applications, including association, sequential patterns, clustering, prediction, classification, and regression (Han et al., 2011). Figure 6.3 represents the major DM techniques.

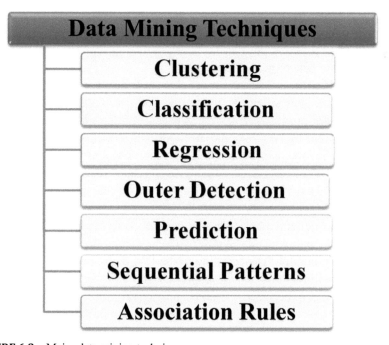

FIGURE 6.3 Major data mining techniques.

1. **Clustering:** This approach generates significant item groups with comparable characteristics. Often people misinterpret it with categorization; however, if they comprehend how these two strategies work, this is not a problem. Clustering allocates objects to classes it produces, contrary to categorization, which assigns items to predetermined classifications. Identical data elements are grouped in the same clusters in this technique. Using metrics to aid maximal data association based on dissimilarities is a main ingredient of groupings. The data is described by a few clusters, which sacrifices some details but improves the overall quality. It uses clusters to model data. Clustering is historically anchored in statistics, mathematics, and numerical analysis, as demonstrated by data modeling. Clusters are related to hidden patterns in ML, the search for clusters is unsupervised learning, and the framework follows a data paradigm. Therefore, clustering is quite useful in DM applications from a practical point of view.

   Medical diagnostics, computational biology, Web analysis, spatial database applications, information retrieval, text mining, forensic analysis, and scientific data exploration are just a few examples of clustering in real-time. In other words, clustering analysis is a DM technique for identifying data that is similar. This technique aids in recognition of data differences and similarities. Clustering is similar to classification in that it incorporates grouping data fragments together based on similarities.

2. **Classification:** The origins of this technology may be traced back to ML. It classifies elements of a data source into predetermined classes or groups. Classification uses techniques like statistics, linear programming, ANNs, and decision trees, among others, in DM. Classification is a design and development approach that allows a program to categorize components in a data set into various categories. This method is used to extract vital and relevant data. This method facilitates the classification of data into various groups. They are: (a) dependent on the sort of data sources mined; (b) depends upon the data type involved; (c) depends upon the type of knowledge discovered; and (d) according to the DM technique used.

3. **Regression:** It is a DM approach for analyzing and identifying correlations between variables that occur due to the existence of another factor. It's being used to express the probability of a certain variable occurring. It is a modeling and planning methodology. We may use it, for example, to predict costs based on other factors like

competitiveness, consumer demand, and data availability. It is mostly used to determine the exact link between two or more variables in a data set.

4. **Outer Detection:** This DM technique is concerned with identifying data elements within a data set that do not conform to expected behavior or pattern. This technique applies to various fields, including fraud detection, intrusion detection, and surveillance. Furthermore, it is referred to as Outlier mining or Outlier Analysis. Outliers are data points that deviate significantly from the remainder of the data set. An outlier exists in the vast majority of real-world data sets. Thus, in the realm of DM, outlier detection is crucial. Outlier detection is useful in various sectors, including recognizing outliers in wireless sensor network data, detecting credit or debit card fraud, detecting network interruptions, and so on.

5. **Prediction:** This persuasion DM technique, as the name implies, assists investigators in matching patterns based on current and past data records in order to forecast future studies. In addition, this approach predicts the relationship between dependent and independent variables. Prediction, in general, involves a combination of various DM techniques, including classification, trends, clustering, trends, and so on. While some solutions rely on ML and AI, others can be completed using simple algorithms. To forecast a future event, it examines prior events or instances in the correct order.

6. **Sequential Patterns:** It is a technique that examines the database to find sequential patterns. This method encompasses discovering unique subsequences inside a set of sequences. The value of a sequence is quantified using various characteristics such as length, occurrence frequency, time, etc. In other words, this technique aids in the discovery or recognition of similar patterns in historical transaction data.

7. **Association:** Out of most of the techniques in DM, it is one of the most frequently employed. The relationship and transaction between its elements are employed to find a pattern in this technique. That's why it is often referred to as a relationship technique. It also aids in the discovery of a connection between two or more items. In the data set, it uncovers a hidden pattern. Association rules are if-then statements that help to indicate the likelihood of data item interconnections in large data sets across different databases. It is a widely used technique to enhance sales correlations in data or medical data sets, and it serves a multitude of purposes.

6.4.3 *DATA MINING (DM) ARCHITECTURE*

The actual data sources are text files, the World Wide Web, data warehouses, databases, and other documents. To be successful, DM necessitates a significant volume of historical data. Organizations commonly store data in data warehouses or databases. Spreadsheets, text files, one or more databases, and other data repositories may be found in data warehouses. The data must be selected, integrated, and cleansed before being sent to the data warehouse or database server. Because the data originates from various sources and in a variety of formats, all of that can't be used immediately for DM because it's not always complete and accurate. As a result, the initial data must be homogenized and cleaned. The necessary amount of data will be acquired from a large number of sources. Then just the most relevant and appropriate data is chosen and sent to the server. As a result, in response to a user request, the server is responsible for retrieving appropriate data from the database based on DM.

Any DM system must include a DM engine. It contains modules for DM tasks such as time series analysis, prediction, clustering, classification, characterization, and association. To put it another way, DM is the bedrock of our DM architecture. It is made up of software and tools that are used to extract information and insights from data gathered from various sources and stored in a database. The Pattern evaluation unit is primarily in charge of quantifying the pattern using a given threshold. It collaborates with the DM engine to reduce the results down to distinct and unique patterns. Stakeholder analysis is frequently used in conjunction with DM modules in this industry to narrow the search to important trends. For example, it may employ a stake threshold to eliminate previously observed patterns.

The DM system and the user communicate through the graphical user interface (GUI) module. It allows the user to use the system quickly and effectively without grasping the process's intricacies. This module interfaces with the DM system to send the results when the user submits a task or a query. The knowledge base is beneficial throughout the DM process. The knowledge base also contains user opinions and data from user experiences. It could be useful for directing the search or determining the significance of the pattern results. To enhance the trustworthiness and accuracy of the outcome, the DM engine may leverage inputs from the knowledge base (Gampala et al., 2020). Figure 6.4 depicts the architecture of DM.

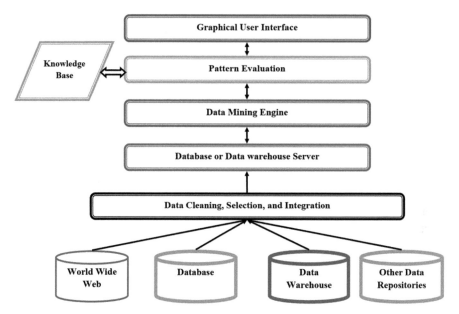

FIGURE 6.4 Architecture of data mining.

6.5 MACHINE LEARNING (ML) TECHNIQUES FOR DIGITAL FORENSIC

The process of discovering algorithms that have improved as a result of data-driven experience is known as ML. The process of creating, examining, and developing algorithms that allow machines to learn with no need for human involvement is known as algorithm development. ML and DM are both part of data science, making sense considering that they both use data. Many people unintentionally use the phrases interchangeably because both approaches are used to address complex problems, given the fact that ML is occasionally used to undertake useful DM (Mohri et al., 2018). Data from DM, on the other hand, can be utilized to train machines.

Furthermore, both processes employ the same fundamental data pattern detection algorithms. ML teaches a computer how to learn and interpret the parameters that are given to it, whereas DM is used to derive rules from massive amounts of data. Figure 6.5 shows the relationship between AI, statistics, DM, and ML.

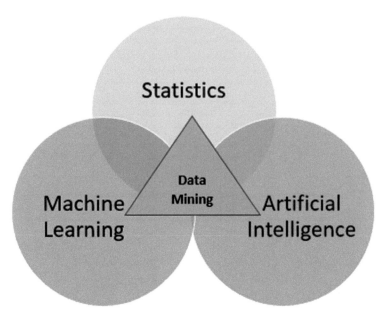

FIGURE 6.5   Relationship between AI, DM, ML, and statistics.

Without being explicitly designed, ML allows a machine to learn from data, improve performance based on previous experiences, and predict future events. ML algorithms create a mathematical model using past data samples, referred to as training data, which helps them make judgments or predictions without being explicitly programmed. The life cycle of ML is a cyclical method for designing and implementing a successful ML project. The primary goal of the life cycle of ML is to find a solution to the problem (Ashmore et al., 2021). Figure 6.6 depicts the seven major steps in the ML life cycle.

Collecting data is the first phase in the life cycle process of ML. Data could be collected from different sources, including mobile devices, the internet, databases, and files. Therefore, we must first find the various data sources. The output's efficiency will be determined by the quality and quantity of the data collected. The more data there is, the more accurate the prediction will be. This process entails: (a) identifying multiple data sources; (b) collecting data; and (c) combining data from diverse sources. Next, we must prepare the data for further processing after it has been collected. Data preparation comprises storing our data in the right place and preparing it for ML training. That whole phase will get categorized into two different sections. They are Data exploration and data pre-processing.

Data wrangling is the stage of converting and cleaning unstructured and raw data into a useful format. It is the method for extracting data, picking a variable to use, and converting the data into a format appropriate for analysis in the next phase. The data has been cleaned and formatted and is now ready for analysis. This process entails: (a) choosing analytic approaches; (b) creating models; and (c) analyzing the results. The goal of this step is to create a ML model that will analyze data using various analytical approaches and then assess the results. The model is then trained, which involves increasing the model's performance in order to reach a better solution to the problem. Next, the model is trained to utilize data sets, which are then utilized for training the model using various ML approaches. Finally, we test the developed ML model once it has been trained on a specific data set. The test data is feed into the developed model and make it ensure that it is rightly working.

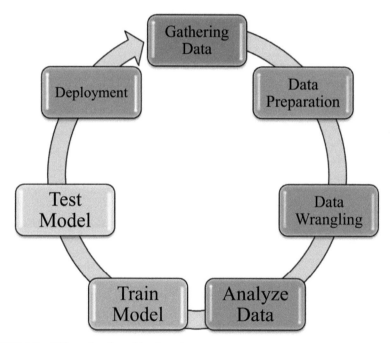

FIGURE 6.6   Life cycle of machine learning process.

6.5.1   *CLASSIFICATION OF MACHINE LEARNING (ML) ALGORITHMS*

The ML algorithms are categorized into four different groups. They are (a) regression under supervised learning; (b) classification under supervision learning; (c) dimension reduction under unsupervised learning; and (d) clustering under unsupervised learning. Regression falls under the topic of supervised learning and includes techniques such as linear regression, decision trees, and so on. The classification under the supervised learning category includes various algorithms such as Linear SVM, Naive Bayes, and decision tree types. The category of unsupervised learning-dimension reduction methods includes techniques such as principal component analysis and latent Dirichlet analysis. Finally, there are numerous algorithms described under the Unsupervised learning – clustering domain. They are primarily utilized in applications that make use of clustering. Among these are K-means, K-nodes, and Gaussian mixture models (Raj et al., 2021; Priyadarsini et al., 2020). Figure 6.7 shows the detailed classification of various ML algorithms.

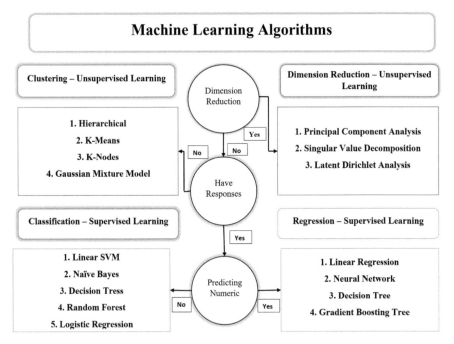

FIGURE 6.7   Detailed classification of different machine learning algorithms.

In the algorithms mentioned above, a few important algorithms like K-NN algorithm, support vector machine (SVM) are mainly used in digital forensics.

6.5.2 K-NN ALGORITHM

The K-NN (Nearest Neighbor) algorithm is a simple and effective ML algorithm based on supervised learning (Khan et al., 2018). This approach considers that the new and old data are similar, and it assigns the new case to the category that is closest to the current categories. Thus, the K-NN approach saves all available data and categorizes new data points depending on their similarity to the current data. It means that by utilizing the K-NN approach, fresh data can be swiftly sorted into a well-defined category. Although this method can be used for both regression and classification, it is most typically employed for classification. Figure 6.8 explains the K-NN algorithm.

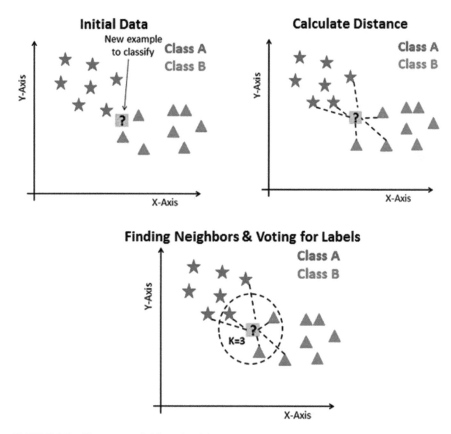

FIGURE 6.8 K-nearest neighbor algorithm.

The working of the K-NN algorithm is explained in the following steps:

- Choose the number K of the neighbors;
- Calculate the Euclidean distance between K of your closest neighbors;
- Based on the Euclidean distance, select the K closest neighbors;
- Count how many data points there are in each category among the k neighbors;
- Assign the newly acquired data points to the category with the most neighbors;
- Finally, our test machine learning model is getting ready.

The K-NN approach is simple to implement and robust to noisy training data. As a result, it may be more successful if the training data is large. However, the main downside of this method is that it requires continual determination of the value of K, which might be challenging at times. Furthermore, the calculation cost is high because it takes time to calculate the distance between data points for all training samples.

6.5.3 SVM ALGORITHM

The support vector machine (SVM) algorithm is another important and most frequently used algorithm in many recent machine-learning-based applications. It's a supervised ML algorithm that can handle regression and classification tasks. However, it is mostly used to tackle classification problems. The value of each feature is the value of a specific coordinate in the SVM methodology, and each data item is represented as a point in n-dimensional space (Zhao et al., 2020), where n is the number of distinct features. Then we locate the hyper-plane that clearly separates the two classes to complete classification. Figure 6.9 shows the graphical representation of the SVM algorithm.

Both linearity and non-linearity input can be classified using the SVM method. It starts by mapping each data point through an n-dimensional attribute vector, with n denoting the total number of attributes. The hyperplane that divides these data components into two groups is then found, with the minimum separation for both categories maximized and classification errors reduced. The selection hyperplane's length and the class's neighboring occurrence define the least closeness for a grouped class. Each feature vector is first displayed as a point in an n-dimensional space, with each feature's value corresponding to a specific coordinate. To finish the categorization, we must first find the hyperplane that divides the two categories by the greatest distance.

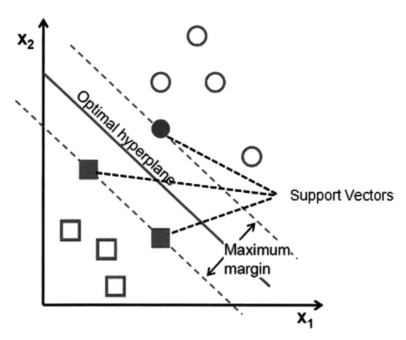

FIGURE 6.9　SVM algorithm.

SVM works well with a clear separation margin and is effective in high-dimensional spaces. It uses a subset of training points (called support vectors) in the decision function. Therefore, The SVM is memory efficient. When the number of dimensions exceeds the number of samples, this method works well.

6.6　CONCLUSION

The current work discusses the importance of the adoption of ML techniques and DM approaches in digital forensics. From the literature, it is inferred that the requirement of advanced approaches to handle the vast volume of data in digital forensics. The present study starts with discussing various digital forensic tools employed in recent digital forensics, and it also explains the different types of data collected using these digital forensic tools. Various important DM techniques that can be adopted in digital forensic are explained with DM architecture. Then, the detailed classification of ML techniques used in digital forensics is explained. The life cycle of the ML process and the important ML methodologies like K-NN and SVM algorithms are also getting explained. From this, we conclude that the inclusion of DM techniques and ML Techniques in modern digital forensics makes it more precise and accurate. It will help investigators retrieve meaningful information from a large volume of data and interpret and use it effectively.

KEYWORDS

- **artificial intelligence**
- **artificial neural network**
- **clustering**
- **data mining**
- **digital forensic**
- **machine learning**
- **pattern recognition**

REFERENCES

Ageed, Z. S., Zeebaree, S. R., Sadeeq, M. M., Kak, S. F., Yahia, H. S., Mahmood, M. R., & Ibrahim, I. M., (2021). Comprehensive survey of big data mining approaches in cloud systems. *Qubahan Academic Journal, 1*(2), 29–38.

Ashmore, R., Calinescu, R., & Paterson, C., (2021). Assuring the machine learning lifecycle: Desiderata, methods, and challenges. *ACM Computing Surveys (CSUR), 54*(5), 1–39.

Casey, E., (2009). *Handbook of Digital Forensics and Investigation*. Academic Press.

Ch, G., Jana, S., Majji, S., Kuncha, P., & Tigadi, A., (2021). Diagnosis of COVID-19 using 3D CT scans and vaccination for COVID-19. *World Journal of Engineering*.

Chopade, R., & Pachghare, V., (2021). Evaluation of digital forensic tools in mongoDB database forensics. In: *Progress in Advanced Computing and Intelligent Engineering* (pp. 427–439). Springer, Singapore.

Chouhan, A. S., Purohit, N., Annaiah, H., Saravanan, D., Raj, E. F. I., & David, D. S., (2021). A real-time gesture based image classification system with FPGA and convolutional neural network. *International Journal of Modern Agriculture, 10*(2), 2565–2576.

Costantini, S., De Gasperis, G., & Olivieri, R., (2019). Digital forensics and investigations meet artificial intelligence. *Annals of Mathematics and Artificial Intelligence, 86*(1), 193–229.

Gampala, V., Kumar, M. S., Sushama, C., & Raj, E. F. I., (2020). Deep learning-based image processing approaches for image deblurring. *Materials Today: Proceedings*.

Han, J., Pei, J., & Kamber, M., (2011). *Data Mining: Concepts and Techniques*. Elsevier.

Khan, M. K., Zakariah, M., Malik, H., & Choo, K. K. R., (2018). A novel audio forensic data-set for digital multimedia forensics. *Australian Journal of Forensic Sciences, 50*(5), 525–542.

Kim, S., Jo, W., Lee, J., & Shon, T., (2021). AI-enabled device digital forensics for smart cities. *The Journal of Supercomputing*, 1–16.

Kohn, M. D., Eloff, M. M., & Eloff, J. H., (2013). Integrated digital forensic process model. *Computers & Security, 38*, 103–115.

Mitchell, F., (2010). The use of artificial intelligence in digital forensics: An introduction. *Digital Evidence & Elec. Signature L. Rev., 7*, 35.

Mohri, M., Rostamizadeh, A., & Talwalkar, A., (2018). *Foundations of Machine Learning*. MIT Press.

Nirkhi, S. M., Dharaskar, R. V., & Thakre, V. M., (2012). Data mining: A prospective approach for digital forensics. *International Journal of Data Mining & Knowledge Management Process, 2*(6), 41.

Priyadarsini, K., Raj, E. F. I., Begum, A. Y., & Shanmugasundaram, V., (2020). Comparing DevOps procedures from the context of a systems engineer. *Materials Today: Proceedings*.

Qadir, S., & Noor, B., (2021). Applications of machine learning in digital forensics. In: *2021 International Conference on Digital Futures and Transformative Technologies (ICoDT2)* (pp. 1–8). IEEE.

Quick, D., & Choo, K. K. R., (2014). Data reduction and data mining framework for digital forensic evidence: Storage, intelligence, review and archive. *Trends and Issues in Crime and Criminal Justice*, (480), 1–11.

Rafique, M., & Khan, M. N. A., (2013). Exploring static and live digital forensics: Methods, practices and tools. *International Journal of Scientific & Engineering Research, 4*(10), 1048–1056.

Raj, E. F. I., & Balaji, M., (2021). Analysis and classification of faults in switched reluctance motors using deep learning neural networks. *Arabian Journal for Science and Engineering, 46*(2), 1313–1332.

Raj, E. F. I., & Kamaraj, V., (2013). Neural network based control for switched reluctance motor drive. In: *2013 IEEE International Conference on Emerging Trends in Computing, Communication and Nanotechnology (ICECCN)* (pp. 678–682). IEEE.

Razaque, A., Aloqaily, M., Almiani, M., Jararweh, Y., & Srivastava, G., (2021). Efficient and reliable forensics using intelligent edge computing. *Future Generation Computer Systems, 118*, 230–239.

Richard, I. I. I., & Roussev, V., (2006). Digital forensic tools: The next generation. In: *Digital Crime and Forensic Science in Cyberspace* (pp. 75–90). IGI Global.

Shen, Z., (2021). *Deep Learning on Image Forensics and Anti-forensics*. Doctoral dissertation, New Jersey Institute of Technology.

Sikos, L. F., (2021). AI in digital forensics: Ontology engineering for cybercrime investigations. *Wiley Interdisciplinary Reviews: Forensic Science, 3*(3), e1394.

Sindhu, K. K., & Meshram, B. B., (2012). Digital forensic investigation tools and procedures. *International Journal of Computer Network and Information Security, 4*(4), 39.

Soltani, S., Seno, S. A. H., & Budiarto, R., (2021). Developing software signature search engines using paragraph vector model: A triage approach for digital forensics. *IEEE Access, 9*, 55814–55832.

Suaib, M., Akbar, M., & Husain, M. S., (2020). Digital forensics and data mining. In: *Critical Concepts, Standards, and Techniques in Cyber Forensics* (pp. 240–247). IGI Global.

Tallón-Ballesteros, A. J., & Riquelme, J. C., (2014). Data mining methods applied to a digital forensics task for supervised machine learning. In: *Computational Intelligence in Digital Forensics: Forensic Investigation and Applications* (pp. 413–428). Springer, Cham.

Tan, P. N., Steinbach, M., & Kumar, V., (2016). *Introduction to Data Mining*. Pearson Education India.

Ulloa, C., Ballesteros, D. M., & Renza, D., (2021). Video forensics: Identifying colorized images using deep learning. *Applied Sciences, 11*(2), 476.

Usman, N., Usman, S., Khan, F., Jan, M. A., Sajid, A., Alazab, M., & Watters, P., (2021). Intelligent dynamic malware detection using machine learning in IP reputation for forensics data analytics. *Future Generation Computer Systems, 118*, 124–141.

Zhao, L., Chen, Y., & Sheng, V. S., (2020). A real-time typhoon eye detection method based on deep learning for meteorological information forensics. *Journal of Real-Time Image Processing, 17*(1), 95–102.

CHAPTER 7

Digital Forensic Investigation: Ontology, Methodology, and Technological Advancement

SUSMITA SINGH[1] and VINEET KUMAR SINGH[2]

[1]*Assistant Professor, Department of Chemistry, Amity Institute of Applied Science, Amity University, Kolkata, West Bengal, India*

[2]*Marine Faculty (Marine Engineer), Sensea Maritime Academy, Kolkata, West Bengal, India*

ABSTRACT

From the genesis of a Digital World, there has always been a genuine concern regarding the exponential and uninhibited mutation of a multi-faceted digitalized domain. The unconstrained interconnectivity has bred numerous digital infections and viral platforms which aim to infiltrate our safety and jeopardize the system. The dangers lurking within the uncontrolled digital realm of the "www" has influenced a dexterous methodology to fortify the security, inhibit crime and facilitate deft investigations, by effortless identification of the digital illegitimate footprints, thereby giving birth to "digital forensics." It is the application of methodical and precise fact-finding techniques to digital felonies and incidents. An appreciation and exercise of OSINT (open-source intelligence) and Ontology in Digital Forensics, to determine morsels of the so called "unidentified sources of the digital domain" which can infiltrate even the most secure-of-sites, has gifted us with the leverage to sketch an amassed giant depiction, thereby further exposing and exploiting frailties of the system. Digital investigation

Advancements in Cybercrime Investigation and Digital Forensics. A. Harisha, Amarnath Mishra, & Chandra Singh (Eds.)
© 2024 Apple Academic Press, Inc. Co-published with CRC Press (Taylor & Francis)

of such a nature, incorporates an algorithm which commences from retrieval and analysis of the data to a probable source of proof presentable in the Judiciary. In this requisite topic, the essential algorithm to be implemented for successful and formidable investigation, is being detailed. "Necessity is the mother of inventions"–this illuminative proverbial truth has led to breath-taking technological inventions like Rydberg Quantum Receiver, RAVEN, Axiom Cyber, ArTHIR, and many more, which will be elaborated in this work. The new normal, since the fall of 2019 has made the world completely rely on digital modes ranging from confidential government data pertaining to Nation's security to our bank accounts. Findings gathered from both government and commercial sites infer that there is a potential risk of pilferage. New digital viral vectors are the "need of the hour." The prime intention of the chapter is to give an analytical and scientific style pertaining to Methodology and Technological Advancements in Digital Forensic Investigation.

7.1 INTRODUCTION

From the Genesis of a Digital World, there has always been a genuine concern regarding the exponential and uninhibited mutation of a multi-faceted digitalized domain. The unconstrained interconnectivity has bred numerous digital infections and viral platforms which aims to infiltrate our safety and jeopardize the system (Halder & Jaishankar, 2011). The dangers lurking within the uncontrolled digital realm of the "www" has influenced a dexterous methodology to fortify the security, inhibit crime and facilitate deft investigations by effortless identification of the digital illegitimate foot-prints. This gave birth to "Digital forensics."

It is the application of methodical and precise fact-finding techniques to digital felonies and incidents (http://archive.ncpc.org/). An appreciation and exercise of OSINT (open-source intelligence) in "digital forensics" to determine morsels of the so called "unidentified sources of the digital domain" which can infiltrate even the most secure-of-sites, has gifted us with the leverage to sketch an amassed giant depiction, thereby further exposing and exploiting frailties of the system. "Necessity is the mother of inventions." This illuminative proverbial truth has led to breath-taking technological inventions and advancements like "Rydberg Quantum Receiver," "RAVEN," "Axiom Cyber," "ArTHIR" and many more (Singh et al., 2020). Let us get started…

7.2 DIGITAL FORENSICS

Digital forensics provides means to deal with cybercrimes using technically proven systems to exhume "potential digital data" that can be presented in the Judiciary. Besides this, it delivers an analytical configuration that empowers criminal detectives to bridge a bond between the perpetrator and the offense. Digital forensics is the fastest growing scientific discipline which stems from many branches. The different branches (https://recfaces.com/articles/digital-forensics#:~:text=Digital%20forensics%20tools%20can%20be%20divided%20into%20several,Mac%20OS%20analysis%20tools%3B%20 8%20Database%20forensics%20tools) of digital forensics are discussed in subsections.

7.2.1 COMPUTER FORENSICS

'Computer forensics' specializes in compendium, recognition, conservation, and evaluation of records from an individual's computer, laptop, and storage computing devices. A computer forensic expert is primarily implicated in the analysis of computer violations. However, their services are often required in civil cases pertaining to data recovery.

7.2.2 MOBILE DEVICE FORENSICS

'Mobile Devise Forensics' specializes in retrieving information's from cellular devices, SIM cards, tablets, GPS devices, game consoles and PDAs. The specialist retrieves audial and graphical information's, call lists and contacts from the devices to be served as evidence.

7.2.3 NETWORK FORENSICS

'Network forensics' specializes in monitoring, registering, and analyzing network activities. A network professionals analyzes activities in case of defense contraventions and cyber-attacks.

7.2.4 FORENSIC DATA ANALYSIS

A division of forensics which specializes in analyzing organized and defined statistics. An FDA (forensic data analyst) is principally concerned in probing into business violations and frauds.

7.2.5 DATABASE FORENSICS

A DFA (database forensic analyst) investigates any unauthorized access to a data bank and accounts for the alterations made in the data. This division of forensics can be used to authenticate business-related treaties and to scrutinize substantial monetary misdeeds.

7.2.6 E-MAIL FORENSICS

An EFA (e-mail forensics analyst) expertise in retrieving pertinent records from electronic mail. The retrieved data ranges from the 'despatchers' and 'acceptors' identification to the substance of the messages, metadata, time-stamps, and sources. "E-mail forensics" tools are extensively applied on a company suspected of e-mail forgery.

7.2.7 MALWARE FORENSICS

A specialist in this area detects, analyzes, and investigates various malwares to hunt down accused and motives behind the incident. He/she also evaluates the harm caused by the attack and determines the malware code.

7.2.8 MEMORY FORENSICS

'Memory Forensics' or 'Live Acquisition' specializes in retrieving data from the Random-Access Memory. At present criminals abstain from any digital footprints on the hard drives. In cases such as these, 'Memory Forensics' aids in tracking down the attack and the hacker.

7.2.9 WIRELESS FORENSICS

'Wireless Forensics' employs specialized instruments and procedures to analyze and scrutinize traffic in a wireless world. This branch of Forensics helps in analysis when digital violations are committed through the infringement of safety conventions in wireless networks.

7.2.10   DISK FORENSICS

A Disk Forensics Analyst specializes in retrieving and recovering records from the physical storage devices, such as servers, memory cards, external USB sticks and flash drives. He/she works in recovering any files pertinent to the investigation, analyzes it, and presents as a testimony in the court.

7.3   ONTOLOGY

It has been observed that during an ongoing "Digital Forensic Investigation," investigators often must deal with new technical and non-technical terminologies. This therefore requires an ontological (thefreelibrary.com) point-of-view to determine the significance of digital forensic jargons in present digital forensic instruments. In the words of Uschold & Gruninger (1996) ontology refers to a mutual appreciation and knowledge of some area of concern which is used as a coalescing structure in unraveling challenges. Appearing in the journal of Karrie & Venter (2014) the concept of Ontology is extensively applied in various areas as a procedure for demonstrating and rationalizing about the field information. Thus, it implies that the field of ontology has predictably matured to cover disparate fields of interest among various stages. Despite, pervasive ontology applications in several spheres, the growth and advancement of ontology and their functions in digital forensics is an ever-growing need. In a supportive statement Noy & Musen (2004) states that ontology has become a semasiological network spine and a universal proposition in evidence collection systems that accelerates varied functions. It comprises of c an action or process of forming a concepts or ideas of that are targeted at coding particular fragments of expertise simultaneously so that easy and uncomplicated conclusions can be drawn for solving problem. Besides, as per Benjamín, Chandrasekaran & Josephson's, ontology can be utilized to deliver evaluation of the composition of intelligence and understanding beyond the establishment of the main frame of any system of information interpretation for that sphere (Chandrasekaran, Josephson, & Benjamins, 1999). For reasons such as the ones rendered above, it is essential to create ontologies that will classify and distinguish the standard expressions in which the combined expertise in this subject can be exemplified (Gruber, 1993). The ontological tactics for deciding the meaning of "Digital Forensic" terminologies has been divided into four parts exhibited in Figure 7.1.

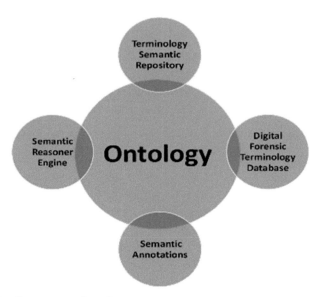

FIGURE 7.1   Components of ontology.

7.3.1   DIGITAL FORENSIC TERMINOLOGY DATABASE (DFTD)

A DFTD or "term-base" is a catalog comprising of model-tailored terminological records and associated info in polyglot design (Wright & Budin,
2001). It is a potent means of promoting uniformity and reliability in the use
of terminology. The jargons of the term base may incorporate, among other
specifics, the meaning, origin, context-of-use, and domain area. The growth
and progress of an ontological method to solve the connotation of digital
forensic lingoes commences with a lexicon archive. With the evolution in
technological innovations in digital forensics, creative, and new terms are
regularly proposed into the domain and new connotations are assigned to
existing terms. Establishment of a lexicon archive will thereby ensure uniformity and precision of terminology in a digital forensic analysis procedure.
An expertly designed lexicon archive will assist detectives to reduce costs,
save time, maintain consistency and improve quality during examinations.

7.3.2   SEMANTIC ANNOTATIONS

"Semantic annotation" is a process of articulating information about a
specific source, phrase or terminology (Karrie & Kebande). The process

includes affixing names, qualities, and traits, descriptions, comments, etc., to domain terminologies. Semantic annotation serves as an excellent means for delivering all the knowledge (together with added metadata) about an active domain lexicon which will start from the glossary-archive.

A typical "Semantic Annotation" exercises a tri-gear mechanism shown in Figure 7.2.

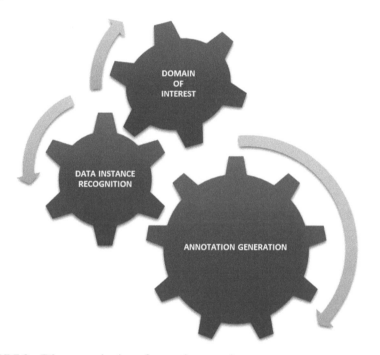

FIGURE 7.2   Tri-gear mechanism of semantic annotation.

The inter connectory gear system makes it conceivable to relate one domain lingo to another. It even aids in annotating several cyber-space forensic terminologies that goes through the cognitive engine and saved in the "semantic repository" for use by any digital forensic tool.

1. **Domain of Interest:** Ontology provides an account of the domain of interest (digital forensics).
2. **Data Instance Recognition:** It aids in discovering all the occurrences of significance in an objective document or ontology-centered lexicon.
3. **Annotation Generation:** It creates a semantic sense for each annotated document or terminology. Through this, any ontology-conscious

system can understand the target document or terminology under consideration.

7.3.3   SEMANTIC REASONER ENGINE

A "Semantic Reasoner" is a system that can deduce rational outcomes from a collection of stated facts or truisms, thereby providing a terminology semantic repository. Besides, it also creates inferences from the accessible gloss information using analytical techniques. The engine can make a decision quickly and effectively in the execution of AI and information-based methods. All existing digital forensic tools can be linked to the "Terminology Semantic Repository" for the effortless grasp of the terminologies, during digital forensic investigation.

7.3.4   TERMINOLOGY SEMANTIC REPOSITORY

It is a sizable and defined set of texts deposited in a knowledge-based format which provides an extremely useful foundation for resolving the meanings of various terminologies.

7.4   COMMON TERMINOLOGY

By now it is evident that terminological ignorance hinders the smooth flow of Digital Forensic Investigation (http://www.forensicsciencesimplified.org/digital/glossary.html). The 'scientific working group on digital evidence' (SWGDE) in alliance with 'Scientific Working Group on Imaging Technology' (SWGIT) has created and constantly updating a thesaurus of terms used in the digital realm. SWGDE has implemented ASTM, an established universal organization, to establish international recognition of terminology.

Some common terms include:

1. **Cloud Computing:** Evidence like applications and software that are recovered from the Internet through a web browser, desktop or cellular phone application. Information is collected on servers at a distant site.

2. **Data:** Facts and statistics collected for reference or analysis.

3. **Data Extraction:** It is where data is analyzed minutely to recover relevant information from data bank (like a database) in a specific manner. Further data processing involves data integration, adding metadata and another process in the data workflow.

4. **Encryption:** The process of converting information or data into a code, especially to prevent unauthorized access.

5. **File Format:** It is a structure by which data is structured in a file.

6. **Forensic Wipe:** It is a confirmable technique used for sterilizing a well-defined region of digital media by over-writing each byte with a well-known value. This prevents adulteration of data.

7. **Handheld (Mobile) Devices:** These are handy data storage devices that deliver digital photography, entertainment, navigation systems, etc.

8. **Hash or Hash Value:** It is a function that can be utilized to navigate data of an arbitrary size onto data of a fixed size.

9. **Log File:** It is a record of events, activities, and associated data.

10. **Media:** These are entities on which information can be saved. It includes thumb drives, hard drives, floppy discs, CD/DVD, SIM cards, etc.

11. **Metadata:** A set of data that describes and gives information about other data.

12. **Partition:** It is the user-defined section of electronic media. Partitions can be used to isolate and obscure information in a hard drive.

13. **Source-Code:** A text containing a list of commands to be collected into an executable computer program.

14. **Work Copy:** A replication of a tape or records that can be used for consequent processing and/or evaluation.

15. **Write Block/Write Protect:** Hardware and/or software methods of thwarting alteration of substance on a media storage devise like a CD or thumb drive.

7.5 TYPES OF DIGITAL FORENSICS INVESTIGATION

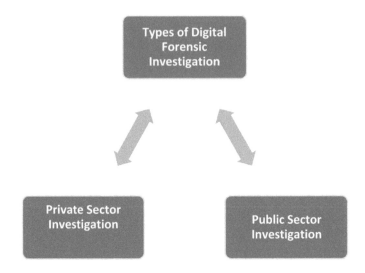

7.5.1   *PUBLIC SECTOR INVESTIGATION*

These are investigations performed against alleged individuals to uncover the scene of crime. Normally, government organizations perform such types of inquiries under criminal investigations.

7.5.2   *PRIVATE SECTOR INVESTIGATION*

These are investigations performed against violations associated to office and corporate. Any breach of the provisions and guidelines of private sectors falls under this investigation.

7.6   OBJECTIVES OF DIGITAL FORENSICS

'Digital Forensics' safeguards and strengthens digital security in the private/ public sector and supports investigators to investigating criminal cases (https:// recfaces.com/articles/digital-forensics#:~:text=Digital%20forensics%20 tools%20can%20be%20divided%20into%20several,Mac%20OS%20anal- ysis%20tools%3B%208%20Database%20forensics%20tools). The essential objectives include:

- Expediting the retrieval, evaluation, conservation, and protection of the information. It helps in preparing digital proof for presentation in the court.
- Ensuring proper implementation of all the essential procedures of assembling proofs as the digital data must not be tainted.
- Recovering erased or concealed data from any digital devices if the data is specifically important for the case.
- Aiding in the identification of a suspect and instituting a purpose for the felony.
- Producing a digital forensic report for swift investigation.

7.7 ALGORITHM OF INVESTIGATION

The inquiry procedure in digital forensics is a stepwise technique, carried out diligently to guarantee that no original evidence is ruined in the course of action while preserving the chain of custody (http://www.forensicscienc-esimplified.org/digital/DigitalEvidence.pdf). Digital investigation of such a nature incorporates an algorithm which commences from retrieval and analysis of the data to a probable source of proof acceptable in the court (Karie & Venter, 2012).

The algorithm for the digital forensics' investigation procedure is as in Figure 7.3.

7.8 DIGITAL FORENSIC TOOLS

Digital Forensic Tools (https://www.guru99.com/computer-forensics-tools. html#:~:text=Best%20Computer%20Forensics%20Tools.%201%20 1%29%20ProDiscover%20Forensic.,4%204%29%20PALADIN.%205%20 5%29%20EnCase.%20More%20items) are the digital modes and method-ologies that assist to protect, detect, obtain, and document digital evidence for legal actions. Besides, making the digital forensic procedure easy and straightforward, they also provide complete reports for legal actions.

The subsequent issues should be considered while selecting a digital forensic instrument:

- Protection;
- Assistance for numerous platforms;
- Customer-friendly application;

INVESTIGATION ALGORITHM

Recognition

Recognition is the first step carried out during any type of investigation process. It involves the recognition of all potential digital sources which have the capacity to store digital information. It could be utilized as a source of evidence. Initially, it is compulsory to be aware of the devices which can be beneficial in the investigation procedure.

Collection

Digital appliances, which are probable sources of evidence are confiscated and appropriately accumulated from the felony scene. Authenticity of evidence is a major priority.

Volatile Data Collection	**Live System Imaging**	**Forensic Imaging**	**Capture Digital Devices Physically**
It comprises the compendium of all the implementing records such as time, date, and RAM data. For the duration of the examination procedure, the system should be powered ON	It comprises of the tomography action achieved on data in the course of the investigation. For the duration of the examination procedure, the system should be powered ON	Precise photo copy of the initial device is designed to execute alterations. Hashing techniques are used to guarantee the confidentiality aspect. For the duration of the examination procedure, the system should be powered ON	When the three techniques stop working, then this method proceeds to the salvage. Here, the digital gadgets are confiscated and then all evidences are gathered. For the duration of the examination procedure, the system may be powered ON/OFF.

Preservation

Preservation is an essential measure to protect the area where the felony has occurred. Besides this, it is crucial to conserve all the 'Electronically Stored Information' (ESI) which can be acquired from the crime scene.

Analysis

Analysis includes in-depth analysis of all the digital proof. Proper tomography is accomplished to guarantee that the original proof do not lose the genuineness. The analysis and research of the criminal situation aids the examiner to determine.

Reporting

It is the decisive phase performed during the scrutiny. Reports are means of displaying all the details crime linking in a methodical manner. These reports are then displayed as the source of proof in the court of law to explain the felony.

FIGURE 7.3 Schematic of investigation algorithm.

- Multiple functions available;
- Assistance for various gadgets;
- Assistance for several file formats;
- Analytics features;
- Plugin's assistance.

7.8.1 RYDBERG QUANTUM SPECTRUM ANALYZER

The Rydberg quantum sensor has the capability to analyze the complete range of real world signals and radio-frequency, thereby leaving a new potential for prompt detection of any digital viral infiltration, spectrum awareness, combat communications and electronic warfare (Meyer, Kunz, & Cox, 2021). It can identify radio frequency band from zero to 20 GHz, Bluetooth, AM and FM radio, Wi-Fi, and other signals of communication. Atomic particles interact firmly with the circuit's electric fields thereby permitting detection, identification, and demodulation of any signal received into the circuit. The sensor utilizes laser beams to create exceptionally excited directly Rydberg atoms above the microwave circuit, to increase the section of the spectrum being evaluated. The Rydberg atoms are extremely responsive to the circuit's voltage thus empowering the device to be utilized as a receptive tool in the Radio Frequency spectrum for the broad array of signals.

7.8.2 RAVEN

RAVEN is an innovative solution for accessing immediate digital evidence procurement and analysis, at the crime scene with the help of a smartphone application developed by MSAB. In forensic technology, it is a world leader for obtaining and analyzing data in confiscated mobile devices (https://www.msab.com/products/msab-platforms/#raven). The company builds superior and user-friendly software for law-enforcement organizations. RAVEN is an extremely portable mobile device toolset for extraction, accurately created and devised for operatives in the field. It is a lightweight solution to instantly obtain information and turn it into a means of actionable intelligence. RAVEN is an assemblage of applications that run on Android-driven devices (mobile phones and tablets). It allows the investigators to take favor of a standard device's capacity and enrich them with forensic data collection qualities, just by downloading all/selected Raven apps. The app allows users to instantly view check watch lists, the extracted results, find the most recent location data.

7.8.3 MSAB TABLET

The tablet offers investigator with instantaneous evidence and information harvesting potential in the most compliant form factor (https://www.msab. com/products/msab-platforms/#tablet). It aids in prioritizing and speeding up the assessment process by granting detectives a rapid access. The tablet can be personalized to company specific systems and configured for a plethora of forensic competence. It proposes control of a streamlined and easy forensic procedure, without questioning the veracity and trustworthiness of the forensic data. It also provides a complete and an integrated software/ hardware solution, with assistance, for extraction of cellular devices through wire or wireless mode. Operatives can execute extractions through an easy extraction wizard, through a touch-based interface. The data obtained can be overviewed, investigated, and acted on, in the "viewer application." It also includes a potent search platform. The search platform along with easy-to-use illustrations aids in the precise assortment of the obtained data to be displayed. The extraction generates a secure. XRY file which contains all the data/information's salvaged from the device in question. The user interface presents an easy control of assignments, being worked on the tablet. When a user is set up, the administrator designates rights to workflow sequences/orders for that user. Examples of the workflow sequences include data extraction, view extraction, SIM id cloner and user-specific multistep workflow configurations. In the "standard user" mode the device is locked down and all user actions are recorded. This facilitates not only in effortless traceability of actions but also makes the extraction procedure transparent. "User Administration" functions also permits the user to expunge and export all user files.

7.8.4 VPER

The VPER kits have been designed to guarantee easy, accurate, and fast forensic data retrieval and analysis, thereby initiating an investigation process in a precise direction (https://digitalintelligence.com/products/ fred_l). Every VPER kit includes a FRED-L laptop. A FRED-L is an AMD Ryzen 7, 3700X, 8-core. About 3.6 GHz (4.4 GHz boost) processor with 32 MB Cache. Besides, each VPER includes an exceedingly valuable "Digital Intelligence Forensic Card Reader," cables, adapters, power supplies and multifarious instruments.

7.8.5   *iVE−VEHICLE FORENSIC SYSTEM*

BERLA's iVe Ecosystem is a vehicle forensic system tool built to recognize, obtain, and analyze significant evidence stored within a vehicle's system (https://www.msab.com/products/ive-vehicle-forensics/). It unearths vital evidence that answers issues like "what happened, where it happened, and who was involved." The Vehicle system stores a substantial amount of data related to recent destinations, frequently visited destinations, contact directories, call records, SMS records, electronic mails, pictures, videos, etc., of the vehicle. The system also logs time-based happenings. Evaluation of these data answers vital questions for examiners and changes the course of an investigation. The iVe Ecosystem is an anthology of tools that aids detectives during the complete vehicle forensics steps with a mobile application for detecting vehicles, for obtaining systems, a hardware kit, and forensic software for scrutinizing records. iVe currently supports many premium vehicles like AUDI, BMW, Chevrolet, FIAT, Ford, GMC, Mercedes-Benz, HUMMER, Jeep, Hyundai, Kia, INFINITI, Lincoln, etc.

7.8.5.1   *KEY FEATURES*

1. Recovery of deleted data.
2. View and analyze results.
3. Logical and physical acquisition.
4. Binary file import and parsing.
5. Generate reports.
6. Search and filter data.
7. Data export.
8. Map GPS and geo-coded data.

7.8.6   *DISK SENSE 2*

Atola Insight Forensic (a hardware and software development company based in Vancouver, Canada, that specializes in creating hard drive imaging tools for the digital forensic post-mortem) has introduced a new hardware imager known as Disk Sense 2. It supports real-time imaging of three drives with other concurrent forensic tasks such as hashing and wiping (https://www.forensicfocus.com/news/atola-insight-forensic-gets-new-hardware-unit-and-supports-3-parallel-imaging-sessions/). This hardware unit is equipped with six source ports:

- 3 SATA/SAS;
- USB;
- IDE;
- Extension port (SAS, M.2NVMe/PCIe/SATA SSD, Thunderbolt, Apple PCIe SSD).

The three concurrent imaging sessions are backed by server-grade motherboard and CPU. The device's ECC RAM aids in safeguarding the data integrity. Its two built-in 10 Gbit ports allow imaging drives into the network at high speeds.

7.8.7   *AXIOM CYBER*

'AXIOM Cyber' facilitates investigators to obtain available Instagram information by a direct procurement, by handling any Instagram or Facebook username and password to substantiate the targeted Instagram account (https://support.magnetforensics.com/s/axiom-cyber). It is available in both browser and mobile application. The default crate for the obtained data is fixed for AFF4-L and could be changed to a standard .zip file by selecting 'EDIT' option. Besides, the user can opt to acquire the data from a particular user account or posts involved with a specified hash tag. A user account can be acquired by searching on Instagram and selection of the account. The account that needs to specify in the Process of AXIOM, would be at the first of the user's data that will display after selecting the specific account. After the username is entered, AXIOM will 'CHECK' to authenticate whether the account is fulfilling the criteria and accessible with the credentials given. AXIOM Process will then construe the main part of the acquisition where the user can do more interpretation in "AXIOM Examine." This ability to get artifacts and Instagram data both from the Cloud and a device of user's is extremely beneficial in generating a comprehensive overview of a user's activity. Procuring Cloud data will permit examiners to gather the most recent activities, but artifacts on the device may exist if data is successively deleted.

7.8.8   *ArTHIR*

'ATT&CK remote threat hunting incident response' (ArTHIR) Windows tool is an integrated and flexible framework that can be used tenuously against one or multiple targets to accomplish threat-hunting, compromise-assessments, incident-response, configuration, containment, and many other actions by applying in-built PowerShell (all versions) and Windows remote management (WinRM) (https://securityonline.info/remote-threat-hunting-incident-response/). This is an upgradient to the popular "Kansa" tool, but with more abilities and facilities than just performing PowerShell scripts. ArTHIR makes it simpler to push and implement any binary program locally or remotely and recover back the output. It intends and plans to map the 'Threat Hunting and Incident Response' portions to the 'MITRE ATT&CK' structure.

7.8.9 VELOCIRAPTOR

Velociraptor is a digital forensics and incident response (DFIR) tool for stalking and tracking at scale (https://www.hackingarticles.in/threat-hunting-velociraptor-for-endpoint-monitoring/). It contains an important language for query called VQL. This VQL language allows swift alteration to fluid DFIR attacks. Velociraptor delivers consummate power in the hands and flexibility of the user. Unlike conventional "Remote Forensic Tools" which accumulate enormous quantity of raw data for offline handling, VQL allows users to perform and execute analysis precisely on the endpoint. This new freedom grants users to amass only high-value, tactical evidence to affect their response, and influence recent state of the art digital forensic analysis processes into detection. To install Velociraptor in Windows, the essential requirements are.

- Minimum 4 Gb Ram and 4 CPU cores in windows 10 platform;
- Admin privileges;
- CMD with admin privilege.

7.8.10 SLEUTH KIT + AUTOPSY

Sleuth Kit + Autopsy is a graphical interface digital forensics tool. It can be used to recover camera photos from memory cards. It is Windows-based (https://www.sleuthkit.org/).

> **Features:**
> - Detection and classification of activity using a graphical interface;
> - Evaluation for electronic mail;
> - Sorting files;
> - Exhibits a thumbnail for easy view;

- File cataloging with arbitrary tag names;
- Removal of data from call history, contacts, etc.;
- Marking of files/folders on the basis of pathname.

7.8.11 PRO-DISCOVER FORENSIC

This is an important computer security tool that studies data at the sector level and helps revive deleted files, examine flapping space and access Windows alternate data streams (ADS) (https://www.prodiscover.com). Pro Discover Forensic permits dynamically a preview and image-capture of the 'hardware protected area' (HPA) of the disk and supports to locate all data on a computer disk, protecting evidence and creating reports.

> ➢ **Features:**
>   - makes a bit-stream copy of the disk (including the hidden HPA section) for interpretation;
>   - finds files on the entire disk; this encompasses slack space, HPA section, and Windows NT/2000/XP alternate data streams;
>   - without altering data on disk previews files;
>   - examines data at the file or cluster level.

7.8.12 CAINE

CAINE or computer-aided investigative environment offers strong security and built-in forensic investigation instrument. It makes a complete investigative environment which is arranged to integrate existing software tools and to give a friendly graphical user interface (GUI). CAINE Linux gives a variety of software tools which can be utilized for database, memory, forensic analysis and network (https://www.caine-live.net). The File Image System analysis of File Systems like FAT/ExFAT, NTFS, Ext2, Ext3, HFS, and ISO 9660 is possible using command-line mode as well as GUI mode. Disk images might be obtained applying the tools that built-in the CAINE or using third-party tools like EnCase, or Forensic Tool Kit.

> ➢ **Features:**
>   - Helps digital detective during the different phases of investigation;
>   - Easy user interface;
>   - Customizable.

7.8.13 PALADIN

PALADIN is Ubuntu centered tool that facilitates to simplify a range of forensic tasks. It gives hundreds of useful tools for investigative any malevolent evidence (https://sumuri.com/software/paladin/). It also aids in streamlining forensic tasks quickly and efficiently.

> ➢ **Features:**
> - Offered in 64-bit and 32-bit versions;
> - Can be run through an USB drive;
> - It facilitates searching of the required information effortlessly;
> - It has 33 classifications that assists in achieving a cyber forensic assignment.

7.8.14 ENCASE

Encase is a computer application that helps to salvage evidence from the hard drives (https://www.guidancesoftware.com/encase-forensic). It conducts a detailed analysis of files to accumulate evidence.

> ➢ **Features:**
> - Data can be acquired from numerous devices, including mobile phones, tablets, etc.;
> - It generates a comprehensive report for maintaining evidence veracity;
> - Enables swift hunting, detecting, recognizing, and prioritizing evidence;
> - Can solve encrypted data;
> - Can operate deep and triage analysis.

7.8.15 SANS SIFT

SANS SIFT, an Ubuntu-based division (https://www.sans.org/tools/sift-work-station/) caters a digital forensic and case reply to analysis capability.

> ➢ **Features:**
> - Compatible on a 64-bit OS;
> - Aids in memory space optimization;
> - Be able to be installed through SIFT-CLI installer.

7.8.16   FTK IMAGER

'FTK Imager' by access-data is FTK used to get evidence (https://access-data.com/productsservices/forensic-toolkit-ftk). It generates duplicates of the concerned data without tampering with the original data. The tool grants user to identify criteria like picture element size, file size, and data type, to decrease the quantity of extraneous data.

- ➢ **Features:**
  - • Offers a scientific approach to expose crime in the cyberspace;
  - • Provides superior image of data by the help of chart;
  - • Aids in password retrieval;
  - • Controls recyclable profiles for different analysis requirements.

7.8.17   CROWDSTRIKE

CrowdStrike is a forensic computer program that delivers endpoint security, intelligence on threats of cyber hacking, etc. (https://www.crowdstrike.com/endpoint-security-products/falcon-endpoint-protection-pro/). It can instantly identify and retrieve from cybersecurity incidents. CrowdStrike aids to find and stop interlopers in actual time.

- ➢ **Features:**
  - • Handles system susceptibilities;
  - • Can automatically analyze malware;
  - • It guarantees all virtual, physical, and cloud-based information center.

7.8.18   WIRESHARK

Wireshark analyzes packets of network for analyzing and fault-finding (https://www.wireshark.org). The tool facilitates the operator to check numerous stream of traffic going through one's computer system.

- ➢ **Features:**
  - • Supports efficient VOIP analysis;
  - • Can decompressed files effortlessly;
  - • Supports instantaneous information reading from ATM, etc.;

- Assists in deciphering various protocols that include IPsec (internet protocol security), SSL (secure sockets layer), and WEP (wired equivalent privacy);
- Allows operator to read/write file at the format of his/her choice.

7.8.19 XPLICO

Xplico is a forensic evaluation application that supports HTTP and IMAP Protocols (https://www.xplico.org).

➢ **Features:**
- Data output is in the SQLite or MySQL database;
- Permits real time cooperation;
- Infinite data entry and filing;
- Supports IPv4 and IPv6;
- Provides PIPI.

7.8.20 X-WAYS FORENSICS

'X-ways forensics' provides an atmosphere for digital (computer) forensic expert (http://www.x-ways.net/forensics/). The program can execute disk replicating and imaging.

➢ **Features:**
- Can read segmentation and file structure configurations inside '.dd' image files;
- Can access disks and RAIDs;
- Inevitably recognizes missing or erased compartments;
- Effortlessly recognizes NTFS (new technology file system) and ADS (alternate data streams);
- Can investigate distant workstations.

7.9 CONCLUSION

The ever-escalating popularity of technology in the modern era, especially since the new normal (fall of 2019, onset of the pandemic) has increased the probability of digital devices being relevant to a criminal/civil litigation. As a direct consequence, the number of investigations and analyzes requiring digital forensic proficiency is resulting in a massive "Digital Evidence" logjam. It is estimated that the number of cases needing digital forensic analysis will significantly rise in the future. It is also possible that each court case will require the evaluation of an increasing number of devices, including smartphones, computers, tablets, wearables, and cloud-based services. The range of new digital evidence resources pose new and intriguing challenges for the digital examiner from a detection, recognition, procurement, storage, and evaluation perspective. This chapter aims to appreciate the various digital viral vectors (technological innovations) which are been continuously injected into the Digital Forensic Investigation system to facilitate enhanced and accurate outcomes.

KEYWORDS

- **database forensic analyst**
- **e-mail forensics analyst**
- **forensic data analyst**
- **internet protocol security**
- **investigation algorithm**
- **ontology**
- **open-source intelligence**
- **remote forensic tools**

REFERENCES

Building Ontologies for Digital Forensic Terminologies. Free Online Library (thefreelibrary. com) https://www.thefreelibrary.com/Building+ontologies+for+Digital+Forensic+Termin ologies.-a0459894717 (accessed on 10 January 2022).

Chandrasekaran, B., Josephson, J. R., & Benjamin's, V. R., (1999). What are ontologies, and why do we need them? *IEEE Intelligent Systems, 14*(1), 20–26.

Gary, P., (2021). *A Road Map for Digital Forensic Research*, Technical Report DTR-T001-01, DFRWS.

Gruber, T. R., (1993). A translation approach to portable ontology specification. *Knowledge Acquisition, 5*(2), 199–220.

Halder, D., & Jaishankar, K., (2011). *Cybercrime and the Victimization of Women: Laws, Rights, and Regulations.* Hershey, PA, USA: IGI Global. 10.4018/978-1-60960-830-9.

http://archive.ncpc.org/ (accessed on 10 January 2022).

http://www.forensicsciencesimplified.org/digital/DigitalEvidence.pdf (accessed on 10 January 2022).

http://www.forensicsciencesimplified.org/digital/glossary.html (accessed on 10 January 2022).

http://www.x-ways.net/forensics/ (accessed on 10 January 2022).

https://accessdata.com/productsservices/forensic-toolkit-ftk (accessed on 10 January 2022).

https://digitalintelligence.com/products/fred_l (accessed on 10 January 2022).

https://recfaces.com/articles/digital-forensics#:~:text=Digital%20forensics%20tools%20can%20be%20divided%20into%20several,Mac%20OS%20analysis%20tools%3B%20 8%20Database%20forensics%20tools (accessed on 10 January 2022).

https://securityonline.info/remote-threat-hunting-incident-response/ (accessed on 10 January 2022).

https://sumuri.com/software/paladin/ (accessed on 10 January 2022).

https://support.magnetforensics.com/s/axiom-cyber (accessed on 10 January 2022).

https://www.caine-live.net (accessed on 10 January 2022).

https://www.crowdstrike.com/endpoint-security-products/falcon-endpoint-protection-pro/ (accessed on 10 January 2022).

https://www.forensicfocus.com/news/atola-insight-forensic-gets-new-hardware-unit-and-supports-3-parallel-imaging-sessions/(accessed on 10 January 2022).

https://www.guidancesoftware.com/encase-forensic (accessed on 10 January 2022).

https://www.guru99.com/computer-forensics-tools.html#:~:text=Best%20Computer%20Forensics%20Tools.%201%201%29%20ProDiscover%20Forensic.,4%204%29%20PALADIN.%205%205%29%20EnCase.%20More%20items (accessed on 10 January 2022).

https://www.hackingarticles.in/threat-hunting-velociraptor-for-endpoint-monitoring/(accessed on 10 January 2022).

https://www.msab.com/products/ive-vehicle-forensics/ (accessed on 10 January 2022).

https://www.msab.com/products/msab-platforms/#raven (accessed on 10 January 2022).

https://www.msab.com/products/msab-platforms/#tablet (accessed on 10 January 2022).

https://www.prodiscover.com (accessed on 10 January 2022).

https://www.sans.org/tools/sift-workstation/ (accessed on 10 January 2022).

https://www.sleuthkit.org/ (accessed on 10 January 2022).

https://www.wireshark.org (accessed on 10 January 2022).

https://www.xplico.org (accessed on 10 January 2022).

Karie, N. M., & Venter, H. S., (2012). Measuring semantic similarity between digital forensics terminologies using web search engines. In: *The Proceedings of the 12th Annual Information Security for South Africa Conference.* Johannesburg, South Africa, IEEE Xplore [R].

Karrie, N. M., & Kebande, V. R. (2016). *Building Ontologies for Digital Forensic Terminologies.* http://www.deg.byu.edu/ding/research/SemanticAnnotation.html (accessed on 10 January 2022).

Karrie, N. M., & Venter, H. S., (2014). Towards a general ontology for digital forensic disciplines. *J. Forensic Sciences, 59*(5), 1231–1241.

Meyer, D. H., Kunz, P. D., & Cox, K. C., (2021). Waveguide-coupled rydberg spectrum analyzer from 0 to 20 GHz. *Physical Review Applied, 15*, 014053.

Noy, N. F., & Musen, M. A., (2004). Ontology versioning in an ontology management framework. *IEEE Intelligent Systems, 19*, 6–13.

Singh, S. K., Azzaoui, A. E., Salim, M. M., & Park, J. H., (2020). Quantum communication technology for future ICT: Review. *Journal of Information Processing Systems, 16*(6), 1459–1478.

Uschold, M., & Gruninger, M., (1996). Ontologies: Principles, methods and applications. *Knowledge Engineering Review, 11*(2), 93–136.

Wright, S. E., & Budin, G., (2001). *Handbook of Terminology Management: Application-Oriented Terminology Management*. John Benjamin's Publishing Company.

CHAPTER 8

Data Breach Fraudulence and Preventive Measures in E-Commerce Platforms

RAJA RAJENDRAN

Department of IT, Canara Bank, India

ABSTRACT

Cybercrime is a moderately new worry for legislation regulating authorities. With the expanding figure of computer users associated with the web, the open door for cybercrime is increasing enormously. Additionally, with the increased utilization of social networking and the capability to make more online purchases, users can submit their personal information for the respective site through the Internet. Cybercrime is a severe menacing issue, and it is difficult to make government officials' laws and challenges monitoring the issues on the web for information security analysts. Cybercrime detection strategies and classification techniques have concocted different levels of success for preventing and protecting data from cybercrime attacks. However, the study shows that many countries are facing this problem even today with maximum damages. In this work, an attempt was made to provide an overview of cyberattacks in India, especially in online shopping forums and data breaches experienced in the past years. Furthermore, it analyzes the preventive measures against these sorts of cyberattacks. From these analytical perspectives, organizations, and governments can develop the infrastructure to minimize cybercrimes and take efficient preventive measures against data breaches. This will ensure the secrecy of organizations from severe cybercrime attacks. E-commerce refers to the interest in buying and promoting matters over the net. Simply, it refers to the economic transactions which might be performed online. E-commerce can be drawn on many technologies,

Advancements in Cybercrime Investigation and Digital Forensics. A. Harisha, Amarnath Mishra, & Chandra Singh (Eds.)
© 2024 Apple Academic Press, Inc. Co-published with CRC Press (Taylor & Francis)

including cell trade, Internet advertising and marketing, online transaction processing, digital budget transfer, supply chain control, electronic data interchange (EDI), inventory control systems, and automated statistics collection systems. The E-commerce security breach is happening by using the net for unfair means with the goal of stealing, fraud, and protection breach. There are diverse types of e-trade threats. Some are unintended, a few are helpful, and some of them are because of human blunders. The most common safety threats are digital payments, e-cash, statistics misuse, credit/debit card frauds, and many others.

8.1 INTRODUCTION

The Internet, computers, mobile phones, and different types of inventions have changed each aspect of human life in the past decades, notably how we communicate around the globe, banking, shopping, acquiring the news, and engage ourselves. Technology changed people's way of thinking, way of acting, and way of responding. The countries also show a keen interest in upgrading technology for social and economic strength. These advancements are an essential reason for the development of unlawful internet-based activities, ordinally known as cybercrime. Since its creation, an unprotected approach to the web and its data has been described as stealing data and information, misuse of privacy, data, and information distortion, and related digital vices leaving the violator barely took note. It is progressively contended that hackers' essential and significant reason to develop computer vulnerability is financial gain.

Notwithstanding the straight financial effect of cyberattack, the computerized data and the information conveyed through the web can have extra content value to the hackers. Cybercrime presents unique and troublesome challenges for law authorization who are charged with countering such activities. Offenses demonstrated as cyber-related crimes are identity theft, money laundering, cyber-bullying, pornography, cyberstalking, cyber-espionage, online grooming, theft of intellectual property rights, pedophilia, etc.

8.2 CYBERSECURITY IN E-COMMERCE

In general, E-commerce is referred to as purchasing and selling items/products through electronic mediums, principally the Internet, the worldwide commercial center. Hence it is featured as virtuality and considered a major milestone in the global marketplace. The benefits of the digital market, such as high interaction, ease of use, transparency, lead to acquiring high popularity of E-commerce among customers and suppliers. Additionally, the increasing profit, business coherence, reduced functional cost, customer services result in a digital marketplace worthwhile to organizations. However, the popularity of E-commerce leads to several security threats, which results in an incredible concern to the customers as well as suppliers. The recent investigations show that the major factor retarding the success of E-commerce is the lack of security. E-commerce security is termed as a mechanism that protects E-commerce assets from illegal access, modification, or destruction. The major security concerns in E-commerce platforms are privacy, authentication, integrity, non-repudiation, and availability. E-commerce cybersecurity framework applies to the component which affects the digital interaction, such as security from viruses, worms, intrusion, and network communications.

When you take into account the fee—to your bottom line and your popularity—defending yourself, your customers, and your data have to be a number one challenge. Educating yourself and your employees about capability threats and understanding how malicious actors reap their goals places you one step in advance of the sport.

If you're unsure which to start, scheduling a risk assessment with a cybersecurity professional is constantly a very good concept. They will go through your structures, applications, and dependencies to become aware of potential hazards, examine the extent of danger associated with them, and put measures in the area to remove or at the least manipulate the chance.

8.2.1 PRIVACY

Privacy is the fundamental right for any customer, irrespective of the marketplace. In general, privacy ensures control over the user-submitted information. From an E-commerce perspective, ensuring privacy is related to safeguarding user data such as payment information, contact number, address, and, most predominantly, user search interest. Any unauthorized third party attempting to read or copy the user information will lead to a serious security risk for the customer. Thus, ensuring privacy in E-commerce is the major security concern for the digital marketplace.

8.2.2   AUTHENTICATION AND INTEGRITY

Authentication is the process of ensuring the entities which are claimed to be. For example, during online payment, user authentication is verified by checking the user-submitted details with the details that are stored in the server database. The simplest way for identifying the authenticated user is by sending the OTP to the verified mobile number. Similarly, integrity in an E-commerce platform is referred to as ensuring the trustworthiness of the user information. Electronic fund transfer and financial information can be protected by preserving the trustworthiness of the information.

8.2.3   AVAILABILITY

The major security concern for the E-commerce service provider is the availability of services. The success of the E-commerce marketplace depends on the high availability of services irrespective of time, i.e., the server should be reliable and provide access to the customer and supplier on a 24×7 basis. Thus, E-commerce service providers should take necessary steps to ensure the services by preventing the platform against service disruptions such as Denial of Service attacks.

8.2.4   RESEARCH CHALLENGES FOR SECURE E-COMMERCE

The literature provides insights into major security concerns for E-commerce. There are still some security issues that arise with respect to the service provider (E-commerce platform), customer, and owner.

8.2.4.1   ELECTRONIC FUND TRANSFER IN E-COMMERCE

E-commerce's online payment security is denoted as secure and systematic genuine transfer of funds or money from the buyer to supplier or merchant via electronic medium. This electronic medium links the supplier and customer by exchanging the funds or money value. Usually, the electronic fund transfer requires bank details or payment card details of the buyer and supplier. Thus, securing these confidential details is the major security concern of E-commerce applications. Hence, transaction transparency is required to protect this information, i.e., after completing the transaction, the

E-payment gateway should not store any information related to the payment card. Nowadays, to attain ease of use for their customers, the E-payment gateway stores the information for future access and is also open to security risks. This results in major security breaches with respect to funding transfer.

8.2.4.2 CUSTOMER AND SUPPLIER PRIVACY IN E-COMMERCE

In the digital marketplace, buyers are required to provide their confidential information such as mobile number, contact address, payment card details to the third-party vendor, i.e., E-commerce vendor, which results in a high risk of sensitive information leakage to the digital environment. There are two major privacy concerns for the customer. First, customers are generally concerned about the reuse of their data, such as their search interest. Usually, customers feel discomfort when they get frequent suggestions related to the recently purchased product. Thus, protecting their search interest is the major concern for the customers. Secondly, customers are concerned about unauthorized access to their confidential information due to security breaches. Hence it is the E-commerce vendor's responsibility to protect the customer's confidential information.

8.2.4.3 SOFTWARE SECURITY IN E-COMMERCE

The software security in E-commerce issues related to the availability of service, confidentiality, and trustworthiness of the information. The availability of service is achieved by setting up the backup server with necessary rescue actions. On the other hand, to ensure security in E-payment or webpage, the E-commerce vendor should design the system equipped with higher-end firewall technology. The firewall should ensure user authorization while accessing the webpage or E-payment.

8.2.4.4 CYBERCRIME IN E-COMMERCE

Cybercrime in E-commerce platforms primarily considers computers as a target model that can intentionally cause security breaches, financial loss, service unavailability by performing virus attacks, worms, trojan horses, phishing, etc. The intruders break the E-commerce web server and cause service interruption for the customer. This will lead to the loss of transactions or personal information provided by customers, which affects the revenue of the E-commerce platform. Intruders or attackers also perform traffic routes which result in customer reach different sites irrespective of the search interest. This routing traffic will cause severe issues to the E-commerce marketplace. Table 8.1 shows the summary of security threats in E-commerce sites.

TABLE 8.1   The Summary of Security Threats in E-Commerce Sites

Security Issues		Concern/Description
Electronic fund transfer security	i.	Protection against leakage of confidential information
	ii.	Ensure transaction transparency
Customer and supplier privacy	i.	No reuse of personal data
	ii.	Prevention of unauthorized access
Software security	i.	Service availability
	ii.	Database threat
Cybercrime	i.	Fraudulence attack
	ii.	Virus
	iii.	Traffic reroutes
	iv.	Phishing

Thus, these investigations indicate that among all the security issues, cybercrime in E-commerce causes severe loss for vendors and customers. Hence, necessary action should be taken to protect the E-commerce platform against cybercrime.

8.3   NEED FOR CYBERSECURITY IN WORLD WIDE WEB

The business world isn't new to data breaches and cyber threats. We have already witnessed the Yahoo data breach in 2016, which turned into a massive catastrophe. The changes introduced by means of the pandemic, including far-off operating, extra agencies moving online, and switching to virtual fees, have given rise to even additional digital threats.

For many small companies, the switch to virtual turned into surprising and supplied little time to reinforce current security features or enforce new processes. Even larger brands determined it difficult to guard statistics even as managing remote groups and using new software programs, including Trello, Slack, Flock, and extra. In 2020, Microsoft suffered a malicious attack that compromised over 280 million Microsoft client records that protected electronic mail addresses, IP addresses and assist case info. A major incident happened in April of 2020 while websites hosted by San Francisco International Airport suffered a cyber-attack in which hackers tried to benefit login credentials.

The recent growth in data breaches has caused a marketplace surplus in stolen identities. Normally, this will result in marketplace saturation; however, because of the quantity of ability fraud schemes, the usage of stolen identities has endured developing, older identities still have resale value. With the stolen statistics traded mainly over the Internet, these progressions have been pondered in changes to the web marketplace for stolen identities. The online market for stolen identities has evolved over the path of three generations. First-generation statistics breaches were exceptionally small, with the early Internet used primarily to facilitate the purchase instead of the sale of the identities. The 2nd generation saw the improvement of an Internet-based totally wholesale marketplace and, with the advent of larger breaches, specialization acting in the fraud cells. The 0.33 and present-day generation is showing marketplace adulthood similarly, with the differentiation of wholesale and retail markets. The darkish net and cryptocurrencies have caused direct-to-purchaser sales of identities, which has had a widespread effect on the marketing and pricing of stolen identities.

The fast shift to virtual has uncovered many vulnerabilities that firms currently face. Based on the World Economic Forum's Global Risks Report (2021), the cyber danger is many pinnacle international threats that need to be countered urgently.

Digital transformation has not most effectively evaporated certain bodily areas but has also merged boundaries among countries. While this is ideal for companies seeking to operate in a global market, it has also created more threats of move-border cyber-assaults. While ransomware, phishing, and social media hacking assaults keep to upward thrust, organizations need to prioritize data security. This is essential because it's miles lawfully required as general data protection regulation (GDPR) asks businesses in Europe to shield private statistics.

Moreover, any breach of protection isn't most effective unfavorable to the corporation's image and has a monetary implication. A commercial enterprise that has a nicely-covered cyber protection device could have the consumer's trust as well. This ensures the protection of any touchy facts, durability, and smooth functioning of the organization.

If you believe you studied that it's far the most influential countries and governments that need to be looking for cyber-assaults, then assume once more. The company area is similarly at threat and needs to ensure that it isn't prone to hazards. This is why corporations need to take appropriate steps to place special protection take a look at in location. Also, it's no longer simply the massive organizations but corporations of all sizes that need to position strong defenses in the face of cyber threats. While destiny is unpredictable, it's miles usually sensible to have a course of action to guard the organization's sensitive facts earlier than any considerable harm happens.

When most of the transactions are happening electronically, they're additionally turning into the source of crime. Lots of data shared, cash transferred are the goal of cybercriminals. All organizations are growing their technical base, increasingly depending upon the Internet for transactions without studying the threat related to the technology. Cybercriminals manipulate financial facts, can flow the electronic possession, interrupted communications with personnel or commercial enterprise companions, scouse borrow intellectual assets, harm a company's reputation, or carry e-trade (or an entire business) to close down.

E-trade enterprise is distinctive in operations, so it has specific challenges and risks as the absence of bodily presence boom the possibilities of frauds and crimes. There are many more motives in India for out-of-control E-commerce cyber frauds. The value of cybercrime will increase as more business functions move online and as consumers and groups around the world are amassing on cyberspace. The risk of highbrow property theft will increase, ensuing in reflect products. The use of technology theft causes enormous loss to the discern organizations because it lessens the rate of return on the improvements. The government must take the extreme initiative to combat the demanding situations of cybercrime; otherwise, an era will negatively impact the enterprise. Globally cybercrime losses are more than 400 billion dollars every year.

8.4   DATA BREACH FRAUDULENCE AND ITS PREVENTIVE ACTIONS FOR E-COMMERCE

8.4.1   OVERVIEW

Cybersecurity is coined to provide prevention against these cyber-attacks through online services to ensure the privacy of the information on the Internet. Cybersecurity is the assortment of devices, strategies, security ideas, security safeguards, guidelines, and innovations that can be used to safeguard the cyber environment and organization and user's assets. The common security objectives in information systems contain confidentiality, integrity, and availability of the data or resources of the organization (or) user's assets include digital devices, telecommunication systems, and information in the digital environment. One of the most troublesome components of cybersecurity is security risks. The conventional approach concentrated most assets on the crucial system components and protect against the threats, which need to leave some less remarkable system segments unprotected and some less risky dangers, i.e., not assured. The various security issues in the cyber environment are IP spoofing, e-mail spoofing, trojan, worm, phishing, and spyware. Among these security issues, cybercrime entailed in an online forum is a serious one like money laundering, e-mail spoofing, etc. Recently, online transactions paying more attention to digitalization like card payment, UPI, and QR scanning. These advancements of digitalization lead to security breaches in payment gateways, say QRishing. Quick response (QR) codes are two-dimensional matrix look-alike barcodes utilized to encode the information. A default camera in a smartphone will act as a scanner for accessing the data that resides in QR. This section provides an overview of cybercrime that exists mainly on E-commerce and analyzes existing prevention measures for these attacks.

8.4.2 RELATED WORKS ON CYBERCRIME INVESTIGATION

As the Internet keeps developing and suddenly transforms into many aspects of modern life, the advantages and opportunities afforded by a globalized digital society are ostensibly huge. Though networked communication advances evolve and increasing dependence is put on the broad scope of capabilities and services on offer, people, associations, and governments are progressively presented to the risks and threats of the hackers. As a result of this substantial computerized freedom, the Internet also offers new and innovative chances for inspired and sorted out hackers to perpetrate a considerable scope of repeatable illegal exercises against a worldwide network close to secrecy. The first detailed cybercrime accompanied the development of Loom, a gadget created by Joseph-Marie Jacquard in 1820, which was utilized to permit the rehash of fabricating the textures induced in compromise for the employees their conventional work and subsistence. Because of which they perpetrate demonstration of counterattack to dampen the use of innovation furthermore. With the continuation of this, a computerized environment encounters many cyberattacks also.

Consequently, research focused on prevention and analytic measures of those cyberattacks, specifically on the banking and e-commerce sector

(Hunton & Paul, 2012). The world of cybersecurity and cybercrime is addressed by describing the various types of cybercrimes and the need for cybersecurity space. The author investigates different problematic elements of cybersecurity such as viruses, worms, hackers, malwares, trojan horses, and password cracking. Additionally, the author describes Jacking, which derives from hijacking where a hacker controls fraudulently (Buch et al., 2017). In 2012 and 2014, an article is presented as a study of cybercrime on the Internet. The authors discussed various cybercrimes that happened in India online. The article gives the basic notion of cybercrimes like cyberstalking, hacking, phishing, cross-site scripting, and vishing. Cross-site scripting is a computer vulnerability that allows code injection into web pages. The author suggests that Cyber Space Security Management is an essential module to monitor and control these types of cyberattacks in India by specifying the roles of independent agencies with unified architecture (Gandhi, 2012; Jain, Neelesh, & Vibhash, 2014).

The outline of cybercrime and its impact on legislation in the United States is given in another article that describes the difference between initial crime legislation and cybercrime legislation. The author suggested that police agencies engage computer shrewdness detectives specializing in cyberattack offenses (Kimberly, 2017). In the year 2019, a report surveyed cybercrimes encountered in India. It briefs all types of cybercriminals and misdeeds with its development, variations, and preventive solutions. According to the published article of TOI by S. S. Gole, the author pointed out that Pune is on the third situation among the urban communities influenced by cybercrimes per the cases enrolled in 2011. Bangalore is the most influenced city with 117 instances, trailed by Vishakhapatnam where 107 crime was registered while the articles of the National Crime Records Bureau. There is two legislation under which the cybercriminal can be a complaint in India: Information Technology Act (2000) and the Indian Penal Code (IPC). IPC, if applicable for conventional misdeeds such as mischief, defamation, theft, fraud, forgery, etc., and IT Act (2000) is for records connected to hacking, unauthorized access, plagiarism, etc. (Gunjan et al., 2013).

Similarly, a study on cybercrimes and cyber-law in India is presented in 2018, which describes cyberlaw as part of the legal system which deals with the web and cyberspace model. Also, the author classifies cybercrimes into various notions depending on the motivation behind the crime. The author suggests safety solutions in cyberspace by ensuring two-step authentication, the additional security level that requires a client's username and password. In India, the author highlighted various cybercrime cases; the Bank NSP,

Bazee.com, Parliament attack, Andhra Pradesh tax, Sony.Sambath.com, are notable cases (Sarmah et al., 2018).

Many research articles are produced with these references, and these provide a new dimension towards the cyberattack. Notable, a research article that suggests a cyber user identification model for detecting and controlling cybercrime. The author utilized an object-oriented paradigm and hardware like GPS, webcam, and fingerprint to identify hackers or cybercriminals. This study assumes that every cybercriminal or hacker usually leaves or creates some impression at the attack time. The uniqueness of the individual is identified with the assistance of biometric sensors (Agana, 2015). An article about the social implications of the cashless economy is presented to describe a simple equilibrium model for demonstrating how a cashless economy leads to the misallocation of resources in the format and informal sector. The article also suggests the source of income in both the industries and the need to move towards a cashless economy. Yet, the article has less focus on investigating the opportunities for cybercrime in online transactions. The major cyber-attack includes malleability attacks to drain the designated account (Cohen et al., 2020). In E-commerce, perspective an article that is presented about cybercrime threats in E-Commerce. The author listed various cyberattacks in an E-commerce environment and safeguard solutions (Sourabh, 2016). In 2013, a report about the cybercrime ecosystems was conferred, specifically focused on credit card fraud, identifying theft in a digital ecosystem. The author describes credit card fraud as the illegal use of the stolen card or card details, such as card skimming without the knowledge of card owners.

As stated by the report, the frequently advertising category in economy servers holds card information as people are move from traditional payment methods [cash] to digital modes of payments (Kraemer-Mbula et al., 2013). Theory of planned behavior (TPB) is proposed to explore the study of cyberattacks and integrate cybercrime perceptions, the trust of sellers with TPB. The premise made in this chapter is that the perception of customers on the presence of cybercrime isn't just behavioral but also deliberate. Thus, customers build a disposition towards cybercrime and become aware of falling victims to such crimes on the web. TPB provides a statistical model for analyzing cybercrimes by considering static factors (Apau, Richard, & Felix, 2019).

Meanwhile, the report analyzes the impact of cloud computing on the cyberattack, which incorporates the investigation of recent advances in cybercrime and subsequent changes in Australian legislation to handle these cyberattacks. The author inferred that cloud computing acquaints various complications with traditional digital forensics processes. Existing methods have centered upon holding a tangible approach to the media that contains information about possible interests. Still, it is frequently impossible or sensible to get the physical media that keeps the user information in a distributed environment. It is because of various features intrinsic to cloud computing. For instance, outsourced data are stored overseas from the investigating agency as the use of the Internet gets more extensive by permitting the customer to share or outsource their information to the web, resulting in colossal prospects of the cyberattack (Christopher, 2013).

As the demand for cellular phones is impressive in the market, the potential date breaches have commonly occurred in portable devices. The chapter analyzes various methods to intercept cybercrime's impact on electronic gadgets. The author suggested the permission-based security model and behavior-based security model to control data breaches. From the analysis, the author founds that device loss or theft, phishing, spoofing, inadvertent declaration of information are the reasons for a data breach (Safavi, Seyedmostafa, Zarina, & Rozilawati, 2013). As a result of the cashless economy and immense growth of portable devices, say mobile phones, the world is experiencing a new cyberattack QRishing. A research article) investigated the viability of QR codes which are utilized for online payments.

The QR codes are primarily popular in small-scale business people because of ease in disposition and utilization. But this technology has the risk factor of allowing the user/customers to scan unauthorized/unauthenticated information from the digital poster, bills, etc., resulting in a new cyberattack for the business people. The chapter carried out two experiments for analyzing the security risk in QR code scanning. The outcome demonstrates that most of the customers who scan QR for payment will reach the associated URL. Providing security controls in desktops or mobile phones may result in a secure transaction on both sides (Vidas et al., 2013). In connection with this, a report was presented as a study on attacks and challenges in QR security. QR codes have discovered their way from automated into our day-to-day mobile phone consumption. QR codes are utilized for publicizing, authentication, and even for money transactions where sensitive information is transferred. The chapter describes two sorts of QR-associated attack vectors as the attacker displaces the original QR code and the attacker tampers the individual modules of QR codes. Also, the author presents an overview of research challenges from a usability perspective. The security issues in the usability of QR include a content display, Content preprocessing, Anti-Phishing tools, and content verification (Krombholz et al., 2014).

Although the general public felt secure, 45% is still a massive part of the general population, an untapped market that may be helped with an extra feeling of safety. Without statistics security, the e-trade enterprise could suffer from first-rate mistrust problems. "If sensitive information is leaked, it may affect the recognition of an enterprise." (Liu et al., 2020). Johansson et al. (2021) focused on the controls of the information resources, i.e., accessing the information like by whom, by how, for what purpose, usage like terms an business enterprise can learn how to follow those controls and use them as assist. USD 700 million was loss in ecommerce sales was found by study which cost 19 times more than the traditional retails methods (Encyclopedia.com, 2021). Small form factor computers or a small farm of servers do not have much power or scalability to process many customer accesses. Compute power and storage ability is more in Cloud computing. Cloud Technology is used to store and exchange data for an excellent result (Hussam & Mohammad, 2018). Nowadays, DoS attacks, malware, debit or credit card fraud, phishing, and other threats have put the E-commerce security at a greater risk. SSL certificates play a major role in reducing this risk in E-commerce (Toapanta et al., 2020).

8.4.3 ANALYTIC REPORT OF CYBERCRIME IN E-COMMERCE

Although major cyberattacks are purposefully designed for exposure, they underline that cyberattack is an absolute risk to billions of online users. The advent of information communication technology (ICT) allows various types of cyberattack behavior such as computer viruses, trojan horses, hackers, etc. The proposed chapter focused on cyberattacks, specifically on E-commerce sites, Banking sectors, and the webserver where users are supposed to submit their personal information.

8.4.3.1 DATA BREACHES

A breach is commonly defined as an incident that confirms information disclosure to an unauthorized unit. The social attacks that were considered for an analytical study in phishing and data breaches in E-commerce sites. The contrast between pretexting and phishing is the absence of dependence on malware establishment in the previous for the assailant to meet their objective. Figure 8.1 depicts the percentage of data breaches encountered in the last four years.

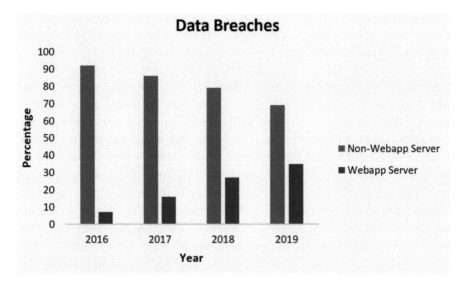

FIGURE 8.1   Data breaches – statistics report.

Table 8.2 shows the detailed data breaches report from the duration of 2016 to 2019. According to data breach reports in the retail sector, card breaches in the sale terminal are low and continue to decline when comparing 2016 and 2019. Attacks against e-commerce payments and web applications are continuously rising in those days because of numerous users and availability. Among these, most data breaches are financial motives, and only a tiny percentage of data breaches have other reasons like fun in that life, espionage, etc. Most of the attackers or intruders are target payment-related activities (almost 64%). More than 20% are credential breaches, and 16 more percentage breaches are in personal data. Due to those breaches, the company or brand reputation will decline among the investors and people who use the product or services of those companies.

Data breaches may cause damages in multiple ways like workforce changes, salary, and other expenses for sealing the security breaches. Customers may move on to the following competitive companies for a better experience, which may bring unwanted media coverage and decline brand value. Business continuity may get affected due to this data or information breach. Also, financial losses might occur by tampering sources where some intruders and hackers change the authenticated data like toll-free numbers or mail IDs for customer support. They publish the fake data at the public forums. The persons eager to contact customer support are misleading to the tampered IDs, leading to financial loss.

TABLE 8.2   Data Breach Report

SL. No.	Activities	2016	2019
1.	Point of sale (POS)	63%	6%
2.	Crimeware	10%	2%
3.	Card skimmers	6%	3%
4.	Web applications	5%	63%
5.	Lost and stolen assets	0%	2%
6.	Miscellaneous errors	1%	8%
7.	Cyber espionage	0%	1%
8.	Privilege misuse	3%	10%
9.	Everything else	12%	5%

Figure 8.2 depicts the basic workflow of cybercrime. The proposed chapter focused on cyberattacks, specifically on E-commerce sites, Banking sectors, and the webserver where users are supposed to submit their personal information. Generally, the user accesses the cloud server for acquiring services like shopping, banking, etc. Consider the scenario of shopping sites which is wide as a part of E-commerce. This is operated by a company that can offer items publicly regardless of no actual shopping mall. The notion of E-commerce is to use the Internet and perform business activities more accessible and faster. A company's website/shopping platform will allow the customer to find their suitable products. They may search through the list and add the product to their shopping cart or remove it from their coach at any time. Upon selecting a suitable product, the customer will navigate to payment, where the online transaction will occur. At this time, a cybercriminal who resides on the Internet will track the customer workflow with the shopping site and established a virtual connection between the cybercriminal (Intruder) and the website. Whenever the user is submitting the card information for payment, the hacker/cybercriminal will be redirecting the link to their account as money laundering. Hence by monitoring the workflow activities on the Internet, a cybercriminal can quickly mock the destination.

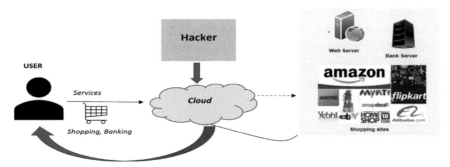

FIGURE 8.2 Basic workflow of cybercrime.

8.4.3.2 E-COMMERCE CRIMES

Any organization that can choose to begin a web-based commercial enterprise or shift towards a multi-channel method, making its commercial to be had online, will adapt to new troubles and threats. Online frauds are par with ones ordinarily known in business corporations. The main fundamental distinction is that you were unable to view your transactions. This phenomenon marks it increasingly complex to identify the user's identity who is purchasing on the site. In any trading venture, the user has to take measures to protect themselves from extortion. There will generally be individuals trying to get something for not anything, i.e., unscrupulous attackers/hackers whose functional purpose is to rip off your enterprise or its customers. As a service provider, it is an endeavor's commitment to teaching themselves what the various E-commerce fraud styles appear to be and take measures to safeguard your business venture.

Meanwhile, innovation has made cyber-attacks reach the hundreds through group e-mailing techniques. Fraudster's strength generation is not functional to masks identity. It generates fake e-mails and sites yet, besides, to reach a massive scope of people all around the international at a meager worth. Additionally, technology made most simpler for the hackers/scammers to play on the sensitive information and speed up our urgent responses. Crimes in E-commerce have numerous flavors, i.e., the essential includes unauthorized purchases. Most retailers offer their customers a method for storing personal information, including purchase records, order history, and shipping details. Hackers who get admission to these obligations can capture/hack them to net a couple of free stuff. Nowadays, numerous deceitful activities are encountered, and different sorts of fraud E-commerce sellers post for it. The various crime activities are credit card fraud, refund cheat, merchant deceit, card testing, friendly deceit, identity theft, Phishing, affiliate deceit, etc.

8.4.3.2.1 Credit Card Deceit

Scammers frequently make use of the credit card information of others during their online shopping. Scammers will get access to the credit card from the card owner by phone calls or by phishing sites to the authenticated mail IDs. Sometimes scammers might be in substantial responsibility for the card. On different occasions, they could have obtained indisputable facts about the card electronically. When the transaction is completed and affirmed, the enterprise is accountable for guaranteeing customer identification. The card proprietor might also be trying to find compensation to the company identical to the transaction amount.

8.4.3.2.2 Refund Deceit

In refund fraud, cybercriminal intentionally utilizes stolen credit cards to make an over-payment on motive. After this, the scammer communicates with the enterprise to notify a random over-payment and asks for compensation. Cybercriminals might ask to refund the extra amount, guaranteeing his credit card is closed with the intention that the merchant transfers the transaction amount in an alternative way. This implies that the original charge of the credit card is not refunded, and the commercial enterprise is accountable for repaying the total amount to the cardholder.

8.4.3.2.3 Merchant Deceit

Generally, in a mart-based business, the retail store is accountable for each customer and the seller. Fundamentally, the items are provided at reasonably-priced charges; they are not consigned in any case. This strategical deceit is additionally applicable in trade. It is not explicit to some specific charge approach. However, that is the path in which the no-chargeback remission method proceeds into its own. In the predominant case, the scammer sold the product and got the charge for imaginary objects. Even in this situation, it is the enterprise that is accountable for the compensation.

8.4.3.2.4   *Card Testing*

The fundamental principle of card testing deceit is that a theft bank card is utilized to initiate a purchase. Yet, the negotiation is then controlled in any such manner that crime identification features are evaded. Additionally, realize how it is needed at that moment apart from friendly deceit, in which the least purpose is to drop the fee when shopping has been fabricated. In card testing deceit, criminals utilize sound analyzes of the fraud detection systems, along with an outstanding deal of understanding approximately the rightful proprietors of theft credit cards. Card testing deceit is the activity of making and examining the legitimacy of a bank card number. Fraudsters target websites that offer an unusual reaction for all types of rejected transactions. For instance, while a card is denied an erroneous expiration date, extraordinary feedback is given to recognize the essential requirements to discover the cessation date.

8.4.3.2.5   *Friendly Deceit*

Friendly deceit portrays more amicable than it is utilizing this methodology. Customers request products or contributions and pay for them ideally using a "pull" installment strategy like a bank card or direct debit. At this point, hackers purposefully provoke a charge-returned, asserting that their credit score card or ledger features were taken. The transaction amount will be reimbursed; however, they maintain the merchandise or administrations. This deceit method is mainly general by services, which incorporate those within the betting or personal background. Besides, friendly deceit leans to be blended with re-delivery. This is where hackers/scammers utilize liberate payment information to reimburse their shopping and not send their products to domestic addresses. Instead, they use agents whose data is being used for purchasing and then in front of them.

8.4.3.2.6   *Identity Stealing*

Identity stealing is the most not uncommon sort of deceit. For instance, a hacker performs an online purchase by utilizing specific identification. This empowers the cybercriminal to buy items online under a fake call and use an individual's bank card.

8.4.3.2.7 Phishing

In this type, an e-mail request for buyer ID, passwords, credit score card information, and diverse individual data. The dealer appears to be a credit group that desires verification of some records because of some modifications in the machine. Phishing permits hackers to get a section to the bank or other cash owed, and it can be utilized for data fraud.

8.4.3.2.8 Affiliate Fraud

There are two sorts of associate misrepresentations. The two of which have a similar objective: collecting extra cash from an associate application with monitoring traffic or login credential information. This can be finished using a completely automated manner or getting actual human beings to log into merchants' websites through counterfeit ledgers. This sort of deceit is charge-method-impartial, however extraordinarily broadly allotted.

8.4.3.2.9 Fake E-Commerce Sites

Generally, our search engines process the data for analyzing and present the results to the users. Some scammers might create fake E-commerce websites with the names and images of the products and describe the products at a very low rate compared to the authenticated stores or popular E-commerce websites. These fake stores are accepting prepaid orders only. To process the order, the customer/client has to pay the amount initially. After payment, they won't send the order ID or the product to the customers. By Search Engine Optimization technologies and other optimization techniques, scammers perform these kinds of frauds.

8.4.4 PREVENTION METHODOLOGIES

The proposed chapter highlights some of the preventive measures of E-commerce crimes. E-commerce fraud management is an essential phase of securing customer's information. Luckily, E-commerce fraud prevention strategies exist to help customers and enterprises proceed further in transactions with safety. Guaranteeing all frameworks are Payment Card

Industry compliant where it builds clear rules for preserving and securing the card information. The most straightforward approach for achieving this is through an E-commerce platform or outsider apparatus-also, the proposed chapter analyzes the usage of address verification to avoid fraud chance. The location verification requires zip or postal code for all transactions, which will probably be sufficient to safeguard against a considerable portion of cybercrime.

Additionally, the proposed chapter examines security code-based preventive measures for all credit card purchases. The backside of the modern credit card holds three sets of figuring numbers, and these numbers are requested to enter for all transactions. But if the card has been stolen, this method is not enough to prevent fraudsters from unauthorized transactions. Hence for every purchase, credit card enterprise must validate their transactions. Meanwhile, customers also need to aware of the banking process, and no one is from the bank will ask for the credit card details like CVV, PIN of the credit card, etc., from the bank customers.

Another prevention method that the proposed chapter examines is, tracking the customer/client's purchase history and behavior. The customer may be tracked if they purchase items that are irrelevant to them. For instance, the specific location-based customer who usually buys shaving lotion and beard oil might suddenly order female cosmetics to a different location irrelevant to their area. This indicates that some fraudulence is going on. Also, some E-commerce site prefers signing procedure at the time of delivering the products. This measure is taken towards preventing criminals from receiving the products. But cybercriminals may spoof the customer's signature also hence this alone could not prevent fraudulence. The cybersecurity forum can also track suspicious session activity. The E-commerce site is continuously monitored and identifies multiple orders for an account in a particular session, massive purchases in a short period, and rapid modifications in the customer's shipping address. The proposed chapter analyzes the prevention measure like keeping track of past fraud attempts, which will assist the enterprise or cybersecurity forum in studying other frauds and building a clear idea in all likelihood, even identifying the vicinity or segment where the ones attempt most often occur. Apart from these prevention measures, the government and private agencies should create awareness about information security frauds, information security laws, punishments enforced for the data breaches, and data tampering like fraud activities. At the time of making IT laws, the resources are more petite. Now numerous resources are available and crimes also. So, the IT laws and punishments must be updated from time to time to cut the crimes.

8.4.4.1 WAYS TO CHECK FAKE WEBSITES

1. **Checking Website Owners Information:** Anyone can check the website registers information publicly available on some third-party websites like https://www.whois.com/whois/.Here we can refer the company details, phone number address details of the company. If there is no information available, we have to move far away from those sites.

2. **Checking Company Address:** To find out if a site is proper or no longer, just test the contact information. If the details they offer are absent, then deal with it as a warning flag and do now not buy whatever from it, even supposing the site looks very attractive. If there is genuinely no information, then it's nice to keep away from this website. There may be info now and again; however, don't be taken in. Though e-mail identification and phone may be communicated back through fraudulent people, they can't sincerely get away from Google's street view. You can test any signs for the office, like a signboard. Check a worker's/proprietor's LinkedIn profile to find out the authenticity.

3. **Read the Policy Details:** The first rule of purchasing online is to examine and understand the trade and go back policy. If the phrases/conditions are clear, then the corporation might be genuine. You can purchase whatever from an internet site that no longer mentions and trades or goes back policy. Also, if the policy appears that it has been just copied and pasted from another internet site, then it's best now not to consider this web page together with your cash. This is one of the approaches to spot a rip-off, as scammers could be dodgy about returning items.

4. **Social Media Checking:** One of the great methods to spot a fake E-commerce website is to look at whether or not they have got a sturdy presence on social media, including Instagram, Facebook, Twitter, LinkedIn, Pinterest, etc. Check that the posts/feedback are similar and if the website name is referred to in all of the social media sites.

   See how many followers they've, read the feedback or remarks, read the corporation's response, and whether the products posted on the website are similar to those on the media websites. If there may be no reply from the employer, and lots of human beings have posted bad feedback, then the site might be a fraudulent website.

5. **Website Design and Languages:** Good websites typically have an appealing design, and English is ideal. There are no mistakes in grammar and spelling, and the enjoyment of buying from a real website online is polished and trouble-free. A faux E-trade website could have many mistakes, even on the important pages. When you notice language errors all over the website online, you need to be cautious about the site.

6. **Ads on the Website:** There would be commercials on every web page, as that is one of the ways to make money online. But if commercials outnumber the content material, then it's time to be careful. Be wary if you have to get beyond numerous redirects and popups. To get to wherein you need to be. To find out whether or not a website is real or no longer, you have got to be very alert, and seeing too many commercials is one of the giveaways.

7. **Based on Payment Mode:** A genuine E-commerce website will receive all predominant debit and credit score playing cards and produce other non-card alternatives also. If a website asks you to transfer cash via PayPal or Western Union or use the only cryptocurrency for the charge, that needs to be an exact pink flag. If you have to pay through a way in which the payment cannot be reversed, you definitely absolutely must not purchase from this site.

8.5 CONCLUSION

The merits related to E-commerce technologies are undoubted. Relevant studies have shown that e-commerce enables price reduction, mass customization, and a high price of aggressive benefits. Cybercrime is and will keep on being one of the most significant topics in information security in the following years. Because of the continuing progress in the digitalization of the nation, an expanded number of assaults and respective misfortunes are expected. Organizations are persistently moving an ever-increasing number of exercises to the Internet and, thus, the economic effect of cybercrime on the economy will increment further. The professionalization pattern of ongoing years concerning the conduct and activities of cybercrime criminals features the significance of further research. By applying the fraudulence analysis in the digital world as presented in the proposed chapter, an individual can be aware of existing cyberattacks and the necessary steps to prevent those cyberattacks. The proposed chapter presented an analytic report on possible cyberattacks in an E-commerce environment. Also, the chapter analyzes the effect of existing prevention measures on these cyberattacks. Organizations and governments must spend more money on infrastructure development to minimize cybercrimes and losses due to data breaches from these analytical perspectives. It may protect organizations from severe cybercrimes.

KEYWORDS

- **cybercrime**
- **data breaches**
- **electronic data interchange**
- **general data protection regulation**
- **information communication technology**
- **phishing**
- **QRishing**
- **security**

REFERENCES

Agana-Correspondence, & Moses, A., (2015). *Cybercrime Detection and Control Using the Cyber User Identification Model.* Retrieved from: https://www.researchgate.net/profile/Agana_Moses/publication/324774119_Cyber_Crime_Detection_and_Control_using_the_Cyber_User_Identification_Model/links/5b81259b299bf1d5a7264b55/Cybercrime-Detection-and-Control-using-the-Cyber-User-Identification-Model.pdf (accessed on 10 January 2022).

Apau, R., & Felix, N. K., (2019). Impact of cybercrime and trust on the use of e-commerce technologies: An application of the theory of planned behavior. *International Journal of Cyber Criminology, 13*(2). Retrieved from: https://www.proquest.com/openview/c6a7719ec954a323 13e60def276a2928/1?pq-origsite=gscholar&cbl=55114 (accessed on 10 January 2022).

Buch, R., Dhatri, G., Pooja, K., & Nirali, B., (2017). *World of Cyber Security and Cybercrime* http://computers.stmjournals.com/index.php?journal=RTPL&page=article&op=view&path%5B%5D=1109 (accessed on 10 January 2022).

Cohen, N., Anna, R., & Labib, S., (2020). Towards a cashless economy: Economic and socio-political implications. *European Journal of Political Economy, 61*, 101820. https://doi.org/10.1016/j.ejpoleco.2019.101820.

Encyclopedia.com, (2021). Encyclopedia.com. [Online] Available at: https://www.encyclopedia.com/books/educational-magazines/security-e-commercesystems (accessed on 10 January2022).

Gandhi, V. K., (2012). An overview study on cybercrimes in Internet. *Journal of Information Engineering and Applications, 2*(1), 1–5. https://www.iiste.org/Journals/index.php/JIEA/article/view/1201 (accessed on 10 January 2022).

Gunjan, V. K., Amit, K., & Sharda, A., (2013). A survey of cybercrime in India. In: *15th International Conference on Advanced Computing Technologies (ICACT)*, (pp. 1–6). IEEE, 3. 10.1109/ICACT.2013.6710503.

Hooper, C., Ben, M., & Kim-Kwang, R. C., (2013). Cloud computing and its implications for cybercrime investigations in Australia. *Computer Law & Security Review, 29*(2), 152–163. https://doi.org/10.1016/j.clsr.2013.01.006.

Hunton, & Paul, (2012). Data attack of the cybercriminal: Investigating the digital currency of cybercrime. *Computer Law & Security Review, 28*(2), 201–207. https://doi.org/10.1016/j. clsr.2012.01.007.

Hussam, H., & Mohammad, A., (2018). Cloud computing: Legal and security issues. In: *8th International Conference on Computer Science and Information Technology (CSIT)* https:// www.researchgate.net/publication/328243752_Cloud_Computing_Legal_and_Security_ Issues (accessed on 10 January 2022).

Jain, N., & Vibhash, S., (2014). Cybercrime changing everything–an empirical study. *International Journal of Computer Application.* http://www.rspublication.com/ijca/ijca_index.htm (accessed on 10 January 2022).

Johansson, S., Kullström, M., Björk, J., Karlsson, A., & Nilsson, S., (2021). Digital production innovation projects – the applicability of managerial controls under high levels of complexity and uncertainty. *Journal of Manufacturing Technology Management, 32*(3), 772–794. doi: 10.1108/JMTM-04-2019-0145.

Kraemer-Mbula, E., Puay, T., & Howard, R., (2013). The cybercrime ecosystem: Online innovation in the shadows? *Technological Forecasting and Social Change, 80*(3), 541–555. https://doi.org/10.1016/j.techfore.2012.07.002.

Krombholz, K., Peter, F., Peter, K., Ioannis, K., Markus, H., & Edgar, W., (2014). QR code security: A survey of attacks and challenges for usable security. *International Conference on Human Aspects of Information Security, Privacy, and Trust*, 79–90. https://link.springer. com/chapter/10.1007/978-3-319-07620-1_8 (accessed on 10 January 2022).

Liu, J., Yuan, C., Lai, Y., & Qin, H., (2020). Protection of sensitive data in industrial Internet based on three-layer local/fog/cloud storage. *Security and Communication Networks, 2020.* doi: 10.1155/2020/2017930.

Munjal, S., (2016). Cybercrimes–threat for the e-commerce. *Journal of Maharaja Agrasen College of Higher Education, 3*(1). https://papers.ssrn.com/sol3/papers.cfm?abstract_ id=2767443 (accessed on 10 January 2022).

Pavlik, K., (2017). Cybercrime, hacking, and legislation. *Journal of Cybersecurity Research (JCR), 2*(1), 13–16. https://doi.org/10.19030/jcr.v2i1.9966.

Safavi, S., Zarina, S., & Rozilawati, R., (2013). Reviews on cybercrime affecting portable devices. *Procedia Technology, 11*, 650–657. https://doi.org/10.1016/j.protcy.2013.12.241.

Sarmah, A., Roshmi, S., & Amlan, J. B., (2018). A brief study on cybercrime and cyber laws of India. *Int. Res. J. of Eng. and Technol., 4*(6). https://www.irjet.net/archives/V4/i6/ IRJET-V4I6303.pdf (accessed on 10 January 2022).

Toapanta, S. M. T., Zamora, M. E. C., & Gallegos, L. E. M., (2020). Appropriate security proto-cols to mitigate the risks in electronic money management. In: *Smart Trends in Computing and Communications* (pp. 65–74). Springer. https://link.springer.com/chapter/10.1007/978-981-15-0077-0_7 (accessed on 10 January 2022).

Vidas, T., Emmanuel, O., Shuai, W., Cheng, Z., Lorrie, F. C., & Nicolas, C., (2013). QRishing: The susceptibility of smartphone users to QR code phishing attacks. *International Conference on Financial Cryptography and Data Security*, 52–69. https://link.springer. com/chapter/10.1007/978-3-642-41320-9_4 (accessed on 10 January 2022).

CHAPTER 9

An Approach to Data Recovery from Solid State Drive: Cyber Forensics

HEPI SUTHAR and PRIYANKA SHARMA

Rashtriya Raksha University, Gandhinagar, Gujarat, India

ABSTRACT

SSD hard disk is a high-speed storage disk, which is widely used in all filed of technology for data storage. This chapter analyzes relevant principle structure and storage characteristics of file data of SSD hard disk propose the recovery possibility and existing challenge of SSD hard disk from three aspects: logic layer, physical layer, and firmware layer. With the in-depth development of computer technology and communication technology and the advent of the era of big data, people are increasingly dependent on high-speed storage. In recent years, solid state drives (SSD for short) is a block device composed of flash memory, flash conversion layer, and controller, which are widely used in various fields, such as aviation, electric power, military, medical, Navigation equipment, etc., as an important medium for data storage, has ushered in a blowout development in recent years. SSD hard disks have developed at a high speed from the high capacity of storage particles, the stability of the main control chip, and the efficiency of data reading and writing. SSD comparative faster than traditional HDD. SSD controller gives processing with data read and write capacity higher. However, many SSD hard disk data cannot be read or written in the application, causing user data damage. SSD data recovery is different from traditional HDD. Data store and retrieve methods different in SDD. Here with tools like FTK imager and Winhex try to find the data. Therefore, how to recover SSD hard disk data after the damage is of important research value in real work.

Advancements in Cybercrime Investigation and Digital Forensics. A. Harisha, Amarnath Mishra, & Chandra Singh (Eds.)
© 2024 Apple Academic Press, Inc. Co-published with CRC Press (Taylor & Francis)

9.1   INTRODUCTION

9.1.1   *COMPOSITION STRUCTURE OF SSD*

The solid-state drive is mainly composed of the main control chip, flash memory particles, a cache chip, and a SATA interface chip (Chang, 2007). Solid state disk (SSD) is a large-capacity memory that uses solid-state semiconductor chips as storage media. According to the semiconductor chip used, it can be divided into flash memory (NAND flash) and volatile storage (DRAM)-based solid-state drives (Hepi & Priyanka, 2021). The latter requires an independent power supply and can only be used in very special equipment. It is beyond the scope of this chapter (Figure 9.1).

FIGURE 9.1   Solid state drive.

9.1.1.1   *MAIN CONTROL CHIP*

The main control chip is the core device of the entire solid-state hard drive. Its role is to accept instructions from the system and rationally allocate the data load on each flash memory chip; the other is to undertake the entire data transfer, connecting the flash memory chip and the external SATA interface (Micron Technology (2008). The main control chip is connected to the flash memory chip and the external interface and controls the entire hard disk work by reasonably distributing the data storage on each flash memory chip (Hepi & Priyanka, 2021). Excellent main control chip has the characteristics of fast data processing speed and advanced algorithm.

9.1.1.2   FLASH PARTICLES

In solid-state hard drives, flash memory particles replace mechanical disks as storage units. According to the difference of electronic unit density in NAND flash memory, it can be divided into SLC (single-level memory cell), MLC (double-layer memory cell), and TLC (three-layer memory cell) (Takeuchi, 2009). These three types of memory cells have obvious life and cost. The difference SLC (single-layer storage) is a single-layer electronic structure, with a small voltage change range when writing data, long life, and more than 1,00,000 reads and writes (Rizvi & Chung, 2010). It is expensive and is mostly used in enterprise-level high-end products. MLC (multi-layer storage) uses a double-layer electronic structure built with high and low voltages. It has a long life and a medium price. It is mostly used in civilian high-end products (Chang, Lin, & Chen, 2016; Guide.net.). The number of reads and writes is about 5,000. Compared with SLC, the write speed and number of times are reduced. The chip adopts a wear-leveling algorithm to meet the requirements of long-term use (Micheloni, Marelli, & Commodaro, 2010). TLC (three-layer storage) has the highest extended storage density of MLC flash memory (up to 3 bit/cell), with a capacity of 1.5 times that of MLC, the lowest cost, low mission life, and the number of reads and writes is about 1,000 to 2,000. TLC is the flash memory particle of choice for mainstream manufacturers. The structure of a solid-state hard disk based on flash memory is much simpler than that of a mechanical hard disk (HDD), which is composed of a shell and printed circuit board (PCB). The shell only plays a protective role, and the core is the PCB. There are main control chip cache chips (some low-end products without cache chips) and flash memory chips for storing data on the PCB (Kim et al., 2013).

9.1.1.3   CACHE CHIP

The cache chip is mainly used for random reading and writing of commonly used files and fast reading and writing of fragmented files. The cache chip is generally set next to the main control chip, and its function is similar to that of a mechanical hard disk. The cache chip can perform functions such as data pre-reading, cache writing, and storing recently accessed data. It is worth noting that some low-end products omit the cache chip to save costs, and its performance will inevitably decrease. Flash memory chips take on the important task of data storage, so they have the largest number on the circuit board and occupy the largest space. The capacity of the solid-state drive is mainly determined by the number of flash memory chips mounted. The capacity of common solid-state hard drives in the market is between 16 GB and 1.6 TB (Figure 9.2).

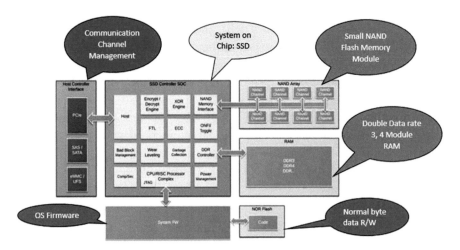

FIGURE 9.2   SSD internal structure block diagram (Hepi & Priyanka, 2021).

9.1.1.4   WORKING PRINCIPLE

The working principle of the solid-state hard disk is shown in Figure 9.1. NAND Flash refers to flash particles (Ko et al., 2019). The SSD controller operates these Flash particles in parallel through several main control channels, just like RAID 0, which can improve the parallelism and efficiency of data writing (Wang et al., 2019). Each Flash particle is further subdivided into multiple blocks (blocks), and each block contains multiple pages. Inside the SSD, the smallest access unit granularity between the SSD controller and Flash is the page. Generally, the size of a page is 4 k, and a block includes 16 pages. When writing data, like the working mechanism of RAID 0, the data is simultaneously written to the available pages in the block of each Flash particle in parallel. When a block is full, another block will be written. Flash memory (Flash Memory) is essentially a long-life non-volatile (the stored data information can still be retained in the case of power failure) memory. A block is a unit, and because of the working principle of the MOS tube, the writing speed of flash memory is slower than the reading speed (Lee et al., 2013; Ahn & Lee, n.d.).

9.2 THE CHARACTERISTICS OF SOLID-STATE DRIVES

Compared with solid-state hard drives and mechanical hard drives. Its main features are as follows:

1. **Fast Reading and Writing Speed:** The solid-state drive uses electrical signals to read and write flash memory chips. Compared with the mechanical hard disk that requires the head to move back and forth, the seek time of the solid-state hard disk is only a few tenths of a millisecond. It has obvious advantages in indicators such as continuous reading and writing of big data and 4 K random reading and writing.

2. **Anti-Shock and Anti-Drop:** Since the all-solid-state flash memory chip is used as the storage medium, there is no need to worry about the collision between the head and the disk surface like a mechanical hard disk. The solid-state drive can work normally when moving at high speeds and shocks, and it has much stronger shock resistance and shock resistance.

3. **Low Power Consumption and No Noise:** The solid-state hard disk does not need to drive the head movement, and there is no bearing and motor, so the power consumption is lower than that of the mechanical hard disk and theoretically no working noise.

4. **Low Working Environment Requirements:** Solid-state drives can work between 10°C and 70°C, while mechanical hard drives can only work in the range of 5°C to 55°C.

5. **Small Capacity and High Price:** In the face of TB-class mechanical hard drives, the capacity of weekly hard drives has obvious disadvantages. With the same capacity, the price of solid-state drives is even more than 10 times higher than that of mechanical hard drives.

6. **Write Life Limit:** The flash memory chip used by the solid-state drive has a limit on the number of erasing and writing. The lowest-end TLC chip theoretically has only 500 erasing and writing times. Even the best SLC chip has only about 1,00,000 erasing and writing

times at most. The most important thing is that the data cannot be recovered after the SSD is damaged. This is the main cause of worries for users (Kang, Lee, & Kim, 2018).

9.3 THE MAIN IDEA OF SSD DISK DATA RECOVERY

SSD disk data recovery mainly includes three aspects:

1. **Logic Layer:** Due to the existence of the TRIM (disable delete notification) instruction and the GC (garbage clear, garbage disposal) mechanism in the solid-state drive (the empty block is collected when the system is idle for the system to use when needed, and the empty block is maintained by establishing a recycling block table method (Gubanov & Afonin, 2014)), so the deleted data is difficult to recover. After deleting a piece of data, the SSD master will first mark the deleted data, that is, trigger the TRIM command. This command will tell the solid-state drive to delete the data location, size, and other information, and then wait for the SSD master to do this part. Data execution GC mechanism. After performing this step, the deleted data that originally existed in the SSD no longer exists. This part of the data has been erased by the GC mechanism, and this erasure cannot be restored. Because some data may be accidentally deleted during use, the SSD cannot be left to GC processing, otherwise, useful accidentally deleted data cannot be recovered. How to recover the data recovery of the SSD with the TRIM instruction and GC mechanism turned on has become the key to the data recovery of the logic layer.

2. **The Firmware Layer:** It is different from the mechanical hard disk. The firmware of the solid-state hard disk is stored in the main control chip, while the firmware of the mechanical hard disk exists in the disk, circuit board, and other places. There is no complete solution to key issues such as how to read firmware information, analyze the firmware structure, and repair damaged firmware.

3. **Physical Layer:** The physical damage of the solid-state drive generally includes damage to the main controller, damage to the circuit board, and damage to the memory chip. For memory chip damage, flash damaged data cannot be recovered. As for the damage of the main control and the circuit board, the current common handling method is to disassemble the Flash chip, and then read and reorganize

the internal data through the chip reading device. This method not only consumes material and financial resources but also requires very high technical personnel's data reorganization technology. Therefore, it is necessary to study how to directly read and extract data without reorganizing data through a chip reading device.

9.4   LITERATURE REVIEW

When against the law and an illegal pastime has been dedicated withinside the bodily global an additionally in digital platform, the proof can regularly be discovered on a suspect's virtual gadgets, social media or at the net. The net expands with extra sensors overseeing the actual global on a everyday basis, including site visitors cameras, ATMs, and webcams. People additionally generally tend to submit extra messages on social networking websites or chatting in IRC rooms, in which IP addresses monitor area and conversations are being registered. Whenever research is ongoing and there's a danger of virtual proof, a virtual forensic research desires to be conducted. This normally consists of apprehending a suspect's virtual gadgets, including a non-public computer, mobile phone, navigation, reminiscence gadgets and trying to find possible proof or clues. When a virtual medium is tested with the aid of using forensic experts, the proof need to every so often be recovered from damaged or deliberately destroyed reminiscence, deleted or misplaced data (Dennon, 2016). Regardless of the tool and records status, a totally vital step must be taken first: create an photo that could be a virtual reproduction of the sate while the tool became gathered. This photo is vital to show the chain of custody, the integrity of the proof likely determined at some point of the investigation, consequently it is able to be demonstrated that the records withinside the way have now no longer been modified via way of means of the investigator or a 3ʳᵈ celebration from the time the tool became gathered till a probable courtroom docket appearance. The step of verifying the integrity typically consists of an evaluation of the fingerprint among the preliminary photo and the proof presented. This Digital proof particularly includes a hash cost of the photo, because of this that a calculated checksum of the records. This checksum it's miles maximum typically calculated via way of means of an MD5 or SHA-1 set of rules. All hash set of rules produce an nearly specific fingerprint, that allows you to constantly be the equal given the equal enter (Dennon, 2016). For example, the MD5 hash set of rules produces a 128 bit checksum of any enter with arbitrary length. Therefore, an precise reproduction or photo of a tool could have the equal fingerprint than the original. A minor addition might make a finger-print distinct from the original. After obtaining a tool that includes virtual memory, the investigator will attempt to take a virtual photo of the gathered tool earlier than seeking out proof. Some instances that don't it's miles to begin with viable because of hardware failure of an intentional or unintended nature. Therefore, a hardware or software program primarily based totally restoration need to be performed.

9.4.1   HARD DISK RECOVERY

Hardware failures needn't be intentional within the nature of any hardware device. Physics on records is extremely fragile, as are the browse and write heads and conjointly storage chip semiconductor. In any of the higher than cases, the foremost promising thanks to restore information from a broken device is commutation components that are broken. Most of the time the disc are still intact, solely the mechanisms, to read their information, they're not operating correctly. In these cases, it's very (Harris, 2015).

It's vital to urge constant hardware because the defective one as a result of every marketer and model uses slightly completely different technologies. Essentially three parts will be replaced. If associate arm facet or head is broken, the complete arm must get replaced if otherwise, the electronic board containing chips and microcode can be replaced, further as an engine of the spindle. The complete spindle can be placed in an exceedingly different box containing all other hardware. Here it's crucial that the disks within the spindle don't shift their position to the opposite disks (Rizvi & Chung, 2010). The chances to revive information from a defective drive in the higher than cases are very high if the replacement is completed with care and in a very clean environment.

9.4.2   FLASH MEMORY HARD RECOVERY

Data restoration from flash reminiscence is extra tough than difficult Hard disk drives. All manage and reminiscence chips are soldered to a board. Therefore, we cannot definitely update part of the tool without locating the precise identical version and update the elements via way of means of re-soldering them. Depending on the kind of flash reminiscence, 2 to 20 chips are located on a board. Re-welding them via way of means of hand is a tough and fragile activity and nearly not possible for more than one chip. Another opportunity is to unsolder every reminiscence chip and every one after the other the usage of unique hardware and tools (Rizvi & Chung, 2010). This is feasible for reminiscence sticks with a chip, on multi-chip SSDs, This technique turns into very complicated due to the fact every supplier makes use of distinctive techniques on a way to use chip, a way to carry out put on leveling and gathering rubbish and a way to distribute data. The discern underneath indicates hardware used to examine a unmarried chip of flash reminiscence.

9.4.3   SOFTWARE RECOVERY ON HARD DRIVES

Data Recovery isn't forever exhausting ware connected task. In more cases, the disk analysis that computer code uses is enough to retrieve information from a disk. Like mentioned above, the particular file is not deleted from the hard drives and, eventually it'll be overwritten by a replacement file. This reality is often wont to recover data. Most importantly, once restoring data from disks and gathering evidence, the information originals should stay intact and altered as very little as possible. So, consultants use specific hardware to repeat the data on disk to a picture or alternative file disco. This device referred to as a write blocker that is employed as an association between the disc drive and also the computer and monitors the commands being issued and prevents the PC writes knowledge to disk (Humphries, 2017). Scan commands are passed to the devices, whereas recording commands are blocked.

9.4.4   FORENSIC SOFTWARE TOOL

Various tools have been developed that forensic personnel can use to recover data from hard disk drives and other digital memories like cache, SSD, expensive software suite as well as open-source Forensic tools. However, one of the best evidence gathering tool known and most common forensic investigation tool is EnCase tool. It can copy discs using bit-stream (bit-by-bit copy) technology to create a virtual file system rebuild. FTK (forensic toolkit by access data) and X-ways are two different windows based tools and the special feature of these three tools is the additional data stored with the disk image as MD5 hash values to prove image integrity. Sleuth Kit is a set of open sources with the disk image as on different operating systems (OS) and supports all common file systems. Autopsy is a digital forensic platform and an interface graphics for the Sleuth Kit and other digital forensic tools. Other Freeware tools well known are Recuva rated very well. PCI file Inspector as well rated good. After creating images of the hard drive using streaming technology bit, each bit in the original drive is stored in the image file and can then be examined (Humphries, 2017). The tools mentioned above can help the examiner to gather possible evidence in existing files and are also capable of restoring data from deleted files or formatted partitions. All the tools mentioned are only capable of processing disks. Not encrypted, if the encrypted file System (EFS) is used, an image can be done. But analyzing the data requires much more effort.

9.4.5   FLASH MEMORY SOFTWARE RECOVERY

The same method is utilised with traditional hard disk drives to analyse an SSD's digital evidence and collect electronic footprint evidence from both existing and erased files and data. EnCase or any other forensic tool described is used to capture a forensic image from the middle, not to alter the original data and to collect files of potential electronic evidence. When partitions were formatted or files were deleted before, exam examiners have little chance of recovery of data back again (Chang, 2007). This is because, in contrast to hard drives, flash memory and in particular, SSDs have internal routines that cannot be influenced from the outside, for example, with a write blocker (Micheloni, Marelli, & Commodaro, 2010).

9.4.6   FORENSIC TOOLS FOR FLASH MEMORY

Forensic Tools that can be used to capture forensic images and gather electronic evidence potentials on SSDs are the same as for hard drives. To read individual memory chips from an SSD or other flash memory in case of hardware problem or to avoid internal routines to change the data saved on the memory chips, these four tools can be used:

- PC-3000 flash memory SSD edition (ACE data recovery);
- Dumppicker;
- Flash extractor; and
- Flash doctor.

All of the above tools work similarly. The equipment, read the content of a memory chip. The software then compares the chip maker and model with a bank and helps recover existing files and deleted files. ACE data recovery announced an expanded cooperation between the company manufacturing software of data recovery and SandForce-based SSD data recovery. As mentioned Previously, SSD controller manufacturers face very strong competition and are not willing to share insight into internal routines, encryption, and wear leveling and garbage collection (Lee et al., 2013). Therefore, the cooperation between a large data recovery company and the biggest SSD Controller manufacturer is a huge step and improvement for forensic examiner and data recovery experts and dramatically increased the recovery rate for SandForce-based SSD (Geier, 2015).

Features	SSD	HDD
Mechanism	NAND NOR flash memory	Magnetic rotating platters
Capacity	Up to 1 TB (notebooks)	Up to 2 TB (notebooks)
	Up to 4 TB (desktops)	Up to 10 TB (desktops)
Durability	Shock-resistant	Fragile
Power consumption	Average 2 W	Average 10 W
Endurance	MTBF > 2 million hours	MTBF < 7,00,000 hours
Noise	None	Present
Operating system boot-time	Average of 10–15 seconds	Average of 30–40 seconds
File opening speed	30% faster than HDD	Slower than SSD
Speed	>200 MB/s	50–120 MB/s
Vibration	No vibration	Moving parts cause vibration
Affected by magnetism	No effect	Can erase data
Full drive encryption	Supported	Supported
Cost	Costly	Cheap compared to SSD
Heating	less	High
Size	Compact	Large
Figure		
	SSD	**HDD**

9.5   EXPERIMENTAL WORK

9.5.1   *SSD DATA RECOVERY METHOD*

Aiming at the idea of data recovery in three aspects: the logical layer, the firmware layer, and the physical layer of the solid-state drive, the researchers carried out a series of research and experiments and summarized three data recovery methods.

> ➢ **Experiment 1: Logical Layer Data Recovery:** The data recovery of the logic layer is explained through experiments
>
>   *Experiment Object:* WD SSD of 120 GB solid-state drive, verify that it supports TRIM, and turn on TRIM. As shown in Figure 9.3, when the TRIM command is turned on, the feedback value is 0 to turn it on, and the feedback value is 1 to turn it off or the TRIM command is not supported.

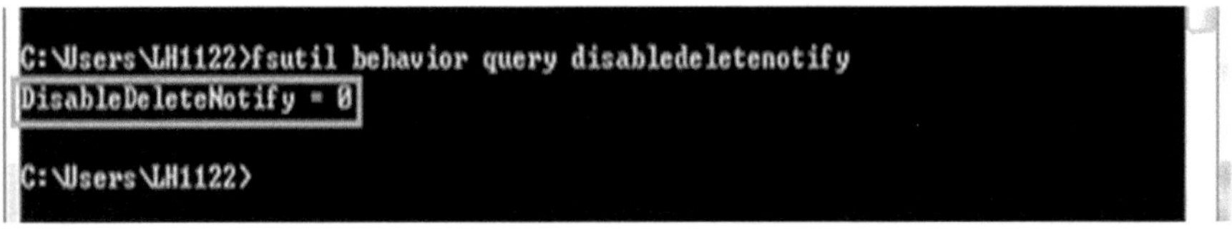

```
C:\Users\LH1122>fsutil behavior query disabledeletenotify
DisableDeleteNotify = 0

C:\Users\LH1122>
```

FIGURE 9.3   TRIM setting.

To add files to the hard disk through win-hex, first, add a small file of 72 K for each file, and keep the remaining memory of the hard disk at about 2G. Select five small files as the files to be deleted, and note the position and the data of the first and last sectors to verify the integrity of the file. Add two large files with a size of 512 MB and write down the location (head sector and tail sector) and data size before deleting to verify file integrity. Delete two large files and five small files. After 2 h, the sector where the original file was located and found that the file still exists, overturning the speculation that "the data is not thereafter 1 h." Add a large file with a size of 1G, and find that the location of the file is in the previously free space of about 1G, and the previously deleted files are all processed by the GC mechanism, that is, GC occurs when writing. As shown in Figure 9.4, after deleting the data for 2 h, it is found that FiLL_10 is still intact.

As shown in Figure 9.5, after writing a 1 G large file, the data in the sector where FiLL_10 is located has been messed up, indicating that the

GC has cleared the data and moved the cleared free space to the new free space.

FIGURE 9.4   File location offset.

FIGURE 9.5   Garbage data location.

As shown in Figure 9.6, the location of the 1G large file is the previous free space. GC is divided into "idle garbage collection" and "passive garbage collection" (Geier, 2015).

Idle garbage collection is to perform garbage collection operations when the solid-state drive is idle. The advantage of this is that it will not occupy additional master control resources and can keep the solid-state drive as high as possible; the disadvantage is that it will increase additional writes. Into zoom. Generally speaking, if new data is being stored or deleted, this mechanism will work automatically. The WD SSD of 120 GB SSD tested above belongs to real-time garbage collection. Therefore, during the test, it was found that the GC trigger was when data was written.

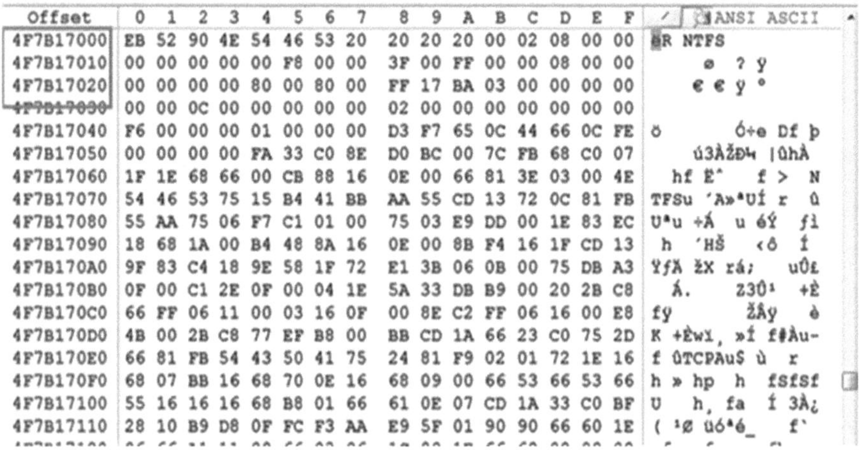

Offset	0	1	2	3	4	5	6	7	8	9	A	B	C	D	E	F	ANSI ASCII
4F7B17000	EB	52	90	4E	54	46	53	20	20	20	20	00	02	08	00	00	ëR NTFS
4F7B17010	00	00	00	00	00	F8	00	00	3F	00	FF	00	00	08	00	00	ø ? ÿ
4F7B17020	00	00	00	00	80	00	80	00	FF	17	BA	03	00	00	00	00	€ € ÿ º
4F7B17030	00	00	0C	00	00	00	00	00	02	00	00	00	00	00	00	00	
4F7B17040	F6	00	00	00	01	00	00	00	D3	F7	65	0C	44	66	0C	FE	ö Ó÷e Df þ
4F7B17050	00	00	00	00	FA	33	C0	8E	D0	BC	00	7C	FB	68	C0	07	ú3ÀŽÐ¼ |ûhÀ
4F7B17060	1F	1E	68	66	00	CB	88	16	0E	00	66	81	3E	03	00	4E	hf Ë�ˆ f > N
4F7B17070	54	46	53	75	15	B4	41	BB	AA	55	CD	13	72	0C	81	FB	TFSu ´A»ªUÍ r û
4F7B17080	55	AA	75	06	F7	C1	01	00	75	03	E9	DD	00	1E	83	EC	Uªu ÷Á u éÝ ƒì
4F7B17090	18	68	1A	00	B4	48	8A	16	0E	00	8B	F4	16	1F	CD	13	h ´HŠ ‹ô Í
4F7B170A0	9F	83	C4	18	9E	58	1F	72	E1	3B	06	0B	00	75	DB	A3	ŸƒÄ žX r á; uÛ£
4F7B170B0	0F	00	C1	2E	0F	00	04	1E	5A	33	DB	B9	00	20	2B	C8	Á. Z3Û¹ +È
4F7B170C0	66	FF	06	11	00	03	16	0F	00	8E	C2	FF	06	16	00	E8	fÿ ŽÂÿ è
4F7B170D0	4B	00	2B	C8	77	EF	B8	00	BB	CD	1A	66	23	C0	75	2D	K +Èwï¸ »Í f#Àu-
4F7B170E0	66	81	FB	54	43	50	41	75	24	81	F9	02	01	72	1E	16	f ûTCPAu$ ù r
4F7B170F0	68	07	BB	16	68	70	0E	16	68	09	00	66	53	66	53	66	h » hp h fSfSf
4F7B17100	55	16	16	16	68	B8	01	66	61	0E	07	CD	1A	33	C0	BF	U h¸ fa Í 3À¿
4F7B17110	28	10	B9	D8	0F	FC	F3	AA	E9	5F	01	90	90	66	60	1E	(¹Ø üóª é_ f`

FIGURE 9.6   Free space location.

Passive garbage collection is doing garbage collection while users are also performing data input/output actions. This mechanism is also called "real-time garbage collection." Since this mechanism occupies a large number of master control resources, it will have a certain impact on the "response time" when there is a data request.

After the solid-state disk is full, delete it all permanently (Shift + delete), and you will find that the file has been marked by the TRIM command. After writing 512 MB of data, it was found that four original files were processed by GC, that is, the data was erased by the SSD master after writing. Explain that when the hard disk is full when the deleted part of the data is written again, the computer will GC process the data larger than the written part, and then take out a part to store the written data, and the rest as free memory; if the hard disk If it is not full, the written part will be stored in the original free memory, and then the deleted part will be GC processed (Cha et al., 2015).

9.5.2   *PHYSICAL LAYER DATA RECOVERY*

Physical faults are generally divided into two types: main control damage and circuit board damage. If the main controller is damaged, it is generally necessary to replace the SSD circuit board with the same circuit board model and the same main control model, and then use a high-temperature blower to disorder the pins of the Flash chip, and then blow the flash chip corresponding to the new circuit board through a high-temperature blower Go to the corresponding location to perform data recovery operations. If the circuit board is damaged, the physical replacement method is also used for recovery, replacing the Flash chip with data on a new circuit board that is intact. For the damage of the Flash memory chip, data recovery is not yet possible.

9.5.3   *FIRMWARE LAYER DATA RECOVERY*

The most intuitive performance that can be observed for the failure of the firmware layer is that the SSD does not recognize the disk. This kind of unrecognized disk has the situation that the model is not recognized, the capacity is not recognized, or the capacity is not recognized. It is necessary to let the solid-state drive enter a specific mode before performing firmware repair, which is recognized by the master, and then short-circuited to enter the specific model. The damaged firmware can be repaired to restore the data normally. There are also some special cases. For example, if there are too many bad blocks or dots, you still cannot recover them despite the above steps, or most of the recovered data cannot be displayed.

9.6   DIFFICULTIES IN THE FORENSICS OF SSD

1. **The Data Storage Method of the Solid-State Hard Disk:** It is completely different from that of a mechanical hard disk. We know that the data storage part of a mechanical hard disk is mainly composed of a disk that stores data, a magnetic head that reads and writes electro-magnetic signals, and an ahead controller. When reading and writing data, the disk rotates at high speed, and the head controller drives the head to move radially on the disk. Under the joint action of the two, the head is finally positioned at the designated position of the disk for operation.

The storage method of the solid-state hard disk is more similar to the U disk. Data storage and reading and writing are performed in the flash memory chip. The flash memory itself is an erasable programmable read-only memory (EEPROM), which can quickly complete three basic operating modes: read, write, and erase. The flash memory in the solid-state drive is divided into blocks, and many pages (Pages) are divided into blocks (Hepi & Priyanka, 2021). When reading and writing data, the computer converts the binary digital signal into a composite binary digital signal (addition of instructions for allocation, verification, stacking, etc.), and sends it to the hard disk adapter interface, and then writes it to the flash memory page after being distributed by the main control chip to realize data storage.

2.   **Data Cannot be Recovered Due to Garbage Collection:** Whether it is a mechanical hard disk or a solid-state hard disk, after receiving the OS write data instruction, it will delete the old data and then write the new data. When receiving the instruction to delete the data, the hard disk only makes a delete able mark at the corresponding position and is deleted when the write operation is to be performed. The problem is that the mass plane is the smallest unit of data writing and the block is the smallest unit of data deletion in solid-state drives. If you want to delete the AI page data in the A block, you must first copy all the data in the block except the AI page to another block, and then clear the A block. This will cause a significant delay in the write operation of the SSD.

To reduce latency, solid-state hard drive manufacturers have introduced a garbage collection mechanism. To put it simply, garbage collection uses the free time of the system to scan all valid pages and merge them into a block containing all valid pages, and those invalid pages and blocks will be completely cleared, and the reclaimed space will be returned. Perform a write operation.

The impact of garbage collection on forensics is that it automatically performs clean-up operations from time to time, resulting in complete data deletion.

Due to the limited number of reads and writes of the flash memory, the data is always stored in a certain block, which will inevitably lead to the early scrapping of this block. To ensure the balanced use of blocks and avoid the premature wear of some blocks and affect the use of the entire hard disk, the weekly hard disk adopts a wear-leveling mechanism. It can evenly distribute files to each block as much as possible to ensure that the number of erasing and writing (P/E) times for each flash block is the same and to avoid excessively repeating the erase operation of a certain part of the block. Effectively prolong the service life of the solid-state drive. The impact of balanced wear on forensics is that each block will be used in a balanced manner, which also means balanced erasure.

The OS has also optimized the solid-state drive. The most representative one is that the TRIM command enables the OS to notify the solid-state drive which blocks are no longer in use so that the solid-state drive can directly perform garbage collection. The close cooperation between pressure RIM and garbage collection and balanced wear mechanism has indeed greatly improved the performance of solid-state drives and effectively extended the service life of solid-state drives. The impact of the TRIM instruction on forensics is that it ensures that garbage collection is completed quickly, that is to say, if one data is deleted in the OS, the solid-state drive will quickly erase all content.

For electronic data forensics, the consequences of garbage collection, equalization of wear and tear, and pressure RIM instructions are disastrous: because of the garbage collection mechanism, after the user deletes the data, the solid-state drive will empty that block instead of just deleting the index and retaining it. Data balanced wear can ensure that garbage collection "equally" erases blocks of the hard disk. The TRIM instruction can ensure the efficient execution of garbage collection. The final result of the combination of these three is: on solid-state hard disks, traditional data recovery cannot be performed, and the feasibility of finding evidence file fragments through unallocated clusters is almost non-existent. Therefore, for solid-state drives, the focus of forensics should be focused on undeleted data.

9.7 CONCLUSION

Solid-state drive data recovery mainly includes three aspects:

1. **Logical Layer:** Hard disks that belong to the real-time garbage collection mechanism can recover a large amount of deleted data without rewriting, but there is no way to restore complete data in the case of secondary writing; hard disks that belong to the idle garbage collection mechanism, files are the initial recovery of the deletion is more likely. After the power-on time is too long, the data will be erased slowly and cannot be recovered.

2. **Physical Layer:** As long as the Flash chip is not damaged or short-circuit damaged, it can be restored by replacing the carrier.
3. **Firmware Layer:** The difficulty is to find a suitable short-circuit point so that the hard drive can enter the factory mode. The important step for writing resource files is to require the hard disk to be in factory mode. After entering the factory mode, write the hard disk resource file, which is the firmware, and then rebuild the hard disk compiler to achieve data recovery.

The practice of digital crimes, which use SSDs as a means or an end to its realization undeniably benefits from the internal mechanisms of an SSD, designed to increase the life and use of the device (Kang, Lee, & Kim, 2018). Even though it already exists concrete methods of safe removal and anti-forensics (AF), just using an SSD can be considered, to some extent, as an advance anti-forensic measure. We expect that factors such as algorithms, mechanism, and the functioning of disks solid-state system will become uniform in the future, as the market expands and as the market expands. As advances are made in the field of reverse engineering leading to a better elucidation of the characteristics of SSDs and consequently advances in practices digital forensics. It is believed, however, that their future lies in the development of studies and techniques focused on the hardware part, instead of the path via software. Past research shows that general digital forensic awareness of the topic is increasing after a period when no one seemed to notice the program. This is a first step important and the basis for further research with the hope of creating standards and new guidelines in the future for forensic investigation of SSD.

KEYWORDS

- **cyber forensic**
- **data acquisition**
- **encrypted file system**
- **forensic toolkit**
- **garbage clear**
- **NAND flash memory**
- **solid state drives**

REFERENCES

Ahn, N.-Y., & Lee, D. H. (2017). Duty to delete on Non-volatile Memory. In *arXiv* [cs.OH]. http://arxiv.org/abs/1707.02842.

Cha, J., Kang, W., Chung, J., Park, K., & Kang, S., (2015). A new accelerated endurance test for terabit NAND flash memory using interference effect. *IEEE Transactions on Semiconductor Manufacturing, 28*(3), 399–407.

Chang, D., Lin, W., & Chen, H., (2016). FastRead: Improving read performance for multilevel-cell flash memory. In: *IEEE Transactions on Very Large Scale Integration (VLSI) Systems* (Vol. 24, No. 9, pp. 2998–3002).

Chang, L., (2007). *On Efficient Wear Leveling for Large Scale Flash Memory Storage Systems.* SAC"07. Seoul, Korea.

Geier, F., (2015). *The Differences Between SSD and HDD Technology Regarding Forensic Investigations.* Sweden.

Gubanov, Y., & Afonin, O., (2014). *Recovering Evidence from SSD Drives: Understanding TRIM, Garbage Collection and Exclusions.* Belkasoft, Menlo Park.

Harris, W., (2015). *How Stuff Works.* Retrieved from: http://computer.howstuffworks.com: http://computer.howstuffworks.com/solid-state-drive.htm (accessed on 10 January 2022).

Hepi, S., & Priyanka, S., (2021). Comparative analysis study on SSD, HDD, and SSHD. *Turkish Journal of Computer and Mathematics Education.*

Kang, M., Lee, W., & Kim, S., (2018). Subpage-aware solid-state drive for improving lifetime and performance. *IEEE Transactions on Computers, 67*(10), 1492–1505.

Kim, J., Lee, Y., Lee, K., Jung, T., Volokhov, D., & Yim, K., (2013). Vulnerability to flash controller for secure USB drives. *J. Internet Serv. Inf. Secure, 3*(3, 4), 136–145.

Ko, J., et al., (2019). Variation-tolerant WL driving scheme for high-capacity NAND flash memory. *IEEE Transactions on Very Large-Scale Integration (VLSI) Systems, 27*(8), 1828–1839.

Lee, J., Kim, Y., Shipman, G. M., Oral, S., & Kim, J., (2013). Preemptible I/O scheduling of garbage collection for solid state drives. *Computer- Aided Design of Integrated Circuits and Systems, IEEE Transactions, 32*(2), 247260,

Martens, S. (2021). *The Phases of Digital Forensics.* University of Nevada, Reno. https://onlinedegrees.unr.edu/blog/digital-forensics/(accessed on 10 January 2022).

Micheloni, R., Marelli, A., & Commodaro, S., (2010). Nand overview: From memory to systems. In: *Inside NAND Flash Memories* (pp. 19–53). Springer.

Micron Technology, (2008). *Wear Leveling Techniques in NAND Flash.* Tech. Rep.

Mike (2018). Safe SSD operating temperature: Is your SSD running too hot? Hard Drive Geek. https://harddrivegeek.com/ssd-temperature/ (accessed on 10 January 2022).

Rizvi, S., & Chung, T., (2010). Flash SSD vs. HDD: High performance oriented modern embedded and multimedia storage systems. In: *2010, 2nd International Conference on Computer Engineering and Technology.*

Rusolut.com. Retrieved January 24, 2023, from https://support.rusolut.com/portal/en/kb/articles/nand-bad-columns-analysis-and-removal-28-10-2019 (accessed on 10 January 2022).

SLC, MLC or TLC NAND for Solid State Drives. By Speed Guide.net. https://www.speedguide.net/faq/slcmlc-or-tlc-Nand-for-solid-state-drives-406 (accessed on 10 January 2022).

Takeuchi, K., (2009). Novel co-design of NAND flash memory and NAND flash controller circuits for sub-30 nm low-power high-speed solid-state drives (SSD). *IEEE Journal of Solid-State Circuits, 44*(4), 1227–1234.

Wang, P., et al., (2019). Three-dimensional NAND flash for vector–matrix multiplication. *IEEE Transactions on Very Large-Scale Integration (VLSI) Systems, 27*(4), 988–991.

CHAPTER 10

A Critical Analysis of the Dark Side of the Dark Web

GOLDA SAHOO

Assistant Professor–Law, FIC Centre for Study in Victimology,
Tamil Nadu National Law University, Trichy, Tamil Nadu, India

ABSTRACT

Cybercrime is a $6 trillion-per-year industry, making it the world's third-largest economy behind the United States and China. As per the FBI's Internet Crime Complaint Center (IC3) examined 7,91,790 fraud and theft reports from 2020 in its annual cybercrime report. To put it another way, there's a lot of money to be made in cybercrime, and cybercriminals are adapting new techniques, business models, and technologies to compete in an increasingly competitive market. It has become clear that criminal organizations are arming themselves with freely available technologies that make their operations simpler while making the lives of their victims more difficult (Ozkaya & Islam, 2019). The dark web is one example of how search engines index the internet. Over a one-month period in the year 2015, two research scholars, named Thomas Rid and Daniel Moore are from London, King's College, categorized of 2,724 contents dark web sites which are live and discovered that nearly 58% of them hosted unlawful materials. Dr. Michael McGuire of the University of Surrey did a study in the year 2019, which was known as the Web of Profit, which found that matter had not become better. The approximate number of dark web since 2016, listings which harm a business has increased by nearly 21%. Among them nearly 62% of all postings (excluding drugs paddling) have the potential to hurt businesses (Retzkin, 2018). It includes the availability of numbers of credit cards, numerous

Advancements in Cybercrime Investigation and Digital Forensics. A. Harisha, Amarnath Mishra, & Chandra Singh (Eds.)
© 2024 Apple Academic Press, Inc. Co-published with CRC Press (Taylor & Francis)

types of drugs, illicit weapons, counterfeit currencies, stolen subscription credentials, raising finances for terrorism, and purchasing software that aids in the hacking of other computers are all available for purchase. In reaction to the COVID-19 lockdowns, the dark web has recently provided options for Mexican gangs to distribute drug shipments and launder money. Mexican TCOs are using the dark web to find buyers for large-scale drug shipments, according to the UN Office on Drugs and Crime (UNODC, 2022). These sites are similar to the now-defunct Silk Road, Alphabay, and Dream Market. Mexican gangs also buy synthetic opioids from China through dark websites. Collaboration is rapidly driving this machine, just as it is in normal corporate circles, with criminals flocking to the Dark Web and other underworld meeting grounds to share thoughts and resources while obtaining more formidable capabilities. Due to the availability of ransomware-as-a-service providers, who sell or lease their ransomware variations to "affiliates" or "clients" who license them to undertake an attack, they can easily pull this off. The Tor network's anonymity makes it particularly vulnerable to DDoS attacks. To escape DDoS attacks, sites are continually changing addresses, creating a very dynamic environment.

The most important factor driving the dark web's rise is legal rather than technological. The fundamental difficulty for national criminal legal systems is the time it takes for possible abuses of new technologies to be recognized and appropriate changes to national criminal law to be made. This can be mitigated by identifying weaknesses in the existing legal system, making appropriate adjustments, and, if necessary, creating new legislation. Furthermore, international cooperation to harmonize various national cyber laws is becoming more crucial.

> ➢ **Objective:** The main objectives of this chapter is to:
>   • To examine the various factors responsible for the increase of the dark web.
>   • To critically analyzes the effectiveness of various existing legal measures in developed countries relating to the dark web.
>   • To identify the legal challenges and suggest measures to combat the issue of the dark web.
> ➢ **Hypothesis:**
>   • Existing municipal law to curb the issue of dark web is inadequate.
>   • Uniformity in the application of cyber law across the globe will control the issue of the dark web.
>   • Methodology: This research is doctrinal and comparative. The chapter is based upon secondary sources like reading down

important case laws, articles published in various international and national journals, reading down the statues and other relevant legislations and referring to various books available in hands, newspaper articles which deal with this concept.

10.1 INTRODUCTION

"One of the main cyber-risks is to think they don't exist. The other is to try to treat all potential risks."

—*Stephane Nappo*

As a result of the tremendous growth in criminal behavior performed via the internet, the scenery of law enforcement has transformed. While huge technological improvements may lead one to consider that track cyber criminals on the internet is a different possibility. But at the same time, the truth is that cybercrime is prospering on the internet in new ways (Gehl, 2018). This is due to criminals using new technology to preserve secrecy on the Internet and allowing them to easily escape from the law enforcement.

10.2 MEANING OF DARK WEB

For a variety of technical reasons, the phrase "Deep Web" signifies as a form of Internet content that is not indexed through any search engines. The Dark Web is a section of the Deep Web which can be deliberately secreted and is hard to find by using any web browser (Gehl, 2018). For content uncovered on the black web, the Tor network is a relatively well-known source. Tor with other related networks allows users to pass through the Web in complete secrecy by way of encryption and transmit them through various networks known as onion routers. Some people, on the other hand, use the dark web's online anonymity to engage in illegal activities like drug trafficking, identity theft, unlawful financial transactions, and so on (Eddison, 2018). Because the Dark Web is different from the visible Web, it is crucial to develop methods to efficiently monitor it. Limited monitoring can be accomplished now using the mapping of the unknown services, customer data monitor, social site monitor, hidden service monitor, and semantic analysis. The Dark Web is the type of the World Wide Web which is hosted on dark nets. These are networks that need a special software to configurations, or authorization to access. Private computer networks can correspond and transact business secretly on the dark web.

10.3   HISTORY AND DEVELOPMENT OF DARK WEB

The message first sent between computers connected by the Defense Advanced Research Projects Agency's was ARPANET network in 1969. Other clandestine networks emerge alongside ARPANET within a few years. The World Wide Web was created by Sir Tim Berners-Lee in 1989, and it was made available to the general public in 1991. The web is an internet service that was created to aid scientists and academics in more effectively exchanging information (Kaplan, 2016). However, by the late 1990s, the internet had become popular and accessible to regular people all over the world thanks to the web. After more than 30 years, Berners-first Lee's proposal for the web has altered everyone's lives. Google has eclipsed Bing and DuckDuckGo as the most used internet search engines in the 21st century (Gehl, 2018). However, there is a lot of web material sent over the HTTP and HTTPS protocols that can't be found via traditional methods (Gehl, 2018). When cyber thieves wish to trade information on the internet, the wise ones stay away from the portions of the internet that are easier to follow. Because of advancements in networking technology, there is now a section of the internet that can only be accessed through completely encrypted anonymizing proxy networks. The standardization of the internet in the 1980s caused the problem of unlawful data storage, and a solution in the form of "data havens" was devised. Then came peer-to-peer data transmission, which led to the creation of decentralized data hubs that could hold illegal material while also being password protected. Ian Clarke created Freenet in 2000, a piece of software that allowed users to visit the deepest parts of the internet anonymously (Islam & Ozkaya, 2019).

The US Naval Research Laboratory released TOR in 2002, a program that hides users' whereabouts and IP addresses. Originally designed for government use to safeguard the identities of American operatives working in oppressive nations such as China, it has since been adopted by the general public. In 2005, Wired magazine claimed that roughly half a million movies were released every day on the Darknet. Everything from Bollywood blockbusters to Microsoft Office was subject to copyright violation. Satoshi Nakamato presented the world of Bitcoin to, an untraceable form of cryptocurrency, in January 2009. Because of a blog post about silk road, which is an popular online bazaar for commercialization drugs on the Dark Web, the value of Bitcoin tripled. Eric Eoin Marques, dubbed "the world's largest facilitator of child pornography" by the FBI, was arrested in the year 2013. The Federal Bureau has shuts down the Silk Road market and arrests the site's creator, and roughly a month later, Silk Road 2.0 emerges. The United States, with 0.4 million daily users, is the top country by relay users who are directly connected to the Dark Web, which is followed by, the United Arab Emirates, Germany, Ukraine, Russia, the United Kingdom, France, the Netherland, Indonesia, and Canada. When it comes to the legality of accessing the dark web from India, it should be noted that there are no explicit regulations prohibiting anyone from doing so. The ability to navigate the Internet anonymously fosters a platform suited for what some countries consider criminal activity, such as "controlled substance marketplaces," "credit card fraud," and "identity theft." The dark web is made up of small, "peer-to-peer" networks, and "friend-to-friend" also largely accepted networks managed by public organizations and individuals, such as "TOR," "Freenet," "I2P," and "Riffle." Silk Road was an underground market for illegal substances, narcotics, and weaponry that operated online (ResearchGate, 2021). The US Federal Bureau of Investigation (FBI) closed the website in 2013. The FBI needed more time to track the company's bureaucrats and servers. Silk Road and other anonymous marketplaces sell anything from everyday items like books and clothing to more illegal items like narcotics and weapons. These sites have the same esthetic as any other shopping website, with a brief description of the items and a photograph to go with them. Therefore, it is accurate to suppose that specialized sites allow the users to exchange both physical and private data such as passwords, right to use to passwords for illegal sites such as pornography which are paid on the surface Web, such as well as PayPal credentials. Another correlation between the dark Web and illegal activity is, terrorist activities which seem tailor-made for one another, with the latter requiring an anonymous network that is both accessible and unreachable. Terrorists would find it difficult to sustain an online presence since these sites may be easily shut down and, more importantly, traced back to their unique poster. Websites like Banker and Co. and InstaCard, which use a variety of methods, enable untraceable bank transactions. In most of the cases, they launder bitcoins by concealing the proper source of transactions or by enabling the users with an unidentified debit card issued by any bank (Dark Web: A Web of Crimes, 2021). The Users are also provided with virtual credit cards that are offered by recognized Dark Web firms (Dean, 2014). Though getting stolen credit card information is a difficult task, this service is provided by the Atlantic Carding website, and the more you pay, the more you get. Accounts linked to ultra-high-net-worth individuals, including business credit cards and even unlimited credit cards, are up for grabs. Personal information about the user, such as name, address, and other details, is available for a price. Dahl (2014). On the hidden Web, the Hidden Wiki is the most important directory. Money laundering, contract murders, cyber-attacks, forbidden chemicals, and explosives-making instructions are also promoted. To avoid discovery, the links to these sites, like those to other dark Web sites, change on a regular basis (Williams, 2011).

10.4   FACTORS RESPONSIBLE FOR THE INCREASE OF DARK WEB

In recent years, the illicit markets on the dark web have seen tremendous technological advancements. Technological improvements, such as the widespread use of cryptocurrencies and secure browser technology like "The Onion Routing," have aided the expanding use of online marketplace for the selling of prohibited items (TOR) (Kaplan, 2016). It's helpful to look at the three components of the Internet to see why law enforcement has become so tough. Then there's what's known as the "Surface Web." "Anything that can be indexed by a conventional search engine, such as Google or Yahoo," according to the Surface Web. It encompasses everything that may be accessed by these engines. The "Deep Web" exists beyond the Surface Web. This expression refers to "the remainder of the Web not covered by the Surface Web." This means that anything that can't be found using the search engine is considered as a part of a Deep Web. The "Dark Web" is the last segment of the Internet. "A small fraction of the Deep Web that has been purposely concealed and is unavailable through regular web browsers." The internet users are unable to access this section of the Web exclusive of the usage of a particular browsers specialized to ensuring the user's complete anonymity. The Dark Web has grown to be a hub for illegal activity and an unrelenting nuisance for the law enforcement as a result of browsers providing users this anonymity. Aside from the ethical issues that arise from law enforcement engaging in illegal activities in order to "catch" cyber offenders, there are other difficulties about this enforcement strategy. Many people regard honeypot traps as a type of entrapment and a breach of civil freedoms, which is a major source of concern. The anonymity that the dark web provides is the most important reason why it thrives as a location where unlawful operations are carried out. Furthermore, transnational cartels that operate as shops selling drugs, guns, and child pornography (CP) movies are very selective about who they allow into their networks. To join a cartel, you must first pay a fee to the syndicate. It is also a lengthy process that can take up to a year before you are permitted access into the inner circle.

Based on research and several years of close monitoring, researchers found 4 key factors for the booming of the dark web. These are as in subsections.

10.4.1 FLOW OF FUND VIA CRYPTOCURRENCIES

The Dark Web's rise has been spurred by Bitcoin's anonymity. This cryptocurrency allows both buyer-seller for conducting secure transactions with anonymity. For bitcoin transaction's wallet IDs are revealed only, keeping the transaction confidential. The ability to keep a bitcoin user's contract private enable them to access illegal markets and buy illegal goods (Blockchain and Cryptocurrencies, 2019). Bitcoin dealings via dark web market have raised from $250 million which was in 2012 to $872 million in 2018, according to a survey published by Chainalysis, a renowned crypto-payment analytic firm. They also estimated that the deals will exceed $1 billion in 2019. Criminals operating on the dark web have started to update their virtual stores with new and attractive items (Johnston, Manheim, & Dion-Schwarz, 2019). One example is cybercriminals exchanging fiat currency for bitcoin at a 10% discount, letting a client to pay $800 in bitcoin and receive $10,000 in their bank account. The expansion of anonymity networks TOR's encrypted browser tools is the largest anonymity network, which is more than 2 million users connecting straight to the service (Mancini & Adrian, 2019). The bandwidth ability of the network has enlarged from more than 60 gigabits per second in 2014 to over 250 gigabits per second in the year 2018 (Seon, 2023).

10.4.2 PAYMENT TO CYBERCRIMINALS

Ransomware attacks have skyrocketed in recent years, and which is one of the biggest for the hackers are getting rewarded. In 2019, for example, cyber-criminals hacked Riviera Beach City's computers, and the city was forced to pay the hackers $6,00,000 in bitcoins as ransom (Handbook of Research on Cybercrime and Information Privacy). The greatest ransomware payment to date was made by Internet Nayana, a Korean web hosting provider, which paid $1.14 million in 2017 (BBC News, 2021). Fraudsters are enticed by these payouts, which leads to the development of the latest ransomware attacks with additional features.

10.4.3 MOUNTING PROFIT OF DARK WEB MARKETS

The Dark Web drug trade in Australia is booming, and it's proving to be very profitable used for the vendors. Ransomware attack is one of the effective ways for thieves to steal large quantities of assets from companies. Several of these organized groups such as Maze are aggressively assigning hackers on the dark web in order to extend their functions and attack new organizations. Cybercrime generates $1.5 trillion in revenue every year, according to a study conducted by Michael McGuire of the University of Surry, including $860 million in drug and weapon sales. Their profit is increasing with the demand.

10.4.4 INCREASING OF ORGANIZATIONS' ATTACK SURFACES

As organizations swiftly migrated to the digitization era, network barriers have dissipated. Integration of systems is now more advanced than it has ever been. As a result of this transition, organizations' attack surfaces have expanded. Because the revenues from such attacks have grown, con artists are pushing for more daring and opportunistic attacks. Businesses can allocate resources to a few areas which has prepared against cybercrime. Cybercriminals are more and more relying on to carry out cyber-attacks to the targets. Despite the fact that major corporations have used TPCRM, their trust in the technology remains poor (Dark Web: A Web of Crimes, 2021). The ability to navigate the Internet anonymously fosters a platform ripe for what some countries consider unlawful activities, such as forbidden substance marketplaces, credit card fraud and identity theft, and sensitive information leaks. Development In the late 1990s, two United States Department of Defense research organizations led efforts for the development of anonymize system. Ordinary internet users would be unaware of or unable to access this hidden network. Creating a foundation dedicated to anonymity for human rights and privacy advocates.

10.5 THE DARK SIDE

What is the extent of illicit behavior through the dark web? Whether browsing for anything unlawful on the dark web? There will be a plethora of options available Dark Web gives people the right to use to information while also protecting them from being persecuted. In more liberal countries, it can be used as an important instrument for whistle blowing purpose and communication, safeguarding citizens from punishment or judgment at work or in the community. Alternatively, it may just give privacy and anonymity for individuals concerned about how firms and governments monitor, use, and potentially monetize their data. Many organizations already have Tor-based hidden websites, including practically every popular newspaper, social media, and even the "US Central Intelligence Agency (CIA)" (https://sgp. fas.org/crs/row/R41576.pdf, 2021). This is due to the fact that a Tor website assured to confidentiality. Both the New York Times and the CIA, are both attempting to increase communication with online interviews who may be able to provide sensitive information. That similar privacy and secrecy that safeguard against tyrants and targeted marketing can serve as a launch pad for illicit activities on the dark web. The reality regarding the dark web is that it facilitates a rising underground market where socio-economic offenders use for traffic drugs, involve in other illegal products and services such as CP, etc., in addition to as long as great privacy and security from strict governments' surveillance. Using Tor to access and browse the dark web is not illegal. It's possible that accessing the dark web is legal. Use of the dark web for purchasing illegal firearms or narcotics is illegal. If anyone using the dark web to anonymously join forums or read unknown blog postings, that person is not violating the law.

   In the year 2015, over the course of five weeks, King's College London academics Daniel Moore and Thomas Rid investigated 2,723 dark web domains (Taylor & Francis, 2021). According to the researchers, unlawful content was found on 57% of the websites. After all, 2015 was a long time ago. The number of illegal operations on the dark web has skyrocketed. The amount of harmful dark web postings has increased by 20% since 2016, according to a study done by the University of Surrey in 2019. This category accounted for 60% of all listings. Some of the areas where the dark side of the dark web is skyrocketing are discussed in subsections.

10.5.1   ILLEGAL PORNOGRAPHY

On the dark Web, pedophilia, or CP (child pornography), is extremely easy to locate. Pornography is tolerated with some limitations on the surface of the Internet (Jenkins, 2003). There are a multitude of sites and communities on the dark Web for people who want to engage in pedophilia. Despite being difficult to uncover, CP is the most popular sort of content on the dark web. A site named Lolita City, which has been closed, had over more than 100 GB of materials in the forms of videos of child pornographic and with nearly 16,000 of members (BBC News, 2021). Websites that provide CP are regularly targeted by law enforcement authorities. The most popular method is to break into a website and track users' IP addresses. In 2015, the Federal Bureaushut down after investigating the PLAYPEN website (Federal Bureau of Investigation, 2021). which was the leading pornography website through the dark web, with more than 2 lakhs subscribers. Websites employ complex systems of manuals, forums, and community regulation. Other material includes sexualized animal abuse and murder, as well as vengeance pornography (Taylor & Quayle, 2004). German police said in May 2021 that Boystown, which is one of the world's largest networks related to CP would be shut down. In raids, four persons were seized, including a Paraguayan, on feeling of controlling the network. Several pedophile private chat sites were also blackout as part of the German-led intelligence operation, according to Europol (Clark, 2002).

10.5.2 TERRORISM

Terrorist groups began using the internet in the 1990s, but due to the rise of the dark web lured them in due to its obscurity, lack of require of control, community interface, and simplicity of usage (Matusitz, 2014). Anonymous transactions were possible with the debut of Bitcoin, enabling for anonymous donations and funding. Terrorists may now acquire weapons using Bitcoin as payment. In 2018, a man named Ahmed Sarsur charged with attempting to use the dark web to acquire explosive materials and engage to aid Syrian terrorists, by providing financial support to them (ResearchGate, 2021). As technology advances, cyber terrorists have been able to prosper by exploiting holes in the system. Following the November 2015 Paris attacks, GhostSec, an Anonymous-affiliated hacker group, hacked one of these sites and replaced it with a Prozac advertising (Jason Rivera & Wanda Archy, 2019).

10.5.3 USE OF SOCIAL MEDIA

The Social Network in Dark Web is a developing collection of social media platforms which is to those found on the Internet (DWSN) (SAGE Journals, 2021). Members of the DWSN can customize their sites, add friends, like postings, and participate in forums, much like they would on a regular social networking site. In order to address concerns with regular platforms and continue to provide service throughout the Internet, traditional social media sites such as Facebook businesses have begun to construct dark-web version of their respective websites. The DWSN's privacy policy, unlike Facebook's, requires players to remain anonymous and not post any personal information (Data Privacy Manager).

10.5.4 LIVE MURDER

The Assassination Market website is a market where a person can wager on the day of a person's death and receive a payout if the date is correctly "guessed," which encourages assassination because the assassin could benefit by placing an correct wager on the timing of the person's death if he knows when the contract will take place (https://www.thesun.co.uk/news/5815189/dark-web-live-torture-hire-hitmen-hackers/). Because the payment is for knowledge regarding the date rather than carrying out the assassination, assigning criminal liability for the assassination is substantially more complicated. There are other websites like White Wolves and C'thuthlu where a person can hire an assassin (The Sun, 2021). There have been rumors that a crowd-funded project is in the works, although these are thought to be scams. Ross Ulbricht who was the Silk Road developer was in custody by homeland security investigations (HSI) for his conduct for purportedly hire a gunman to murder several person, but the accusations were in the end. There is another barbaric recent trend that person can watch live telecast murder via the dark web. It associate with the word "Red Room" was derived which was based on a Japanese anime and urban legend (Gehl, 2018). According to the data, all of the recorded incidents appear to be hoaxes. On June 25, 2015, YouTubers Obscure Horror Corner reviewed the indie game Sad Satan, which they said they broadcast on the dark web. There are several inconsistencies in that channel's broadcast led to doubt on the reporting of such accounts of these events (Terrorism in Africa: The Evolving Front in the War on Terror, 2010).

10.5.5 HUMAN EXPERIMENT

The Human Experiment is another area where website that reported on medical tests allegedly conducted on unregistered citizens who were homeless and poor. As per the data, they were taken up from the street, experimented on, and then usually died. Though the website has been stopped officially since 2011 still these barbarous activities are still exist (Unhcr.org., 2021).

10.5.6   STEALING

There are several pages on the dark Web, hosted by people who are expert at stealing and will steal whatever thing one can't manage to pay for or refuse to pay for (The Gateway, 2021).

10.5.7   ILLEGAL FIRE ARMS

Euroarms is a website which sells a wide range of firearms and can send them to your door in any European country. These firearms' ammunition is sold separately, and that website must be found on the dark Web on its own. The shadow side there are around 66,000 distinct URLs which are ending in on the TOR network (Kumar & Rosenbach, 2021). The most prevalent uses of these sites, according to a report by computer security firm Hyperion Gray, are communication through various forums, chat rooms, and file and picture hosting, over, and above business through marketplaces.

10.5.8   BITCOIN OPERATION

Bitcoin, the world's earliest decentralized currency and payment network, was founded a decade ago by an anonymous cryptography expert (renowned for his ability to crack passwords) who went by the moniker named Satoshi Nakamoto (Investopedia, 2021). In 2011, Bitcoin, started out as one of the specialized mode of exchange for the technical sector, became the preferred currency for drug paddlers transacting on the Silk Road. The amalgamation of the encrypted network which has been concealed from the world and a transactional currency that is untraceable by law enforcement agencies has resulted with major market for unlawful merchants advertising illegitimate goods, during the last five years. A noticeable crypto-payment expository association. Concurring to the trade, bitcoin exchanges on the dull web are expected to beat $1 billion in 2019. In the event that genuine, this would set an unused record for illicit exchanges within the business. According to the investigation, the number of Bitcoin exchanges tied to unlawful exercises has declined by 6% since 2012, bookkeeping for less than 1% of all Bitcoin movement. Cash washing is assessed to account for 2 to 5% of worldwide GDP in a given year (Bank, 2021), or $1.6 trillion to $4 trillion, agreeing to the Joined together Nations. Despite the truth that the full financial esteem of unlawful dim web action is still small, numerous of today's most perilous perils to society work within the shadows of the Tor organize, requiring the consideration of worldwide controllers, budgetary teach, and law require-ment authorities.

The challenge for controllers and law requirement organizations is to plan approaches that strike a compromise between maintaining generous standards in an age of data control and finding and disposing of the dark web's most vindictive exercises. Over the final few a long time, the universal community has made noteworthy advances in handling these issues by making strides in data sharing, culminating law enforcement's specialized capacity to close down huge unlawful markets, and directing bitcoin transactions. The to begin with step in tending to the dim web's most damaging conduct and monetary teach is made strides data sharing among law requirement associations. The dim web's around the world nearness needs worldwide participation.

Recently, in 2018–2019, in collaboration of both the European Union Interpol and Interpol brought law authorization organizations from 19 countries together to distinguish 247 most important targets and trade the operational insights required to implement them. Individuals of the gather were able to form captures and closed down 50 illegal dark-web locales this year, counting two of the foremost vital sedate markets, Divider Road Showcase and Valhalla (Interpol.int., 2021). The rise of unlawful dull web exchanges has impelled a few governments around the world to move forward residential law authorization agencies' capacity to combat criminal exercises, such as the US Government Bureau of Examination (FBI). Concurring to reports, the FBI has undertaken operations that permit them to "de-anonymize." The FBI accomplishes this by setting up hubs within the Tor organize that permit it to get to the characters and areas of a few criminal Tor-based webpages. The FBI's takedown of the "Silk Street 2.0" site, which was the biggest illicit dull web marketplace in 2014, was the primary noteworthy move. Amid the site's two and a half a long time of operation, thousands of medicate traffickers and other illegal sellers conveyed hundreds of kilograms of unlawful drugs and other illegal things and administrations to well over 1,00,000 buyers, concurring to the examination. As a result of these illegal exchanges, hundreds of millions of dollars were washed through the location. Generally, the location had sold over 9.5 million Bitcoins, which were esteemed around $1.2 billion at the time. AlphaBay and Hansa Showcase, two of Silk Road's most major successors, were closed down in 2017 (https://www.wired.com/story/white-house-market-dark-web-drugs-goes-down/). As outlined by a later Dutch operation to require control of a huge dull web vendor, run it anonymously for a month, and after that utilize the data collected to close down hundreds of other dark web stores, dim web requirement abilities have kept on developing.

10.6   IT IS NECESSARY TO CREATE NEW REGULATIONS

In June 2019, the Monetary Activity Errand Constrain issued suggestions for associations that handle cryptocurrency installments to distinguish both the source and beneficiary of cash exchanges. Taking after the G20 Summit in 2018, pioneers encouraged that universal administrative organizations see into approach reactions for crypto resources, especially within the zones of know your client, anti-money washing, and counter-terrorist financing. In spite of the truth that the start-up biological system of trades, wallets, and other crypto installment facilitators needs the framework required to actualize such financial-sector-like standards, controllers must begin laying the foundation for more control. The dark web has been contended to ensure gracious rights such as "free discourse, security, and secrecy," however it is dreaded by certain prosecutors and government authorities as a sanctuary for illicit conduct. Policing implies concentrating on particular private web practices that are unlawful or vulnerable to web regulation. Authorities ordinarily use an individual's IP (Web Convention) address when examining web suspects; in any case, since Tor browsers ensure secrecy, this can be not practical. As a result, law authorization has to a run of unusual ways in arrange to follow down and secure anybody taking an interest in illegal via such web (Wired, 2021). The Open Source Insights (OSINT) is another sort of information collecting instrument that collects data from open sources legitimately. Both law enforcement and judicial authorities from Australia, Germany, Cyprus, the Netherlands, Canada, Austria, Sweden, the United Kingdom the United States collaborated on this operation. In addition, law enforcement is effective while working together, and today's announcement sends a strong message to criminals selling or buying illicit goods on the dark web that the hidden internet is no longer hidden, and their anonymity is no longer anonymous. By assisting police forces around the world in understanding and investigating digital crimes.

10.7   ROLE OF INTERPOL

INTERPOL aided in the expansion of a blockchain analytics means called GraphSense that simplifies the tracing of bitcoin dealings in collaboration with the European Union-funded Project called "TITANIUM" (Ec.europa. eu., 2021). Investigators can use this tool to "follow the money" by searching bitcoin addresses, tags, and transactions, as well as finding clusters connected with an address. With the facilitate of a dark web monitor, an analytical tool that will collect data on criminal behavior on the Darknet and for use it to give actionable intelligence mean to support investigating around the world. The data and succeeding study will also aid in the identification of recent trends, research, and prevention recommendations.

- PGP (pretty good privacy encryption program) keys;
- Darknet marketplace domains;
- E-mail addresses;
- Usernames and aliases;
- Darknet forums;
- IP addresses.

"The INTERPOL Darknet and Cryptocurrencies Task Force" is another working group on a global cryptocurrency nomenclature that will outline what information should be collected from apprehensive cryptocurrency transactions (Interpol.int., 2021).

The "Practitioner Manual for ASEAN Countries to Counter Terrorism Using the Darknet and Cryptocurrency" (Rand.org., 2021) will provide step-by-step instructions for law enforcement personnel on how to investigate terrorist operations on the Darknet, particularly those involving cryptocurrency. It will go over the various methods, techniques, and procedures for analyzing cyber-terrorism intelligence, as well as the difficulties that investigators may face.

In addition to that, the UNODC also suggested all member states in enhancing the ability of essential crime prevention measures to work more efficiently, with a focus on vulnerable groups. To achieve this purpose, the UNODC collaborates with both national and regional level. UNODC also focuses on developing specific tools and guides to assist policymakers and suppliers of technical assistance. The UNODC fosters multi-sectoral, multi-disciplinary strategies, programs, and activities by recognizing the numerous variables that contribute to crime and functioning as the guardian of UN principles and strategies in criminal justice and crime prevention. The rule of law and those human rights recognized in international instruments to which Member States are signatories must be protected in all aspects of crime.

10.8   POSITION IN INDIA

10.8.1   ORGANIZATIONAL STRUCTURE TO ADDRESS CYBERCRIME IN INDIA

The Chief Ministries and officials who are working to curb cybercrime against have been given in subsections.

10.8.1.1   MINISTRY OF ELECTRONICS AND INFORMATION TECHNOLOGY

This is one of the prime government bodies towards the successful application of the IT Act. The Ministry is accountable to issue suitable directions, rules, and regulations on account of the Central Government for the effective implementation of the IT Act. The Ministry also functions as an appointment authority for the smooth functioning of the IT Act (Meity.gov.in., 2021).

Under Section 69 of the IT Act authorizes the Central Government to:

> "Order or give directions to any agency of the government to interpret, monitor or decrypt any information received or stored through any computer source (IT Act, 2000; Section 69). Issue direction for blocking of access by public any information generated, transmitted, received, stored or hosted in any computer resources (IT Act, 2000; Section 69A(1)). Prescribed procedures and safeguard for blocking access to any information" (IT Act, 2000; Section 69A(2)).

10.8.1.2   INDIAN COMPUTER EMERGENCY TEAM

The IT Acts authorizes the "Indian Computer Emergency Team" to act as the national agency for incident response. They are performed by the team (http://202.62.95.70:8080/jspui/bitstream/123456789/13788/1/Module%20 3%20%26%204.pdf):

"Forecast and alerts of cybersecurity incidents, Collection, analysis, and dissemination of information on cyber incidents, Issue guidelines, advisories, vulnerability notes and white papers relating to information security practices, Emergency measures for handling cybersecurity incidents, procedures, prevention, response, and reporting of cyber incidents."

10.8.1.3   CYBER-CELLS

For the purpose of tackling the issue concerning cybercrime, separate cells have been established within the special police units for investigation. Those cells function within the institutional framework, present inside the appropriate jurisdiction.

The police in investigating cyber offenses under the IT Act:

> ➢ **Section 78:** "Notwithstanding anything in the Code of Criminal Procedure, 1973, any offense under this Act shall be investigated by a police officer not below the level of Deputy Superintendent of Police."

> **Section 76:** Any computer, floppies, compact discs, computer system, tape drives, or any other accessories connected thereto, especially in respect of which any provision of the IT Act, orders, rules, or regulations completed there under has been or is being contravened, shall be liable to confiscation: Provided, however, that where it is established to the satisfaction of the court adjudicating the confiscation that the person in whose possession the computer, computer system, compact discs, floppies, tape drives, or other accessories related there Instead of ordering the confiscation of the computer system, floppies, computer, tape drives, compact discs, or any other accessories related thereto, the court may make any other sort authorized by this Act against the person who violates the requirements of this Act, its rules, orders, or regulations as it sees fit.

> **Section 80(2):** When a person is arrested under Subsection (1) by an official who is not a police officer, that officer must take or send the person apprehended before the magistrate with power in the case or the OIC of a police station without delay. From the foregoing clauses, it is obvious that there are no explicit legal measures in place in India to fight the dark web issue.

In succeeding sections, we discuss some challenges faced by investigating agencies why dealing with cybercrime in general.

10.9 THE CHALLENGES DURING THE INVESTIGATION OF DARK WEB

Cybercrime investigations are a comparatively recent development. Techniques being applied by the investigators are still being an improvement process. Though efforts have been made to develop a method on how to perform this new way of investigations, the procedure for conducting an investigation is still different from the jurisdiction. In reality, cybercrime investigations are still in their teething stages and yet to be improved with the passage of time. There are several challenges in the cybercrime investigation mechanism. Some of the challenges are due to the inadequacy of law, and few are due to the rapid evolution of technologies. Due to rapid technological advancement, cybercrime investigators must persistently upgrade their expertise and knowledge. The present ways to combat crime often nonfunctional in the online world. The development of a holistic manner to address various aspects of cyber-related crimes are new challenges for legislators as well as for investigation agencies.

Some of the major challenges which the cyber investigators are facing are discussed in subsections.

10.9.1   *EXPANSION OF INTERNET USERS*

With the expansion of internet user in people's everyday life is one of the biggest challenges for the cyber world. In the year 2005, the number of people who for the first time experienced the internet in an advanced economy exceeded their number more than in industrialized states. The enhanced number of internet users who were inter-connected through the worldwide connectivity networks signify a challenging task for the cyber investigators as one of the main shortcomings which enable a chance to the criminals is the "lack of the understanding of individual security online along with the application of social engineering techniques (Rash et al., 2009)"; and, secondly, "while identity theft, spam, and phishing activities can be performed automatically (Berg, 2007; Ealy, 2003) without investing much money or effort, it is very hard to automate the process of investigation (Gercke, 2009, p. 65)."

10.9.2   *THE ACCESSIBILITY OF MEANS AND INFORMATION*

The internet was intended as an unrestricted access communication network for obtaining information, but this beneficial function becomes a tool in the hands of criminals by finding an easy way to gather information to commit cybercrime (Richard, 2005). The easy accessibility of various software and equipment are allowing the hackers to hack even protected password and increasing the possibility of utilizing search engines for unlawful means "(Long, Skoudis, & van Eijkelenborg, 2005; Dornfest, Bausch, & Calishain, 2006)."

10.9.3   *CHALLENGES IN TRACKING THE OFFENDERS*

Several possibilities for concealing identity in the Information and Communication Technology and the availability of various methods for unidentified access and browsing make it tough for police to trace offenders. Especially where the criminal activities become inter-states, it is extremely difficult to investigate.

10.9.4 THE ISSUES CONCERNING TO BORDERS OF THE COUNTRY

Criminal law and investigations of a country have always been an issue for national sovereignty, "while the protocols applied for internet data transfers are based on the most optimal routing meaning that data transfer processes go through more than one country (Sofaer & Goodman, 2001, p. 7)." As the cyber world has no specific limits, both the offender and the victims of such cybercrime may be situated in various jurisdiction, which demands mutual cooperation between the countries concerned into an international investigation (Putnam & Elliott, 2001, p. 35 et seq.; Sofaer & Goodman, 2001, p. 1 et seq.). Nevertheless, the authorization of the concerned government is necessary by the principle of national sovereignty which required time to meet basic formalities. The states, which lack proper frameworks for such cooperation become a security zone for cybercriminals and, consequently, that obstruct the investigation process.

10.9.5 STRONG ENCRYPTION

Strong encryption is a fundamental element of the digital world which helps to secure the protection of the basic fundamental rights (https://economic-times.indiatimes.com/wealth/personal-finance-news/cyber-criminals-stole-rs-1-2-trillion-from-indians-in-2019-survey/articleshow/75093578.cms?from=mdr). But at the same time, the effective utilization of the technologies also assists the scope for cyber-criminals. An increasing number of "Electronic Service Providers" apply encryption by default through their services. However, tools which facilitate personal encryption and/or anonymity concerning communications are widely encouraged. As a result, the current investigation mechanism, such as the lawful interception of communication, are becoming least significant and technically unfeasible.

10.9.6 LACK OF EXPERTISE IN LAW ENFORCEMENT AGENCIES

Besides this, non-reporting cybercrime, privacy, anonymity, and the development of emerging forms of cybercrimes, the inadequacy of techno-legal expert, lawyer, and high-tech savvy judicial officers, issues are some of the prominent concerns to beat for the appropriate implementation of cyber laws. The function that the cyber investigators are bound to play in pursuance of combating cybercrime, is impaired with the above-mentioned factors.

Besides investigating the crime in the cyber world is not only complex but also policing of the same in general directions to be hampered. It is hard for the police force to initiate investigations with the lack of the clarity and reporting which could occur due to various reasons, i.e., the disinclination of business enterprise, to report for cybercrime in view of reputation and the lack of detailed information and sometimes lack confidence in law enforcement agencies. The advancement of ICT technologies enables cybercriminals the occasion to enhance revenue with marginal impact on a particular victim. One of the vital challenges for the cyber investigator is the justification for such type of crime and to initiation of the investigation. On the one side, the absence of proper application to control, the primary structure of the internet and the construction of the network need the reform for policing the cyber world. At the same time, the basic purpose of the internet as a means for sharing of knowledge at the same time, the free access to get information should not be discarded. Therefore, the major task is also to maintain the balance between open access of the internet without hamper to the public at large.

10.10   CONCLUDING REMARKS

In the final analysis, the time has come for the INTERPOL and police force in national level to renovate and reform investigating mechanism for the successful trial of cybercrime case. Prompt reaction to the Interpol directions and requests for bilateral treaties, mutual cooperation between the countries regarding cyber forensic technology and more inter-country training programs are some of the suggestive measures to curb the jurisdictional issues. In some situation, when it is tough for the police to investigate cases, they can take the assistance of private agencies having more expertise with cybercrime investigations. This way the burden of the police department can be minimized and better result better performance due to expertise. In the United States, a private agency, "identity ecosystem steering group (IDES)" has been established to support the "National Strategy for Trusted Identities in Cyberspace (NSTIC)" (Lawnn.com., 2021). The outcome of various experimentation lead to the conclusion that, the transnational nature of cyberspace, the strategy calls for global cooperation and state authorities to take responsibility, collaborate, and communicate with industry and academia.

KEYWORDS

- **Central Intelligence Agency**
- **cryptocurrencies**
- **cybercrime**
- **cyberspace**
- **dark web**
- **homeland security investigations**
- **illegal firearms**
- **internet crime complaint center**
- **INTERPOL**
- **pretty good privacy**

REFERENCES

Bank, E., (2021). *Understanding the Crypto-Asset Phenomenon, Its Risks and Measurement Issues*. [online] European Central Bank. Available at: https://www.ecb.europa.eu/pub/economic-bulletin/articles/2019/html/ecb.ebart201905_03~c83aeaa44c.en.html (accessed on 10 January 2022).

BBC News, (2021). *Hackers Take Down Child Pornography Sites*. [online] Available at: https://www.bbc.com/news/technology-15428203 (accessed on 10 January 2022).

BBC News, (2021). *South Korean Firm's 'Record' Ransom Payment*. [online] Available at: https://www.bbc.com/news/technology-40340820 (accessed on 10 January 2022).

Clark, M. D., (2002). *Obscenity, Child Pornography and Indecency*. United States: Novinka Books.

Dark Web: A Web of Crimes, (2021). ResearchGate (PDF). [online]. Available at: https://www.researchgate.net/publication/338878596_Dark_Web_A_Web_of_Crimes (accessed on 10 January 2022).

Data Privacy Manager. [online] Available at: https://dataprivacymanager.net/how-to-protect-your-privacy-on-social-media/ (accessed on 10 January 2022).

Ec.europa.eu. (2021). *Documents Download Module*. [online] Available at: https://ec.europa.eu/research/participants/documents/downloadPublic?documentIds=080166e5cc987392&appId=PPGMS (accessed on 10 January 2022).

Eddison, L., (2018). *Tor and the Deep Web: The Complete Guide to Stay Anonymous in the Dark Net*. (n.p.): CreateSpace Independent Publishing Platform.

Federal Bureau of Investigation, (2021). *Playpen' Creator Sentenced to 30 Years | Federal Bureau of Investigation*. [online] Available at: https://www.fbi.gov/news/stories/playpen-creator-sentenced-to-30-years (accessed on 10 January 2022).

Gehl, R. W., (2018). *Weaving the Dark Web: Legitimacy on Freenet, Tor, and I2P*. United States: MIT Press.

Gehl, R. W., (2018). *Weaving the Dark Web: Legitimacy on Freenet, Tor, and I2P*. United States: MIT Press.

https://economictimes.indiatimes.com/wealth/personal-finance-news/cyber-criminals-stole-rs-1-2-trillion-from-indians-in-2019-survey/articleshow/75093578.cms?from=mdr (accessed on 10 January 2022).

https://sgp.fas.org/crs/row/R41576.pdf (2021). [online] Available at: https://sgp.fas.org/crs/row/R41576.pdf (accessed on 10 January 2022).

Inside the Chilling World of Dark Web Sites Where Users Can Hire Hitmen. [online] Available at: https://www.thesun.co.uk/news/5815189/dark-web-live-torture-hire-hitmen-hackers/ (accessed on 10 January 2022).

Interpol.int. (2021). *INTERPOL and the European Union*. [online] Available at: https://www.interpol.int/en/Our-partners/International-organization-partners/INTERPOL-and-the-European-Union (accessed on 10 January 2022).

Investopedia, (2021). *How Bitcoin Works*. [online] Available at: https://www.investopedia.com/news/how-bitcoin-works/ (accessed on 10 January 2022).

Islam, R., & Ozkaya, E., (2019). *Inside the Dark Web*. United Kingdom: CRC Press.

IT Act, (2000). Section 69. https://indiankanoon.org/doc/1439440/ (accessed on 10 January 2022).

IT Act, (2000). Section 69A(1). https://indiankanoon.org/doc/166979650/ (accessed on 10 January 2022).

IT Act, (2000). Section 69A(2). https://indiankanoon.org/doc/162711216/ (accessed on 10 January 2022).

Jenkins, P., (2003). *Beyond Tolerance: Child Pornography on the Internet*. United Kingdom: NYU Press.

Johnston, P. B., Manheim, D., & Dion-Schwarz, C., (2019). *Terrorist Use of Cryptocurrencies: Technical and Organizational Barriers and Future Threats*. United States: RAND Corporation.

Kaplan, F., (2016). *Dark Territory: The Secret History of Cyber War*. United Kingdom: Simon & Schuster.

Kumar, A., & Rosenbach, E., (2021). *The Truth About the Dark Web: Intended to Protect Dissidents, it Has Also Cloaked Illegal Activity*. [online] imfsg. Available at: https://www.elibrary.imf.org/view/journals/022/0056/003/article-A007-en.xml (accessed on 10 January 2022).

Lawnn.com. (2021). *Article: Cybercrime: A Menace to India*. [online] Available at: https://www.lawnn.com/article-cybercrime-menace-india/#:~:text=E.g.%20In%20U.S.%20a%20privately,science%20departments%20at%20local%20universities (accessed on 10 January 2022).

Mancini, G. T., & Adrian, T., (2019). *The Rise of Digital Money*. United States: International Monetary Fund.

Matusitz, J., (2014). *Symbolism in Terrorism: Motivation, Communication, and Behavior*. United Kingdom: Rowman & Littlefield Publishers.

Meity.gov.in. (2021). *Ministry of Electronics and Information Technology, Government of India | Home Page*. [online] Available at: https://www.meity.gov.in/ (accessed on 10 January 2022).

Ozkaya, E., & Islam, R., (2019). *Inside the Dark Web*. United Kingdom: CRC Press.

Rand.org. (2021). [online] Available at: https://www.rand.org/content/dam/rand/pubs/research_reports/RR3000/RR3026/RAND_RR3026.pdf (accessed on 10 January 2022).

ResearchGate, (2021). *Dark Web: A Web of Crimes.* [online] Available at: https://www.researchgate.net/publication/338878596_Dark_Web_A_Web_of_Crimes (accessed on 10 January 2022).

Retzkin, S., (2018). *Hands-On Dark Web Analysis: Learn What Goes on in the Dark Web, and How to Work With it.* United Kingdom: Packt Publishing.

Richard, W., (2005). *Downing, Drafting Procedural Laws: Empowering Law Enforcement with Legal Tool Needed to Investigate and Deter Cybercrime.* www.cybersecuritycooperation.org/ (accessed on 10 January 2022).

SAGE Journals, (2021). *Power/Freedom on the Dark Web: A Digital Ethnography of the Dark Web Social Network – Robert W. Gehl–2016.* [online] Available at: https://journals.sagepub.com/doi/10.1177/1461444814554900 (accessed on 10 January 2022).

Taylor & Francis, (2021). *Cryptopolitik and the Darknet.* [online] Available at: https://www.tandfonline.com/doi/full/10.1080/00396338.2016.1142085 (accessed on 10 January 2022).

Taylor, M., & Quayle, E., (2004). *Child Pornography: An Internet Crime.* (n.p.): Taylor & Francis.

Terrorism in Africa: The Evolving Front in the War on Terror, (2010). United States: Lexington Books.

The Demise of White House Market Will Shake Up the Dark Web. [online] Wired. Available at: https://www.wired.com/story/white-house-market-dark-web-drugs-goes-down/ (accessed on 10 January 2022).

The Gateway, (2021). *What is Cyber Theft?-The Gateway.* [online] Available at: https://www.frontierinternet.com/gateway/what-is-cyber-theft/ (accessed on 10 January 2022).

The Sun, (2021). *Inside the Chilling World of Dark Web Sites Where Users Can Hire Hitmen.* [online] Available at: https://www.thesun.co.uk/news/5815189/dark-web-live-torture-hire-hitmen-hackers/ (accessed on 10 January 2022).

Unhcr.org. (2021). [online] Available at: https://www.unhcr.org/56bb369c9.pdf (accessed on 10 January 2022).

Vice.com. (2021). *Study Claims Dark Web Sites Are Most Commonly Used for Crime.* [online] Available at: https://www.vice.com/en/article/3daqxb/study-claims-dark-web-sites-are-most-commonly-used-for-crime (accessed on 10 January 2022).

CHAPTER 11

Tools and Protocols for the Analysis of Mobile Phones in Digital Forensics

AKANKSHA,[1] VAISHALI TYAGI,[1] CHINTAN SINGH,[2]
SOURABH KUMAR SINGH,[2] and AMARNATH MISHRA[3]

[1]*MSc Forensic Science Student, Amity Institute of Forensic Sciences, Amity University, Noida, Uttar Pradesh, India*

[2]*Research Scholar, Amity Institute of Forensic Sciences, Amity University, Noida, Uttar Pradesh, India*

[3]*Professor (Forensic Science) & Director, Lloyd Institute of Forensic Science, Greater Noida, Uttar Pradesh (affiliated to the National Forensic Sciences University, Gandhinagar, Gujarat), India*

ABSTRACT

In today's world, mobile devices have become the utmost priority of every individual. A mobile phone or cell phone is a portable telephone with an access of cellular radio system hence it can be used worldwide. Mobile phones are considered as portable computers as they can do all things that a computer can do. But nowadays, it plays a vital role in several crimes and become an important investigative tool in forensics for tracing the crime. Mobile forensics (MF) is just another part of digital forensics which involves collection and preservation of evidence in the digital format in order to proof or disproof a crime. Steps involve in MF are identification, preservation, acquisition, analysis, documentation, and preservation. Prior to examination and analysis of digital evidence, seizure is done. Certain classification tools are present which are being used for extracting information required which is present in the digital format from the devices these are manual, physical, and

Advancements in Cybercrime Investigation and Digital Forensics. A. Harisha, Amarnath Mishra, & Chandra Singh (Eds.)
© 2024 Apple Academic Press, Inc. Co-published with CRC Press (Taylor & Francis)

logical extraction. Various challenges may impact the collection and examination of data from the mobile devices. These can be differences in hardware, legal issues, anti-forensic techniques, alteration of device, accidental reset, lack of resources and dynamic nature of evidence. Android data acquisition and iPhone data acquisition both follows distinct procedure. Both android data acquisition and iPhone data acquisition access logical extraction and creates a backup copy of data.

11.1 INTRODUCTION

In this era of modern science and technology, the use of mobile phones is merging at an alarming rate and at an exponential speed, mobile forensics (MF) develops an important aspect as mobile devices are often found at the places where crime took place (Ayers, Jansen, Moenner, & Delaitre, 2007). The extensive use of forensics widely plays an important role in the criminal investigations, internally from a corporate auditing frauds and accounting case to a criminal investigation carried out at law enforcement agencies. Many crimes took place which one way or other, involve any kind of mobile device or any other form of digital evidence (Bates, 1998). Thus, this field of forensics is emerging as an important tool in tracing crime, criminal, and their reason for criminality in order to make the world a better place.

On the basis of, information given by NSD or commonly known as National Security Database, digital forensics is an emerging branch of computer forensics that deals with the crimes often done digitally or involve any such crime of digital devices which help us to proof or disproof a statement (Carrier, 2001; Casey, 2010). Being a branch of Digital forensics, Mobile device forensics deals with the recovery and gathering of digital evidence or data from mobile phones and similar devices from our daily life. These smart phones these days have pre-installed operating system (OS), software applications thus, make the interaction with the user convenient and very hands-on. In the modern era, certain cybercrimes such as cyber bullying, fraud, espionage, and other cross-border crimes are done through mobiles (Conder & Darcey, 2009). Mobile is a digital device having both internal storage memory and external memory cards are used for the storage of data (Garber, 2001; Garfinkel, 2010; Garner, 2011). External memory storage devices such as memory cards, pen drives, etc. Subscriber identity module or SIM is a major identity module, separates the phones numbers, names of the contacts and network settings from the personal information present in the mobile phone (Halderman, Schoen, Heninger, & Felten, 2008).

Mobile devices can be used as a minicomputer as they can save our personal information such as contacts, photos, notes, chats, passwords, user account credentials, internet browsing history, messages, etc. (Al-Mutawa, Baggili, & Marrington, 2012). It is the easy source for the criminal to attain your personal information and fraud a person. The application of MF is of great use not only in the law enforcement investigations, but also in the corporate investigations and their auditing criteria for information security and managing different kinds of risk involved with the functioning of the organization, private investigations for personal use, criminal, and civil defense, military intelligence and electronic discovery of some confidential information that may have some value (Park, Chung, & Lee, 2012; Jansen, & Ayers, 2007). Numerous businesses may turn to mobile evidence if they believe their intellectual property is being stolen or any of their employee is engaged in fraudulent activity.

11.2 MOBILE FORENSICS (MF)

Mobile devices play a major role in the collection of evidence in the form of digital data. MF is a very interesting and less explored branch of digital forensics that focuses on forensically sound examination of digital evidence collected from a cell phone found at the crime scene. It allows for the systematic gathering and processing of proof stored on a smartphone, such as text messages, personal chats, call recording, MP3 and MP4 files, browser history, location, and so on, without compromising with the integrity of the data (Dasari, Nandagiri, & Satish, n.d.). One of the most important things to be prioritize is to gain access to intentionally erased and deleted data from mobile phones.

11.3 LITERATURE SURVEY

There is not much history associated with MF because it is a newly emerging field and it has still to go a very long route to develop. It all started in 1990s and 2000s when there were a lot of advancement took place in mobile phones and when smartphones were starting to emerge but at that time only few and wealthy people own mobile phones. Later on in 2000s when a large population start using them it become quite popular among people, and everyone used it for storing their important data and documents and mobile phones become the most personal form of storage so when any crime took place the first commodity that meant to checked so in this way MF came into action.

Forensics or forensic science in the branch of science that deals with the use of scientific techniques to solve the criminal cases. In MF we use tools or techniques to analyze the mobile devices found at the crime scene which may have some potential evidences because as there increased number of mobile phones therefore they are found in most of the cases according to the survey of last 10 years it has been found that almost in 45% cases mobile phones have been found at the crime scene (Agar, 2003). Also, almost 60% of cases involve mobile phones because there are cases that involve sharing of uncensored pictures or unethical images of something. Many cases of kidnapping have been solved by tracking the device of the culprit and much more.

> **Importance of Mobile Devices:**
>   - Use to track mobile id and location;
>   - Use to extract information about the important data;
>   - Listen to call recordings;
>   - Recover images or videos;
>   - Help to plot a storyline;
>   - Help to recover the mobile history;
>   - Help to obtain text messages that have been deleted;
>   - Link the culprit with the victim;
>   - Can access to the device which are not accessible;
>   - Recover passcodes.
> **Handling of Mobile Devices:**
>   - **Step 1:** Take notes of everything and photograph each of them individually.
>   - **Step 2:** Determine if the device is power on or off.
>   - **Step 3:** If the device is power off, do NOT turn it on.
>   - **Step 4:** If the mobile phone found at the crime scene is on, take proper care.
>   - **Step 5:** Collection and packaging.
>   - **Step 6:** Transport.
> **Information That can be Recovered from the mobile Devices:**
>   - All information about call log like incoming, outgoing, and missed calls;
>   - Contacts saved in the device;
>   - Text messages and other multimedia and SMS send from the device;
>   - Pictures, videos, call recording;
>   - Notes, calendar marked notices;
>   - Browsing history;
>   - Account linked;

- Location;
- Deleted data;
- Data from apps installed in the device;
- Wi-Fi connection information;
- Presentations, other important documents;
- Passcodes, swipe codes and other personal information, etc.

11.4 CHAIN OF CUSTODY AND THEIR DIFFERENT PHASES

Chain of custody involves all the steps and procedure followed from recovery of an evidence till its presentation in the court of law. In mobile forensic and its related investigation, maintaining chain of custody is the most important aspect to maintain the integrity of an evidence. Chain of custody of MF includes various phases as in subsections (Kostadinov, 2019).

11.4.1 THE EVIDENCE INTAKE PHASE

The initial stage of investigation is evidence intake phase, which includes all the sort of paperwork or documentation of the type of information the investigator is seeking for. It also includes developing objectives for the investigation.

11.4.2 THE IDENTIFICATION PHASE

The investigator should identity certain details before examining any device like:

- The legal issues associated with the investigation;
- Objectives of investigation;
- Identifying information that needs to be yield from the device;
- Data storage both removable and internal;
- Other potential evidence.

11.4.3 THE PREPARATION PHASE

After identification, this phase involves preparation of methods and techniques that will be used in acquiring data from that particular device basically it is the preparatory phase where we identify the techniques that should be used for that particular device.

11.4.4   THE ISOLATION PHASE

The tendency of mobile to obtain other inappropriate data by connecting with Bluetooth, Wi-Fi, or by another means is very high, so in order to get only relevant data for investigation, isolation techniques and instruments are used like faraday bag.

11.4.5   THE PROCESSING PHASE

This method is all about acquisition techniques like which method or which technique should be used to acquire a particular sort of information in a forensic way.

11.4.6   THE VERIFICATION PHASE

After the processing phase is completed comes to the verification phase in which up to what extent the data is accurate has been verified. There are several methods of verifying the data like:

1. **Comparing Extracted Data to the Handset Data:** This method usually involves the comparison of the data to the data displayed on the device.
2. **Using Multiple Tools and Comparing the Results:** Using different techniques to obtain the similar results.
3. **Using Hash Values:** After acquisition techniques are applied all the data like images are hashed so as to ensure there is no any changes in the data and whenever that image is open or any changes are made then hash value become different this is imp to maintain the integrity of the evidences.

11.4.7   THE DOCUMENT AND REPORTING PHASE

The documentation plays a major role throughout the case because it helps to keep a record of all the events that took place during that course of time. It should include the following:

- Date and time of examination;
- Condition at the time of recovering the evidence;

- Photos and individual data;
- Phone power status;
- Model no and name;
- Tools and techniques used for processing.

11.4.8   THE PRESENTATION PHASE

The last step that comes under chain of custody is the presentation of evidence in the court of law. It involves all the evidence linked to the case, forensic reports obtained and other clues so as to reach the final conclusion of the case and provide justice.

11.4.9   THE ARCHIVING PHASE

As cases went on to many years before getting a final decision and in the process of that it becomes quite a difficult to retain the evidences like images, etc., so as to preserve the data all the images and the data is archived and used as and when required.

11.5   PROCESS OF MOBILE FORENSICS (MF)

Our aim of using MF is to recover, preserve, analyze, and to present the digital evidence from a mobile device found at the crime scene. Certain specific rules are set out for collection, preservation, and transporting the data obtained as evidence (Raghav & Saxena, 2009). Steps of collection and preservation of an evidence in MF are as follows:

- Seizure;
- Acquisition;
- Examination and analysis;
- Reporting.

11.5.1   SEIZURE

Prior to the process of acquisition and examination, seizure of digital evidence should be taken as an priority. As this is one of the most important steps which influences the other steps such as acquisition, examination, and reporting (Sannella, 1994). Important points that should be followed while seizing the mobile evidence. They are as follows:

1. **Documentation of the Environment:**
   - Always take photographs of the evidence along with the surroundings.
   - Take all the additional things (cable, charger, hard drive, adapters, docking stations, etc.), which the investigator found at the crime scene.
   - Do not turn off the mobile phone, collect it as you found.

2. **Seizure Isolation Technique:**
   i. **Faraday's Bag:** A faraday bag is made up of a double layer of conductor, and it is the best method to isolate the mobile phones found at the crime scenes. In some cases, mobile found at the crime scene have low battery and in order to keep them on and also to maintain their connectivity they are kept in faraday's bag because it has cord with the help of that it awakes the phone.
   ii. **Jamming:** This technique is basically used to interrupt connection of a mobile phone with other devices. It is done by matching the frequency of radio waves of jammer with the device in order to avoid any unnecessary data to interrupt in the investigation.
   iii. **Airplane Mode:** This mode can be obtained from the mobile setting it doesn't involve any tools and it helps to block any transmission radio signals from the mobile phone.
   iv. **Enable Stay Awake Settings:** If the device found at the crime scene is without a lock we can directly go to the settings and enable stay awake setting in order to keep the device on and it utilizes the minimum battery possible of the device. Document the state in which the investigator found the mobile as running/not running, locked/unlocked, etc.

3. **Documentation of IMEI:**
   - International mobile station equipment identity (IMEI) is used in the smartphones as an identifier.
   - Always try to obtain the IMEI as it is meant to identify a device in a cellular network.

4. **Handling Locked Mobile Phones:**
   - Not to lock the mobile phone, if found to be in unlock state.
   - Try to detect the passcodes.

5. Disconnecting Mobile Device from the Network Connection:
- Put the mobile phone on airplane mode for disconnecting it from a network connection.
- Potential use of faraday bag, it will help in draining out the battery from the mobile phone.
- Remove the sim card if you cannot be able to access the airplane mode and faraday bag (Forman, 2003; Brown, Hua, & Gao, 2003).

11.5.2 ACQUISITION

Retrieving the data from the mobile device is done by using right passcodes, PIN, pattern, etc. Once we have successfully done the seizure of the evidence we never work our original evidence we usually make multiple copies to work on it, a process referred to as Imaging or Acquisition. Various tools and software such as hard-drive duplicating software or software imaging tools, such as FTK Imager, DCFLdd, IX imager, Guy Mager, True Back, encase are used for duplicating the data. To prevent the original file from tampering it is stored carefully (Yu & Lau, 2005). After the proper retrieval of data, analysis of data is done by using appropriate approach (Sathe & Dongre, 2018).

11.5.2.1 ACQUISITION TOOLS VENDORS

1. **MSAB:** It is a Sweden-based company which has created a software named XRY in 2003 which is used to extract data from mobile phones when they are connected to the computers, it can even recover the deleted files. This software is used by the army, law enforcement agencies, forensic labs. Its reliability is also even checked by many government agencies and data produced from this software is also accessible in the court of law.
2. **Cellebrite:** It is an Israeli company which creates software and hardware for phone-to-phone data transfer, it also creates backups. It is accessible to law enforcement agencies and forensic teams in most of the cases. This company was established in 2007 (Sathe & Dongre, 2018).
3. **BlackBag Technologies:** These make software for extraction of data from Apple devices. It has software like:

- BlackBag blacklight-cross platform macOS and windows forensic software.
- BlackBag MacQuisition-for acquisition of iOS devices.
4. **Magnet Forensics:** It is a company started by Adam Belsher. It also creates software to acquire the required data.
5. **Access Data:** It is a company which provides us with the tools for analysis of mobile phones to recovery data.
6. **Oxygen Forensics:** It is a software that is made to extract data from advanced level. It can decode any password and lock easily and helps to recover data by creating a backup, it can also extract data from iCloud and google id.

11.5.3 EXAMINATION AND ANALYSIS

The examination and analysis of recovered mobile phone evidence is done by cyber analysts with the help of various tools and techniques which can acquire the data, store the data, extract the data and various other things that are need to be carried out. Prior use of specialized tools (EnCase, ILOOKIX, FTK, etc.), is done by the investigators or examiners to support with inspecting and recuperating data. Separation of irrelevant information from the relevant information from the obscured data (Dasari, Nandagiri, & Satish, n.d.).

Type of retrieved data differs depending on the investigation, for example, e-mail, chat logs, images, contact numbers, confidential information, order history, financial information, our identity cards and marksheets in Digi lockers and much more.

The data can be recovered and reproduced from different things associated with the devices and also from different types of software and memory (Spector, 1989).

11.5.3.1 MOBILE OPERATING SYSTEM (OS) OVERVIEW

The main problem before applying any acquisition technique is understanding the OS of mobile phones because of advancements in technology different OS have different system access, for example, android devices gives more terminal access which iOS devices does not give (Sannella, 1994).

1. **Android:** Google is the developer of android hence google has free source of access to all android devices. It is a Linux-based OS also

known as Linux-based OS. It is also a free OS hence any manufacture of any mobile phone does not want to develop their own OS then they can use android (Bharati, Ramith, & Amala, 2015).

2. **iOS:** It is basically an OS of iPhone devices which is developed by apple. It is not open for most of companies. It is accessible most in apple devices like iPhones, iPods, iPads, etc. Hence, it is considered to be safer than android because all the activities are monitored by Apple.

3. **Windows Phone:** Windows is the type OS which have been developed by Microsoft for mobile phones and PCs as well. Windows phone is almost similar to Windows present in the PCs but more in a minimizable level.

4. **BlackBerry OS:** Blackberry is an OS which is developed and brought into use by blackberry Ltd. which is also referred to as research in motion (RIM). It also does not provide access to much of the devices only blackberry phones are operable on this OS. This OS is also known for its security reasons.

11.6 DATA RECOVERY SOFTWARE

There are certain software that allows us to recover data from the devices like:

1. **Stellar Data Recovery:** It is ISO9001:2008-certified software that is used to recover data lost from catastrophic situations.
2. **Sys Tools Data Recovery:** Recover windows-based programs.
3. **Iso Buster:** It helps to recover all kinds of media from CD, DVD, etc.
4. **Bit Recover:** It can recover data from windows 10, 8.1, 8, 7, 2000, XP.
5. **Bit Recover Virtual Drive Recovery Software:** This uses to recover data or images stored as .vhd .vmdk format.
6. **Bring Back:** This uses to recover data from Windows and Linux (ext2) operating system.
7. **CnW Recovery:** This uses to recover corrupted data or formatted one.
8. **Deleted File Recovery:** This uses to recover permanently deleted files.
9. **Sys Info Tools Virtual Disk Recovery Software:** This uses to recover data from corrupted VMDK, VHDX, VHD, and VDI files.
10. **DDL Data Recovery Tools:** These are used to recover information from hard drives that have broken or are physically damaged.

11. **Disk Drill:** It recovers data from Mac OS X.
12. **DRS:** It recovers media from HDD storage.
13. **Hetman Partition Recovery:** This recovers data from inaccessible data storage.
14. **GoPro Recovery:** This uses to recover fragmented data.
15. **RAID Reconstruction:** This uses to recover data from level 0 RAID (striping) and level 5 RAID drives.
16. **ReclaiMe Pro:** It is a data recovery toolkit from NAS recovery.
17. **e-ROL:** It has the ability to recover files that have been deleted accidentally over the internet.
18. **Recuva:** This uses to recover data from windows accidentally deleted from the internet.
19. **Restoration:** Same as Recuva.
20. **Undeleted Plus:** It can restore data from solid-state devices.
21. **R-Studio:** This recovers files from NTFS, NTFS5, UFS/UFS2 and so on.
22. **Deep Spar Disk Imager:** This uses to recover data from hard drives which has some kind of bad sectors in it.
23. **Adroit Photograph Recovery:** This uses to recover fragmented photos from Canon and Nikon cameras.
24. **Kernel Data Recovery:** This uses to recover corrupt databases, desktops, and laptops.
25. **Free Recover:** This uses to recover data from NTFS drives.
26. **PC-3000:** This uses to recover data from damaged HDD and SSD.
27. **PC Inspector File Recovery:** This uses to recover files from FAT12/16/32.
28. **Salvation DATA:** This uses to recover data from CCDV and DVR.
29. **Sys Info Tools Data Recovery:** This uses to recover data from FAT file systems.
30. **Mini Tool Power Data Recovery:** This uses to recover data present in the SD card of our device and also from the USB storage.
31. **M3 Data Recovery:** They have been associated with recovering data from windows and MAC operating systems.

11.7   IP AND MAC ADDRESSES

The full form of IP address means internet protocol address. It is basically a number which is used to identify a particular device, printer or any other device. These are unique to every device IP address have two versions IPv4 and

IPv6, where IPv6 uses 128 bit-numbers and IPv4 uses 32 bit-numbers. Four in IPv4 represents the group of numbers, i.e., there are four groups of letters in it, for example, 127.0.0.1. In the case of IPv6 there are six groups of numbers that are separated by colon for example, 2001:odb4:rty6:8876:0000:78rt. Mac address stands for media access control address. It is mostly embedded into the network card. For simple understanding we can take MAC address as the serial number present on the phone and IP address just like our phone number (Sengul & Erhan, 2017).

11.8 FORENSIC TOOL CLASSIFICATION PYRAMID (FIGURE 11.1)

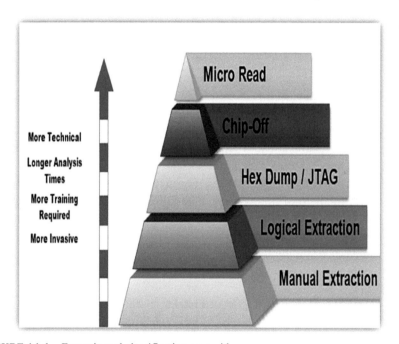

FIGURE 11.1 Forensic tool classification pyramid.

11.8.1 NON-INVASIVE METHODS

1. **Manual Extraction:** This basically includes photographing all the sides of the mobile phone in which condition it was developed at the crime scene and related evidence with the background.
2. **Logical Extraction:** This method basically includes interaction between mobile phone to the forensic labs where its examination is

meant to be done. It includes Bluetooth, USB cable or any other cord or device with which it could be done. In this, deleted data can be recovered.

3. **JTAG Method:** It helps us to recover data from all the different types of software this method is used when the device is damaged or corrupted. This is done via connecting device with TAPs (test access ports) that can obtain raw data from the device.

4. **Hex Dump:** It is similar to JTAG that can recover raw data which is present in the flash memory. But the major difference between the two is that the result obtained in this method is in the binary form which later need to be translated into the original data.

11.8.2   INVASIVE METHODS

1. **Chip-Off:** It is an expensive method of extracting data. In this method first chip is removed from the device and under investigation the data is recovered the data obtained through this method is raw data when which is later on decoded and analyzed. This method is only advisable to use when other non-invasive methods have already been applied to extract data.

2. **Micro-Read:** This method is more technical because in this method with the help of microscope, data is meant to be located on the chip with the help of gates. It requires high expertise to exercise this method. It is highly expensive and time-consuming and is only done in cases of national security.

11.9   MOBILE FORENSIC CHALLENGES

As MF has helped us to solve many a number of cases, but it has its limitations too; in some cases, because of certain factors, it is quite difficult to obtain information, and this is because of certain challenges that every examiner faces:

1. **Hardware Differences:** Because of increasing advancement in hardware, it becomes difficult for the examiner to adapt all the types of hardware present in the market because every hardware has different features and it becomes quite difficult for the examiner to testify them.

2. **Mobile Operating System (OS):** Just like hardware every device has its own mobile OS and these systems has their own security issues which also create hindrance in testifying them.

3. **Mobile Platform Security Features:** Some mobile security features give access to a lot of data, but some mobile platform does not even allow the data to be revealed due to which it becomes difficult for the examiner to obtain data.

4. **Lack of Resources:** Due to lack of tools and proper acquisition techniques sometimes the data can't be recovered.

5. **Generic State of Devices:** In some mobile phone, even when the mobile is switched off the alarm clock still works because device is in its generic state due to which sometimes data can be lost.

6. **Anti-Forensic Technique:** There are many anti-forensics (AF) techniques that hackers use like data hiding, forgery of data, adding of irrelevant data or misguiding data to the important data.

7. **Dynamic Nature of Evidences:** Digital evidences can be altered easily on the mobile devices when browsing into any other websites.

8. **Accidental Reset:** While examining sometimes examiner when end up resetting the complete data which may end up destroying the complete data present in the device.

9. **Device Alteration:** Sometimes changing of device or while transferring of data from one device to another, important data gets lost.

10. **Passcode Recovery:** In some devices, at the time of passcode recovery there are only limited number of chances after which the phone gets locked forever and due to which we can't get the data.

11. **Communication Shielding:** In some cases, if isolation techniques are not practiced properly, then after connecting with Wi-Fi or Bluetooth foreign data may enter the device and ruin it.

12. **Lack of Availability of Tools:** Lack of availability of tools like faraday's bag and other techniques data can be altered.

13. **Malicious Programs:** In some cases, alterations are done to the device by adding the malicious programs that have viruses which may end up crashing the hardware and software of the device due to which important data can be lost.

14. **Legal Issues:** Due to some legal issues involved with the mobile OS it becomes difficult for the examiner to testify.

15. **Laws Under which Mobile as an Evidence is Accepted Under Court of Law:** At the end, it is very necessary to discuss whether all these mobile evidences are even accessible in the court of law or not because if they are not, then all the investigations and analyzation process is of no use for us. In the Indian constitution, there are no specific rules for mobile as evidence, but according to Section 65B of the Indian Evidence Act (IEA), all devices, whether mobile phones,

CCTV cameras, CDs, or any other kind of electronic devices are presented in front of a court if they are compatible enough to prove the crime or defend the innocent person. Also, 65B(4) says that this evidence can only be presented by an expert who has enough training, knowledge, and experience in the required field, which was decided by a bench of U Lalit and Justice AK Goel.

KEYWORDS

- **android data**
- **digital forensics**
- **iPhone data**
- **malicious programs**
- **mobile devices**
- **mobile forensics**
- **operating system**
- **test access ports**

REFERENCES

Agar, J., (2003). *Constant Touch: A Global History of the Mobile Phone.* Cambridge: Icon Books, 172 pp., 20 illus., cloth $20, ISBN 1-84046-419-4.

Al-Mutawa, N., Baggili, I., & Marrington, A., (2012). Forensic analysis of social networking applications on mobile devices. *Digital Investigation, 9*, S24–S33.

Ayers, R., Jansen, W., Moenner, L., & Delaitre, A., (2007). *Cell Phone Forensic Tools: An Overview and Analysis Update.* NISTIR 7387.

Bates, J., (1998). *Fundamentals of Computer Forensics.* Information Security Technical Report, Elsevier.

Bharati, W., Ramith, N., & Amala, N., (2015). *Mobile Operating System: Analysis and Comparison of Android AND IOS.* ISSN: 2348 - 6090.

Brown, L. D., Hua, H., & Gao, C., (2003). A widget framework for augmented interaction in scape. In *Proceedings of the 16th Annual ACM Symposium on User Interface Software and Technology (UIST '03).* Association for Computing Machinery, New York, NY, USA, 1–10.

Carrier, B., (2001). Performing an autopsy examination on FFS and EXT2FS partition images: An introduction to TCTUTILs and the autopsy forensic browser. *Proc. SANSFIRE 2001 Conference.*

Casey, E., (2010). *Handbook of Digital Forensics and Investigation.* Academic Press.

Conder, S., & Darcey, L., (2009). *Android Wireless Application Development.* Addison Wesley.

Dasari, M. S., Nandagiri, R. G. K. P., & Satish, D., (2015). *The Forensic Process Analysis of Mobile Device. 6*(5), 4847–4850.

Forman, G., (2003). An extensive empirical study of feature selection metrics for text classification. *J. Mach. Learn. Res., 3*, 1289–1305.

Garber, L., (2001). Computer forensics: High-tech law enforcement. *IEEE Computer, 34*(1), 202–205.

Garfinkel, S., (2010). Digital forensics research: The next 10 years. *Digital Investigation, 7*, S64–S73.

Garner, (2011). *Sales of mobile Devices in Second Quarter of 2011 Grew 16.5 Percent Year-on-Year; Smartphone Sales Grew 74%.* Gartner, Inc.

Halderman, J., Schoen, S., Heninger, A., & Felten, E., (2008). Lest we remember - cold boot attacks on encryption keys. *Proc. 17ᵗʰ USENIX Security Symposium.*

Jansen, W., & Ayers, R., (2007). *Guidelines on Cell Phone Forensics.* Recommendations of the National Institute of Standards and Technology.

Kostadinov, D. (2019). *The Mobile Forensics Process: Steps and Types.* Infosec Resources, (n.d.). https://resources.infosecinstitute.com/topic/mobile-forensics-process-steps-types/ (accessed January 30, 2023).

Park, J., Chung, H., & Lee, S., (2012). Forensic analysis techniques for fragmented flash memory pages in smartphone. *Digital Investigation, 9*(2), 109–118.

Raghav, S., & Saxena, A. K., (2009). Mobile forensics: Guidelines and challenges in data preservation and acquisition. *IEEE Student Conference on Research and Development* (pp. 5–8). Malaysia.

Sannella, M. J., (1994). *Constraint Satisfaction and Debugging for Interactive User Interfaces.* Doctoral Thesis. UMI Order Number: UMI Order No. GAX95- 09398. University of Washington.

Sathe, S. C., & Dongre, N., M., (2018). IEEE 2018 2ⁿᵈ International Conference on Inventive Systems and Control (ICISC)–Coimbatore, India (2018.1.19-2018.1.20). *2018 2ⁿᵈ International Conference on Inventive Systems and Control (ICISC)-Data Acquisition Techniques in Mobile Forensics.*

Sengul, D., & Erhan, A., (2017). *Analysis of Mobile Phones in Digital Forensics.* MIPRO, Opatija, Crotia.

Spector, A. Z., (1989). Achieving application requirements. In: Mullender, S., (ed.), *Distributed Systems.*

Yates, I. I., (2010). Practical investigations of digital forensics tools for mobile devices. In: *2010 Information Security Curriculum Development Conference.* ACM.

Yu, Y. T., & Lau, M. F., (2005). A comparison of MC/DC, MUMCUT and several other coverage criteria for logical decisions. *Journal of Systems and Software.* (in press).

Zareen, & Baig, S., (2010). Mobile phone forensics challenges, analysis and tools classification. *Fifth International Workshop on Systematic Approaches to Digital Forensic Engineering (SADFE.2010)* (pp. 47–55).

CHAPTER 12

Analysis of Tree-Based Machine Learning Techniques for Credit Card Fraud Detection

JITENDER TANWAR,[1] SHUBHAM SINGH,[2] AKASH KUMAR,[2]
MANDEEP MITTAL,[4] LEENA SINGH,[2] and SUDHANSHU TRIPATHI[3]

[1]*Amity Institute of Information Technology, Amity University, Noida, Uttar Pradesh, India*

[2]*Amity School of Engineering and Technology, Amity University, Noida, Uttar Pradesh, India*

[3]*Amity University in Tashkent, Labzak Tashkent City, Uzbekistan*

[4]*Department of Mathematics, Amity Institute of Applied Sciences, Amity University, Noida, Uttar Pradesh, India*

ABSTRACT

A credit card is an easy and most beneficial target for hackers. On the other hand, it is a huge threat to the financial industry. Thousands of people suffer great losses from this problem every year. Although it is not an easy problem, machine learning (ML) has given us hope and played an important role in the fraud detection of online transactions. Online fraud detection is facing difficulties like having unbalanced data, not having confidential data, non-availability of analyzed data, etc. The processed and analyzed data can significantly improve the performance of ML algorithms. In this chapter, we analyzed and compared the performance of Random Forest, AdaBoost Classifier, XGBoost Classifier and finally ensembled them to get better results.

Advancements in Cybercrime Investigation and Digital Forensics. A. Harisha, Amarnath Mishra, & Chandra Singh (Eds.)
© 2024 Apple Academic Press, Inc. Co-published with CRC Press (Taylor & Francis)

We have evaluated their performances based on their Matthews correlation coefficient (CC) score, AUC-ROC score, and their confusion matrix. The result analysis shows the order of importance of features for selected ML algorithms. The results indicate that the Random Forest has the best score in both the evaluation metrics which comes out to be 0.9981.

12.1   INTRODUCTION

Fraud in credit card transactions is the usage of an account by an unauthorized user or some fraudster (Sahin, 2013). It is a criminal activity that brings serious consequences to financial industries. To avoid fraud activities there are two measures: fraud detection and prevention. Fraud prevention means anticipating and stopping the fraud before it happens. Fraud detection means detecting the fraud and stopping it if the fraud is attempted.

The world is filled with companies, clients, and fraudsters. Whenever a client suffers a fraud, the company must also suffer both financially and reputation-wise (Awoyemi, 2017). To provide quality service to customers, companies keep on hiring new schemes. They use data analysts to understand their customers. They use techniques to detect and prevent fraud and machine learning (ML) is one of the solutions. Using different ML techniques, we can identify fraudulent transactions and avert them from happening (Pozzolo, 2018). As e-commerce is rising the credit card transactions are also rising rapidly (Srivastava, 2008). This rise has made a serious impact on financial industries. The number of credit card users in 2019 in India was around 47 million and rise to 52 million in 2020. Numerous authorization techniques are not much effective in credit card fraud cases. Fraudsters use the internet to hide their identity and location. The global credit card fraud in 2018 reached 24.26 billion USD (Nilson Report, 2018).

In this chapter, the ML techniques are utilized for identifying frauds in credit card transactions, i.e., Random Forest, AdaBoost Classifier, and XGBoost Classifier. The real datasets are occupied from Kaggle, and it is generated by a python simulator PaySim through automation. The dataset has further divided into train, validation, and test datasets. The algorithms are trained over the testing data set, then validated through the validation dataset, and finally tested through the testing dataset. For the assessment of the models, evaluation metrics AUC-ROC and MCC (Matthew's correlation coefficient) have been used, and a comparison has been made between them.

12.2 RELATED WORK

In this section, the efforts of different researcher's has-been reviewed. Credit card fraud is very common nowadays and a lot of research has been done already. Few researchers have made some serious contributions but still fraudsters find new ways to challenge them and therefore we should keep learning and find better ways to stop them.

Bhanusri et al. (2020) have investigated ML algorithms like random forest with boosting, logistic regression, and naive bayes. They detect fraudulent transactions and reduce the number of false alerts. According to their comparative analysis, the random forest with boosting technique outperforms the other techniques used for credit card fraud detection. However, they could not be able to differentiate the fraud and non-fraud transactions.

Researchers Randhawa et al. (2018) have used both single and hybrid models. They have used a publicly available dataset for both types of models and they have added noise to data so that the robustness of the algorithms can be measured. At first, they used some classical single algorithms like naïve-bayes, decision-trees, random-forest, multilayer perceptron, support vector machine (SVM), feed forward neural network, linear-regression, and logistic-regression. After that, they combined these models with adaptive boosting and finally made different combinations of these algorithms making hybrids like naïve bayes combined with gradient boosting and more. They used the MCC score for their final evaluation. The best MCC score they got is 0.823, using majority voting.

Lakshmi & Kavila (2018), have used three ML techniques like Random-Forest, Logistic-Regression, and Decision-Tree for fraud detection. Around 70% of the dataset is used for training and 30% is used for testing and validation for evaluation. The accuracy, error rate, sensitivity, and specificity are utilized to evaluate distinct variables for three algorithms. The suggested system's performance is measured using accuracy, sensitivity, error rate, and specificity. The accuracy of logistic regression, random forest classifiers, and decision tree, respectively, is 90.0, 95.5, and 94.3. Random forest classifier outperformed the decision tree and logistic regression when compared.

Chen et al. (2014) used hybrid classifier ensembling and clusters for fraud handling. Raghavan et al. (2019) have compared various machine-learning and deep-learning models in their research. They got the best results when SVM was combined with CNN. CNN was best in all deep-learning algorithms as a fraud detector. Another deep learning-related work is Habibpour et al. (2021), who researched the value of uncertainty related to the predictions of

DNN (deep neural network). They used three deep UQ techniques (EMCD, MCD, and ensemble).

Domashova & Zabelina (2021) also used ensemble methods for analysis but with dimensionality reduction techniques to examine. They implemented the model on the cloud-based platform SAS (statistical analysis system) Viya, which is a popular anti-fraud system.

A comprehensive analysis of various machine-learning methods was presented in Pooja et al. (2021) paper with pros and cons. They worked on various algorithms like the Hidden Markov model, SVM, Random Forest, ANN, LSTM, and Genetic algorithms out of which ANN gave them the best results but its training process was very expensive. Here, we learned that fraud detection should not be too much costly both in time and money.

Sara Alqethami et al. (2021) showed various optimization methods of first and second-order learning algorithms like Backpropagation, Steepest Descent, Quickprop, and Gauss-Newton. First-order methods use gradients to construct further training iterations and second-order methods use Hessian to calculate further iterations. They used these algorithms to enhance the accuracy of various Neural-Network algorithms and got the highest score of 96%.

Fenila et al. (2021) used BiGRU (bi-directional gated repeated unit) and BiLSTM (long bi-directional memory) and got 80% and 81% AUC scores. They also used undersampling, oversampling, and SMOTE to handle imbalanced data and it also helped in increasing the accuracies of models. As we knew the dataset was highly unbalanced, we also used these techniques to reduce the side-effects of imbalanced data.

Bocheng Liu et al. (2021) created two fraud-detection models using Fully Connected Neural Network and XGBoost and got an AUC value of 0.912 and 0.969, respectively. They also build an interactive online fraud transactions detection system. After all, the purpose of all this research is to detect frauds in real-world transactions systems.

12.3   MATERIALS AND METHODS

12.3.1   DATASET

The problem in this research is the lack of available datasets for credit card transactions. In this chapter, mobile money transactions dataset has been used which can be downloaded from Kaggle. For researching fraud detection, we need a financial dataset, but due to the privacy policies, there is a lack of datasets. Here, a synthetic dataset has been created through a simulator named PaySim. It used a sample of real transactions of an African country total of a month transactions to create a synthetic dataset that looks similar to the normal transactions. The synthetic data is 1/4 of the original data.

The dataset contains 11 features. The variables are as follows step one unit of time, types of transactions, amount of transactions, nameOrig origin of the transaction, oldbalanceOrg old balance of origin of the transaction, newbalanceOrig new balance of origin, nameDest recipient customer, oldbalanceDest old balance of the recipient, newbalanceDest new balance of the recipient, isFraud this is target variable 0 means non-fraud 1 means fraud. PaySim is the simulator which produced the synthetic dataset then preprocessing is done, after processing the dataset, it is given as input to all algorithms to get their results which are then ensembled to get our final predictive model.

Figure 12.1 is the flow diagram of the process of building our model for fraud detection. PaySim2 is the simulator that gave us a synthetic dataset, this data was further preprocessed, and feature engineering was done. After that, this dataset was fed to all three tree-based algorithms to get their results, and then they were ensembled to get a combined result and all four results were then compared.

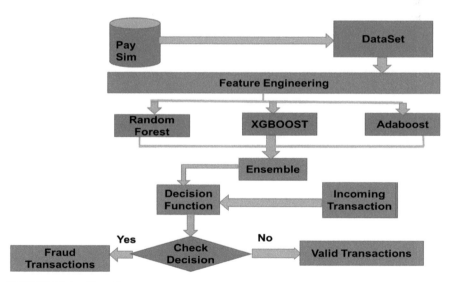

FIGURE 12.1 Flow diagram.

12.3.2 ALGORITHMS

12.3.2.1 RANDOM FOREST

Random Forest algorithm is used ensemble learning that trains several decision trees in parallel; this process is known as bagging. It divides the training dataset into n samples and uses these samples to train n different decision trees (Lin, 2017). This division of dataset into samples is known as Bootstrapping. In simple words, it trains multiple decision trees in parallel on different subsets of the dataset. Finally, it aggregates the decisions of individual trees using the ensemble technique of max voting. The final prediction is an output of voting from these decision trees. This algorithm is really good for structured data.

12.3.2.2 AdaBOOST

AdaBoost is also a type of ensemble modeling in which multiple lower accuracy classifiers are combined to gain higher accuracy classifier. Steps of Ada-Boost are, firstly it selects a training dataset randomly and then trains a model on this random dataset then predictions are made through this model (Cao Ying, 2019) and errors are calculated between actual and predicted values then it assigns weights to observations predicted wrong and the models whose overall accuracy is high and then the process is repeated, for final prediction max-voting is used among all the classifiers you have made.

12.3.2.3 XGBoost

XGBoost is one of the most frequently used ML algorithms in competitions in Kaggle for structured or tabular data. It is also an Ensemble learning algorithm that uses boosting techniques. It is made of gradient-boosted decision trees, which have high speed and performance. It uses an approach where new models are created which can predict the errors of the previous model and then combine them to make final predictions (Mitchell, 2017). Other boosting algorithms increase weights of weak classifiers, but in gradient boosted algorithms they try to optimize the loss function. It uses the gradient descent algorithm to reduce the loss while adding new models. Its optimization techniques give better performance and speed using the least amount of resources.

12.4 WORK PROCEDURE

In this section, the experimental setup is done in the following order, firstly a publicly available dataset is taken, and all the pre-processing techniques are applied to it, secondly, some feature engineering is done (Khurana, 2018), and then the models are created and tested on the final dataset. All the experiments are performed in Jupyter Notebook.

12.4.1 EXPERIMENTAL SETUP AND OBSERVATION

In the credit card dataset, the fraud transactions are very small in comparison to valid transactions, so the dataset is highly skewed, and to tackle this problem a little bit, weights are used while building models (Kirkos, 2007). To evaluate the final predictions three different evaluation metrics are used, to get a better understanding. A consistent score or result of these metrics means better performance.

The confusion matrix is a two-cross-two matrix that contains four possible outcomes of a binary classifier.

$$Confusion\ Matrix = \begin{matrix} TP & FN \\ FP & TN \end{matrix}$$

ROC (receiver operator characteristic curve) can help in deciding the best threshold value. It is generated by plotting the true positive rate (y-axis) against the false positive rate (x-axis).

$$TPR = Sensitivity = \frac{TP}{TP+FN}$$

$$FPR = 1 - Specificity = \frac{FP}{FP+TN}$$

MCC is robust to imbalance. It is related to the chi-square statistic and takes values between −1 and +1. For binary classification:

$$MCC = \frac{TP \times TN - FP \times FN}{\sqrt{(TP+FP)(TP+FN)(TN+FP)(TN+FN)}}$$

12.4.2 PREPROCESSING

Pre-processing is the process of preparing the raw data for model training. It consists of finding missing values and removing or imputing them (Ravisankar, 2011). There were no missing values in the dataset, and all the values with 0 only were replaced with –1. One hot encoding was used for categorical variables. Min-max scaler is used for scaling the features. Furthermore, the features which are not contributing to model results were removed. Some insights into the dataset are discussed in subsections.

12.4.2.1 TRANSACTIONS BY HOURS OF THE DAY

As an added insight, Figure 12.2 shows activity based on hours, the observations from this were substantial and can affect the project flow in any significant manner. The fraud transactions were continuous along with the 24 hours duration, whereas valid transactions were significantly less at night.

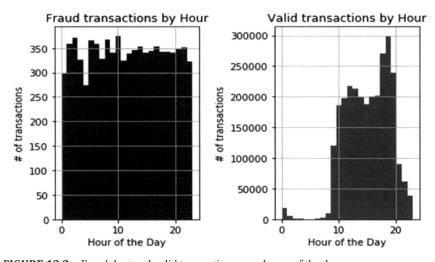

FIGURE 12.2   Fraudulent and valid transactions over hours of the day.

12.4.2.2 AMOUNT OF TRANSACTIONS

Certain trends were seen in the fraudulent activity based on the amount of the transaction as shown in Figure 12.3. The fraud transactions never crossed the 10-lakh line, whereas valid transactions were high up to 1 crore.

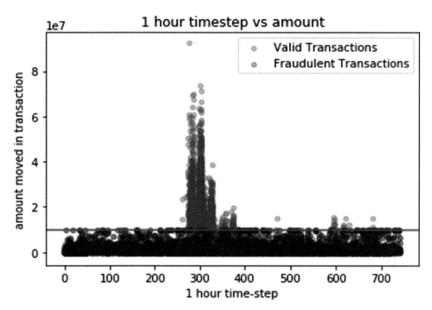

FIGURE 12.3   Scatter plot of transactions based on the amount.

12.4.2.3   TYPES OF FRAUD TRANSACTIONS

Figure 12.4 shows the types of transactions present in Fraud and Valid transactions. From the figure, we can see fraud transactions were of two types only and valid transactions were of five types.

```
Fraud transactions by type:
 CASH_OUT    4116
TRANSFER    4097
Name: type, dtype: int64

 Valid transactions by type:
 CASH_OUT    2233384
PAYMENT     2151495
CASH_IN     1399284
TRANSFER     528812
DEBIT         41432
Name: type, dtype: int64
```

FIGURE 12.4   Types of transactions fraud vs. valid transactions.

12.4.3 FEATURE ENGINEERING

Feature engineering is the synthesis of new features from existing ones or raw data. When these features are preprocessed by models (Coates, 2011), they could improvise the performances of ML models. Also, random forests and gradient boosting algorithms get significantly benefit from synthesized features as cited in (Heaton, 2016). Feature engineering requires more labor than any other process in ML (Bengio, 2013). In our case, it has played the most effective role in the model's results (Duman, 2011).

The Features which were added to the dataset are the first hour of the day. This feature talks about the hour of the day in which the transaction has occurred. Secondly, Error Balances, the dataset has some errors, like the transaction amount was not matching to the amount received or deducted from the destination or origin account, respectively. So, two new columns were introduced using these errors and named as 'errorBalanceOrigin' and 'errorBalanceDestination' as shown in Figure 12.5.

HourOfDay	errorBalanceOrig	errorBalanceDest
1	0.00	181.0
1	0.00	21363.0
1	213808.94	182703.5
1	214605.30	237735.3
1	300850.89	-2401220.0

FIGURE 12.5   The new features in the dataset.

12.4.4 RESULTS

12.4.4.1 RANDOM FOREST

The random forest algorithm is very flexible and easy to implement, making it perfect for the starter algorithm. Simply put, the random forest builds multiple decision trees (Tanwar, 2018) and combines them to get more accurate. Figure 12.6 shows the feature importance considered by the random forest. It can be seen from the figure the random forest is placing the most weight on the newBalanceOrig variable, i.e., it is most concerned with the new balance of the origin of the transaction and will base its prediction mostly on this variable (Figure 12.7).

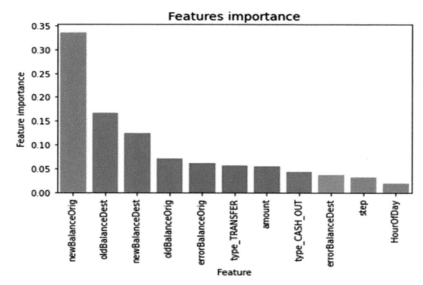

FIGURE 12.6 Feature importance for random forest.

From the confusion matrix, it can be seen that out of 5,55,000 transactions 0 transactions resulted in false positives where the algorithm flagged a clean transaction as a fraudulent transaction, and there were only 6 fraudulent transactions the algorithm failed to detect. This algorithm was able to detect every valid transaction and showed an exceptional performance.

12.4.4.2 ADABOOST

The basic idea of Ada-boost is to line the weights of classifiers and train the info sample in each iteration such it ensures the accurate predictions of negative observations. Figure 12.8 shows the feature importance taken into account by the AdaBoost. It can be seen from the figure that the AdaBoost is placing the most weight on the errorBalanceOrig feature, i.e., it is most

concerned with the error in balance at the origin side and will base its prediction mostly on this variable. Figure 12.9 shows the confusion matrix of
AdaBoost.

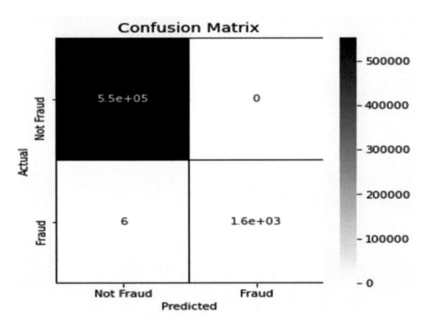

FIGURE 12.7   Confusion matrix of random forest.

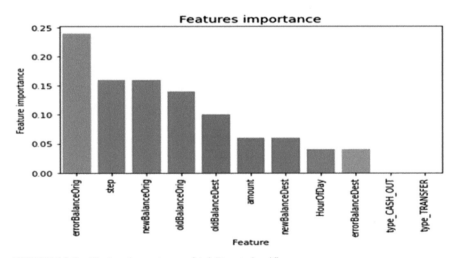

FIGURE 12.8   Feature importance of AdaBoost classifier.

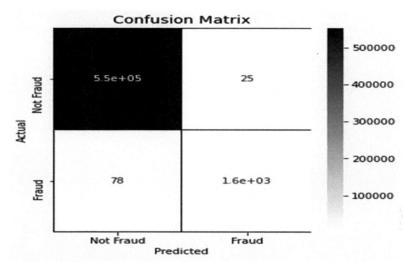

FIGURE 12.9 Confusion matrix of AdaBoost classifier.

From the confusion matrix, it can be seen that out of 5,55,000 transactions 25 transactions resulted in false positives where the algorithm flagged a clean transaction as a fraudulent transaction and there were 78 fraudulent transactions the algorithm failed to detect.

12.4.4.3 XGBoost

Figure 12.10 shows the feature importance of the XGB tree.

As can be seen from Figure 12.10, the XGB Tree is placing the most weight on the errorBalanceOrig feature, i.e., it is most concerned with the error in balance at the origin side and will base its prediction mostly on this variable. Figure 12.11 is the confusion matrix of XGBT.

From the confusion matrix data, out of 5,55,000 transactions only 10 transactions resulted in false positives where the algorithm flagged a clean transaction as a fraudulent transaction, but there were only 6 fraudulent transactions the algorithm failed to detect.

12.4.4.4 ROC-AUC SCORE

Table 12.1 lists ROC-AUC values of each algorithm, in which Random Forest and XGBoost have the approximately same score of 0.998 whereas AdaBoost is slightly behind them by a 0.997 score.

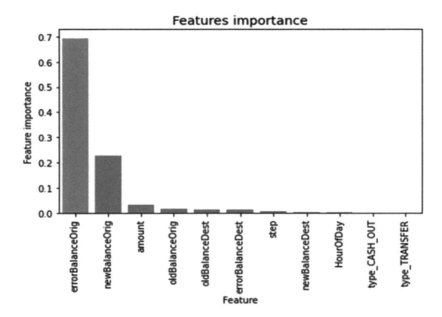

FIGURE 12.10   Feature importance for XGB tree.

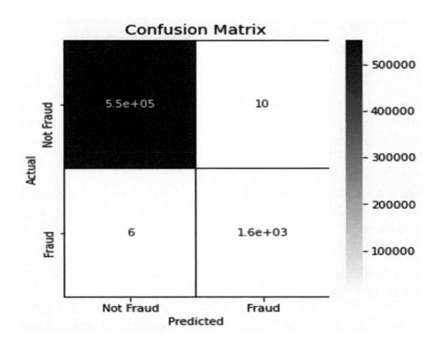

FIGURE 12.11   Confusion matrix of XGBT.

TABLE 12.1   ROC-AUC Score

Algorithms	ROC-AUC Score
Random forest	0.9981740718198417
AdaBoost	0.9762403067253764
XGBoost	0.9981650210468151
Ensemble	0.9981740718198417

12.4.4.5   MCC SCORE

Table 12.2 lists the MCC score of all algorithms where Random Forest and XGBoost are showing consistent results in both AUC-Roc and MCC.

TABLE 12.2   MCC Score

Algorithms	MCC Score
Random forest	0.9981669812672261
AdaBoost	0.9681787197506385
XGBoost	0.9951230367215831
Ensemble	0.9981669812672261

12.5   CONCLUSION

The implementation and testing of three decision tree-based algorithms are presented in this chapter. The motive was to get higher accuracy as possible on this dataset and comparison of their results. Although these three algorithms follow ensemble learning, we have again ensembled these three models to get a better result. The MCC metric is used for the final prediction measure. The best score of 0.99 is achieved using Random Forest, XGBoost, and Ensemble. The accuracy of the Ensembled model was 99% similar to Random Forest and XGBoost. We found that if you spent more time on feature engineering rather than on model training results could be better, which we have done. We have successfully learned and implemented these four different models and got an accuracy of 99%. This chapter can give you a brief introduction to credit card frauds and different ML algorithms. It also tells you about different aspects of ML like feature engineering, evaluation metrics, etc. Credit card fraud is a topic of concern for everybody, and continuous progress has been made in detecting frauds through ML.

KEYWORDS

- **AdaBoost**
- **bi-directional gated repeated unit**
- **credit card**
- **deep neural network**
- **ensemble learning**
- **fraud detection**
- **random forest**
- **XGBoost**

REFERENCES

Awoyemi, J. O., Adetunmbi, A. O., & Oluwadare, S. A., (2017). Credit card fraud detection using machine learning techniques: A comparative analysis. In: *2017 International Conference on Computing Networking and Informatics (ICCNI)* (pp. 1–9). doi: 10.1109/ICCNI.2017.8123782.

Bengio, Y., Courville, A., & Vincent, P., (2013). Representation learning: A review and new perspectives. *IEEE Transactions on Pattern Analysis and Machine Intelligence, 35*(8), 1798–1828.

Bhanusri, A., Ratna, S. V. K., Jyothi, P., Varun, G., Sai, R., & Rohith, S. S., (2020). Credit card fraud detection using Machine learning algorithms. *Journal of Research in Humanities and Social Science, 8*(2), 04–11.

Bocheng, L., Xiang, C., & Kaizhi, Y., (2021). Online transaction fraud detection system based on machine learning. *J. Phys.: Conf. Ser. 2023*, 012054.

Cao, Y., Qi-Guang, M., Jia-Chen, L., & Gao, L., (2013). Advance and prospects of AdaBoost algorithm. *Acta Automatica Sinica, 39*(6), 745–758.

Chen, F. H., Chi, D. J., & Zhu, J. Y., (2014). Application of random forest, rough set theory, decision tree and neural network to detect financial statement fraud taking corporate governance into consideration. In: *Proc. Int. Conf. Intell. Comput.* (pp. 221–234).

Coates, A., Ng, A. Y., & Lee, H., (2011). An analysis of single-layer networks in unsupervised feature learning. In: *International Conference on Artificial Intelligence and Statistics* (pp. 215–223).

Dal, P. A., Boracchi, G., Caelen, O., Alippi, C., & Bontempi, G., (2018). Credit card fraud detection: A realistic modeling and a novel learning strategy. In: *IEEE Transactions on Neural Networks and Learning Systems* (Vol. 29, No. 8, pp. 3784–3797). doi: 10.1109/TNNLS.2017.2736643.

Duman, E., & Ozcelik, M. H., (2011). Detecting credit card fraud by genetic algorithm and scatter search. *Expert Syst. Appl., 38*(10), 13057–13063.

Fenila, N. J., Roshan, J. R., Sakthi, E. K., & Sanjeev, K. N. M., (2021). Intelligent transaction system for fraud detection using deep learning networks. *J. Phys.: Conf. Ser., 1916*, 012031.

Habibpour, M., Gharoun, H., Mehdipour, M., AmirReza, T., Asgharnezhad, H., Shamsi, A., & Khosravi, A., et al., (2021). *Uncertainty-Aware Credit Card Fraud Detection Using Deep Learning. arXiv preprint arXiv:2107.13508v1.*

Heaton, J., (2016). *An Empirical Analysis of Feature Engineering for Predictive Modeling* (pp. 1–6). Southeast Con. 2016. doi: 10.1109/SECON.2016.7506650.

Jenny, D., & Olga, Z., (2021). Detection of fraudulent transactions using SAS Viya machine-learning algorithms. *Annual International Conference on Brain-Inspired Cognitive Architectures for Artificial Intelligence: Eleventh Annual Meeting of the BICA Society, Procedia Computer Science, 190,* 204–209.

Khurana, U., Samulowitz, H., & Turaga, D., (2018). Feature engineering for predictive modeling using reinforcement learning. *Proceedings of the AAAI Conference on Artificial Intelligence, 32*(1). Retrieved from: https://ojs.aaai.org/index.php/AAAI/article/view/11678 (accessed on 10 January 2022).

Kirkos, E., Spathis, C., & Manolopoulos, Y., (2007). Data mining techniques for the detection of fraudulent financial statements. *Expert Syst. Appl., 32*(4), 995–1003.

Lakshmi, S. V., & Kavila, S. D., (2018). Machine learning for credit card fraud detection system. *International Journal of Applied Engineering Research, 13*(24), 16819–16824.

Lin, W., Wu, Z., Lin, L., Wen, A., & Li, J., (2017). An ensemble random forest algorithm for insurance big data analysis. In: *IEEE Access* (Vol. 5, pp. 16568–16575). doi: 10.1109/ACCESS.2017.2738069.

Mitchell, R., & Frank, E., (2017). Accelerating the XGBoost algorithm using GPU computing. *PeerJ Computer Science, 3,* e127. https://doi.org/10.7717/peerj-cs.127.

Pooja, T., Simran, M., Nishtha, S., Jitendra, K., & Ashutosh, K. S., (2021). *Credit Card Fraud Detection Using Machine Learning: A Study.* arXiv:2108.10005v1 [cs.AI].

Raghavan, P., & Gayar, N., (2019). *Fraud Detection Using Machine Learning and Deep Learning,* 334–339. 10.1109/ICCIKE47802.2019.9004231.

Randhawa, K., Chu, K. L., Manjeevan, S. C. P. L., & Asoke, K., (2018). *NandiCredit, Card Fraud Detection Using AdaBoost and Majority Voting, 6.*

Ravisankar, P., Ravi, V., Rao, G. R., & Bose, I., (2011). Detection of financial statement fraud and feature selection using data mining techniques. *Decision Support Syst., 50*(2), 491–500.

Sahin, Y., Bulkan, S., & Duman, E., (2013). A cost-sensitive decision tree approach for fraud detection. *Expert Syst. Appl., 40*(15), 59165923.

Sara, A., Badriah, A., & Al Ghamdi, M., (2021). Fraud detection in e-commerce. *IJCSNS International Journal of Computer Science and Network Security, 21*(6).

Srivastava, A., Kundu, A., Sural, S., & Majumdar, A., (2008). Credit card fraud detection using hidden Markov model. *IEEE Trans. Depend. Sec. Comput., 5*(1), 3748.

Tanwar, J., Sharma, S. K., & Mittal, M., (2018). Secure framework for web-services communication. In: *International Conference on Automation and Computational Engineering.* ISBN: 978-1-5386-5464-4. https://doi.org/10.1109/ICACE.2018.8687009.

The Nilson Report, (2018). [Online]. Available: https://nilsonreport.com/mention/1313/1link/ (accessed on 10 January 2022).

CHAPTER 13

Black Hats: A Major Threat to Your Finances

SUNIDHI JOSHI,[1] SOURABH KUMAR SINGH,[2] and AMARNATH MISHRA[3]

[1]*MSc Forensic Science Student, Amity Institute of Forensic Sciences, Amity University, Noida, Uttar Pradesh, India*

[2]*Research Scholar, Amity Institute of Forensic Sciences, Amity University, Noida, Uttar Pradesh, India*

[3]*Professor (Forensic Science) & Director, Lloyd Institute of Forensic Science, Greater Noida, Uttar Pradesh (affiliated to the National Forensic Sciences University, Gandhinagar, Gujarat), India*

ABSTRACT

Nowadays, criminal approaches have changed from an age-old to a new digitalized method, but somewhere people are not much aware of it. The reason behind the commission of such crimes is the motive, and the intention of the individual. The only aspect on which the investigator need to focus is whether the commission is for legal purpose or illegal purpose. We are using digital means for every possible thing; hence it is necessary to know about this emerging type of cybercrime, that is financial cybercrime, as it has a direct impact on the finance of an individual, institution, or organization. Sometimes it becomes very difficult to even trace the offender of the crime as the person doing this is also a part of the organization who knows the loopholes or the vulnerabilities of the structures. In computer-related crimes, the victim, suspect, and target are the computer itself. The main motive of the criminal is either to defame someone or to get personal profit. In computer-related crimes, there are many tools, techniques, and measures that can help

Advancements in Cybercrime Investigation and Digital Forensics. A. Harisha, Amarnath Mishra, & Chandra Singh (Eds.)
© 2024 Apple Academic Press, Inc. Co-published with CRC Press (Taylor & Francis)

in retrieving data, the functionality of different software, and preventing hacking, virus attacks, and ransomware, with which digital forensics is not limited here has many more applications as well. Many technologies and measures can help in preventing such crimes. This chapter mainly deals with financial cybercrime and how with time, it has taken a turn and gets transformed into a heinous act. We studied the type of criminals, and how one can be a victim, and in the end, we studied how an individual can prevent himself/herself from this heinous crime and some case studies that reside in the category of financial cybercrime.

13.1   INTRODUCTION

Crime is not a new word. It can be heard in our day-to-day life; while defining crime, it can be defined as any illegitimate work that bears a punishment from a court of law (The Institute of Company Secretaries of India, 2016). This punishment can be in the form of a fine or imprisonment. Crime was certainly existing for ages, but with this advancement in technology, and communication emergence of a new mode of crime has occurred, which is termed cybercrime. If anyone wants to know, what cybercrime is, one should first go for the concept of crime and then what cybercrime is? As such, there is no definition of cybercrime that will comprehend every aspect of it. We can say that it is a type of crime that involves internet fraud, hacking, cyber terrorism, etc., or any illegitimate action that is committed by the use of computer or computer devices or networks (Sarmah, Sarmah, & Baruah, 2017).

One of the major categorizations of cybercrime is financial cybercrime. These days with an increase in the use of the internet for various purposes has invited a new way to an existing crime that is having a direct threat to the finance, i.e., money, property of an individual. According to a study by the World Economic Forum in 2018, it was observed that cheating and financial crime is an industry that is to be considered as a trillion-dollar industry and the private companies spent a lot of money for anti-money laundering (AML) which is approximately $8.2 billion (in 2017) (Hasham, Joshi, & Mikkelsen, 2019).

Financial computer-related offenses can be defined as the offenses that bear the inclusion of systems for the commission of offenses against some-one's property, illegally withdrawing the money from someone else's bank account, and many more for their personal use. E-banking, online buying, and other forms of electronic commerce are commonly used to carry out financial fraud. Cards such as credit cards, allow for credit expansion and payment postponement, whereas debit cards do not. Fraud that is perpetrated using banking cards such as credit card or any other payment device as a bogus source of reserve in a transaction charge or debit cards your account at the time of the transaction. It could be to obtain items without paying for them or obtaining illicit money from an account. Identity theft is often accompanied by credit card fraud.

Majorly when we see ATM frauds, they are the result of the various actions committed by the customer, their negligence, or we can say lack of awareness among people regarding frauds, how to use, what to share or not, and many others. When talking about a country like India, we can see here crimes do not occur due to mistakes they occur due to negligence and lack of awareness. Today when everything has been digitalized these financial frauds have also become financial cybercrimes as they started using some known cybercrimes that we used to hear very rare, but now they are prevalent all over the world no matter how much the country is developed or not. Nowadays, a very common term is internet trading, that means when we purchase and sell various securities via the use of the internet. Several strategies are used by the criminals that are listed below (Figure 13.1):

FIGURE 13.1   Some types of cyber-attacks.

1. **Phishing:** It is defined as a technique one uses to get sacred information of the end-user, and use it for identity theft (Hoonakker, Carayon, & Bornø, 2009). An e-mail is sent to the users which will seem to be legitimate but will contain a link that will ask for the bank details and any other personal information and then forward it to another party that will exploit the details. It also involves the usage of spam to get access to the individual's bank. One should keep in mind that e-mails from legitimate institutions will never demand the personal data of an individual.

2. **HTTPS Protocol:** This will appear at the beginning of the URL for any site that implements security encryption. Generally, individuals do not look after it and get into the trap.

3. **Social Engineering:** It is a cybercrime in which the attacking person will contact you via call, social media they will first gain the trust of the person and slowly take all the sacred data such as ATM PIN credit card number, etc., and ultimately loot the person.

4. **Spoofing Websites:** It mean sometimes the fraudsters make the fake sites which look like legitimate ones but in reality, it is a spoof site (Vadza, 2013). These sites are created to get personal information like the bank details of the victim. They will generally ask for your details and sometimes will ask you to do transactions.

5. **E-Mail Spoofing:** It is a technique in which the perpetrator forges the header of the e-mail so that it will look like the original authenticated source e-mail.

6. Sometimes there are also possibilities that while doing some personal work, transaction, etc., from some device that might be the cybercafé one or any other place at that point of time some people forgot to delete their data or there is some bug sort of thing in that computer which records everything and ultimately the person becomes a victim of the crime.

7. **Hacking:** It is a method by which illegitimate access is made into the system with the intent of getting the sacred information of others (Dashora, 2011).

8. **Vishing:** It is a type of social engineering in which the perpetrator communicates over a phone call to acquire the personal as well as financial details of an individual (Nkotagu, 2011).

9. **Denial of Service:** It is sometimes used by the attackers to flood the site with requests or to create network traffic that will ultimately prevent the candidate from accessing the service.

10. **Watering Holes:** It can be classified as an evolved or modified technique known as spear phishing. In this, some malicious code is endorsed into some public sites that a few individuals go for (Kaur, 2017).

13.2 CAUSES OF CYBERCRIME

One of the studies published in 2019 says that there was a work "the idea of law" published by Hart which says that "people are helpless so the standard of law is required to ensure them" (Harshita, 2019). Current study applies the very same principle in terms of the cyber world it says that our devices do not have much power only the law is required to protect the citizens from these offenders of cybercrime. Some of the reasons are:

- Human negligence is the major cause;
- Loss of evidence;
- Easily accessible;
- Lack of awareness;
- Devices can store much information.

13.3 FINANCIAL CRIME VS. FRAUD VS. CYBER BREACHES

Financial crime versus fraud versus cyber breaches there is a clear-cut demarcation that can be seen in the way it is detected, processed, prevented. This modern era has removed this demarcation to nil as the modus operandi has become very similar and has grown to that much extent that's sometimes it is very difficult to distinguish between them. This distinction is not based on any legislative body or any regulatory body, but it is very often the result of the institutional systems. Earlier in past decades, we say that a financial crime means a crime in which there is tax evasion edition, money laundering, bribery. Nonetheless, it's most typically addressed as a compliance difficulty, inclusive of whilst monetary establishments use anti-cash laundering sports to avoid fines. Fraud, on the other hand, refers to an expansion of crimes involving the deception of economic people or services to behavior theft, consisting of forgery, credit score scams, and insider threats. Monetary organizations have constantly dealt with fraud as loss trouble, but in current years, powerful analytics were used to hit upon or even prevent fraud in real-time. Cyber breaches can be defined as an event that really or possibly jeopardizes the secrecy, judgment, or accessibility of a data framework or that constitutes an infringement or inescapable danger of abusing security approaches, security strategies, or worthy utilize approaches. Economic establishments ought to use most of the same techniques to shield assets in opposition to all three sorts of crime because the difference among them will become much less critical.

13.4 FINANCIAL CRIME, FRAUD, AND CYBER OPERATIONS

Nowadays the higher organization are putting efforts to bring hand in hand the cybercrime or cyber operations, financial crime, and fraud. In financial institutions, the public dealing employees, as well as the back-office employees, are working together for this cause. Risk features and regulators are catching on as properly. Risk functions and regulators are catching on as well. Issues such as AML today are inscribed as regulatory issues. In these cases, the primary step is to first elucidate the activities related to risk management and explicate the duties and responsibilities throughout the defense line (Hasham, Joshi, & Mikkelsen, 2019). The following steps will give assurance that the process is completely described coverage by the following and replicating the attempt.

1.	The first line of defense	The business and enterprises.
2.	The second line of defense	threats like financial crime, fraud, and cyber operations.

When it comes to crime associated with finance it includes four types of curative measures:

- To identifying and authenticating the buyer;
- Tracing and detecting transaction;
- Abnormalities in behavior; and
- Action to eliminate hazards and related issues.

Each of those activities, whether or not taken in response to fraud, cybersecurity breaches or attacks, or different monetary crimes, are processed by the alike information and further processes. Indeed, bringing those statistics resources together with analytics materially improves visibility at the same time as supplying an awful lot deeper notion to intensify the utility of detection. In many times it additionally enables prevention efforts.

While giving an extra comprehensive view of the basic procedures. Financial institutions can make efficient enterprises and produce such a system that will help the customer to have a better experience, also help in better quality decision and regulation of better and efficient cost. Accordingly, the re-structuring of the system takes place as per the need.

Each of these activities, whether taken in response to fraud, cybersecurity breaches or attacks, or other financial crimes, are supported by many similar data and processes. Indeed, bringing these data sources together with analytics materially improves visibility while providing much deeper insight to improve detection capability. In many instances it also enables prevention efforts.

13.5   CARBANK ATTACK CASE

This attack was one such which illustrates the complete profile of modern-day frauds and monetary or financial crime. This series of attacks began in the year 2013 which caused a monetary loss which was amounting to be more than $1 billion. The criminals who were part of a planned perpetrator organization gained access to the systems with the help of phishing attacks and shipped balances to their bank accounts, and they programmed banking machines (ATMs) to withdraw cash. The offense was noticeable and it was been notified that this attack is being done on multiple banking companies which took place in a coordinated and simultaneous way. The offenders showed a high level of understanding and knowledge of the computer-related crimes and digital environment which include operations linked with banking systems, their controls, and even the faults (Hasham, Joshi& Mikkelsen, 2019). They used banking machines (ATM), transaction cards (i.e., credit and debit cards), and transferring of wire among other methodologies. This assault proved that there are no longer differences between computer-related crimes, primitive frauds, and criminality against finance.

13.6   TYPES OF CYBERCRIMINALS

Cybercriminals can belong to any of the age groups, it involves children to adolescents to adults (Sharma, Ghisingh, & Ramdinmawii, 2014) (Figure 13.2):

1.   **Children in the Age Group of 9–16 Yrs.:** One will think about how a child can be a criminal but in cyber we come across many of the cases in which children deliberately or in a confused state become the cybercriminal. When it comes to hacking it feels great to a child

that he can hack but his curiosity will bring him to a crime that would be unknown to him but illegitimate in the eyes of law.

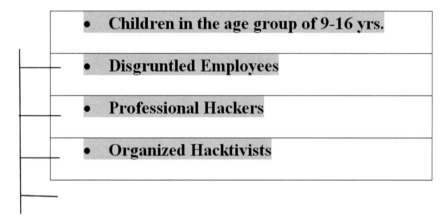

FIGURE 13.2   Types of hackers.

2. **Organized Hacktivists:** These are the ones for whom this is a profession they do the hacking with a purpose which might be on a political basis or social or religious basis.
3. **Disgruntled Employees:** What happens is some of the employees that might not be pleased by the organization or feel disheartened they are being the people of an organization knows the loopholes, vulnerability of the system harms the organization and fellow employees. This results in many financial cybercrimes and sometimes the entire system breakdown occurs.
4. **Professional Hackers:** They are that professionalist that are hired by big organizations to bring down information related to other organizations or sometimes to protect their system from the attack of some sort of cyber-attacks that might be imposed by other hackers.

13.7   VICTIMS

The victims of financial cybercrime may be financial institutions, organizations, or people. In terms of FIs and agencies, the options for prevention and movements are extra normal in comparison to people going through cybercrime. However, the effects of a successful cyber-attack at the FIs and companies may be tremendous and devastating. In addition to the misplaced number of sums which might also affect the monetary balance of the entire organization negatively or even result in financial disaster, the adverse effect on the reputational side in the eyes of commercial enterprise partners, clients, and society may additionally leave irredeemable harm pleasantly (Financial Cybercrime: How the System is Failing the Victims, n.d.).

As for the individuals who are the victims of cybercrimes, having their debts stolen, savings emptied, or IDs stolen and debts taken with the aid of their names, there may be a bigger hole in the device in acknowledging, taking it severely, and appearing upon the cybercrime. Every person can be the goal of the cyber attackers, but in maximum cases, they hit upon senior or older people who have the better opportunity of missing the capability or proficiency in using generation effectively or who are digitally incapable. If people will have to face such circumstances in their day-to-day life, there will be fear and bad feedback of technologies and financial institutions among people so government, along with the institutions must take necessary steps to combat these assaults against people. The modern-day economic and security systems fail in serving the sufferers of cybercrime who stumble upon difficulties in reporting the incidents or receiving strong guide, consequently a reform inside the machine is as urgent as it is important.

13.8   IMPACT OF CYBERCRIME ON THE BANKING SECTOR

Due to the emergence of mobile phones with Internet access, the number of detected cybercrime cases has increased significantly. Today, mobile phones are used to perform many online tasks and exercises, such as creating websites, shopping, and e-commerce, paying service fees, and writers are constantly using them to access proprietary information. Among the various aspects that have inspired the commission of cybercrime, monetary gain for many years remains the invariable factor leading to the emergence of new ways of performing such tasks, including revenge, coercion, and political motives.

The success rate of simple phishing attacks is as high as 45%, which is worrying because people are unsure what steps to take to avoid falling victim to this type of internet and technology scam. A statistic says that the cybercrime that occurs for monetary profit (NTRO) is about 3,855 in number and amount 534 phishing attacks in number, according to 2014 data (Harshita, 2019). This data is only based on the cases about which detailed information was there not including those cases whose information is missing or they are not reported or taken care of. Outside the world, banks deviate from the practical goals of visually perforated time-to-use (DDoS) classes in a blueprint to avoid presumption of the exemption profession more dangerous measures in parallel, such as adding malware or interfering with Computer Resources. Known as the Continuous Violence of State-of-the-Art, this secret plan hack is the newest kid on the board with a well-rounded build and enhanced versatility. When attackers do not want to provide meaningful data, they destroy the bank's website to punish their failed efforts. In addition to the financial benefits of successful digital attacks, proximity to the commonly cited illegal online activities, disclosure of personal data as the "dark web" in typical cybercrime, but more recent efforts and sophisticated hacking units. Sensitive data such as stolen/leaked motherboard numbers, web-based management accounts, transaction logs and proxy access are exchanged for cash in these online extortion networks.

13.9   PREVENTIVE MEASURES

According to a study the average cost of defrauding and cybercrime is around to have more than dollar three trillion globally on a very basis however the focus of tracking system operated by way of the financial establishment is to remain compliant with the policies to a minimal degree and the issue of the identity of a suspicious economic pastime with maximum performance stays unnoticed for the financial institutions to acquire each on the identical time they want to define powerful and analytical goals first and then support there monitoring structures as a consequence with advanced analytical technology and gear along with system studying or machine learning (ML), artificial intelligence (AI) or behavioral biometrics Decreasing the false positives ought to be one of the goals. As there one of the maximum times and efforts taking issues for the financial institutions and consequently an important handicap for spitting extra electricity end time for detecting and combating real threats and attacks. Growing the efficiency with faster identity off suspicious how we can be set as the second intention. Using technologies together with the device getting to know behavior biometrics can use simple person facts to rapidly come across typical activities and alert the tracking teams every other aim to fight monitor cybercrime can be an expansion of protection coverage (Financial Cybercrime: How the System is Failing the Victims, n.d.). The connectivity of the technological landscape has enlarged the areas that the threats may also reach. It has grown to be important for financial institutions to make sure that they cover all of the possible regions which can be at risk of these cyber-attacks. Sooner or later economic institutions are entitled to broaden their predictions and prevention gear. So, one can actualize the above-stated goals the simplest answer for the economic institution is to shape partnership for legal and technological guidance. Getting used to AI and predictive analytics can improve the information business enterprise soda economic institution can merge and examine records extra without difficulty towards cybercrimes. Particular records are tracking and studying strategies which can be helpful for the financial institutions in making sure safety and prevention in opposing two suspicious activities which might be:

1. **Digital Entity Fingerprinting:** A fixed validation assessment to recognize if the tool used, get entry to location, or IP deal with is suspicious or not.

    Session monitoring analysts of user conduct stumble on if there is an actual man or woman or a bot inside this session (https://fineksus. com/financial-cybercrime-how-the-system-is-failing-the-victims/).

    There is a method called behavioral biometric evaluation in this technique the following things are noted that are:

    - Typing speed;
    - Cursor movement;
    - Duration of reaction whether consistent or not.

2. **Multi-Layered Solutions:** The threats cybercrimes pose today are without a doubt extra serious than previous years and they hold evolving at a quick pace. With the advancement of technologies, offenders have also opted for new strategies to find out the loopholes and then ultimately assault many. To combat these types of cyber threats and assaults the organizations need to upgrade themselves with these technologies to effectively stand against them. We used to rely upon the firewalls and antivirus structures for the prevention of cyber assaults; but, these aren't sufficient for safety anymore. The modern arena threats in cyber can be best resolved using multi-layered techniques which includes first of all a type of defense system that has prioritized the safety standards for statistics (https:// fineksus.com/financial-cybercrime-how-the-system-is-failing-the-victims/), it includes:

    - Prevention of statistics loss;
    - Statistics profiling;
    - Data collection.

With the use of these methods of cyber defense, the various cybercrimes can be detected and prevented before they occur.

13.10   MEASURES THAT CAN BE TAKEN FOR DOING ONLINE TRANSACTIONS

- One should always keep in mind that he/she should not disclose his/her internet banking password, one-time password, credit card, and debit card CVV: Card Verification Value number, expiry month and year, and automated teller machine (ATM) PIN to anyone whether he is known or unknown.
- Online banking passwords should be changed simultaneously over time and one must be careful while setting the password it should not be the same password which one is using for other works and should not be easy in such a way that anyone will guess it.
- For setting the internet banking password a set of guidelines is available on the site first carefully read that and then set the password (https://fineksus.com/financial-cybercrime-how-the-system-is-failing-the-victims/). Your password must be the combination of the following:
  o   One upper case letter;
  o   One lower case letter;
  o   One special character;
  o   One numerical value.

- The easiest way these days is that most of the offenders of cybercrime send e-mails to people which will look genuine to an individual and as soon as the person is going to click on it he/she will land on a page that contains the malicious thing that can even hack all the details of the person in seconds. So, one must be careful of such types of e-mails. Do not open any mail that is not authentic.
- Check the transactions and activities in your account on a daily basis and if you encounter any such activity then immediately report it to your respective banks.
- While logging in to your bank site always be careful that the site must contain https protocol in the beginning if it does not have "s" after http don't login into that site. In this protocol "s" means secure.
- You have seen that while logging in your browser asks for a saving password always click on the never option and always be careful that you delete the browsing data after the transaction is done.
- Never do the transaction on any public device as these devices have some trackers in them which can steal your username and password and ultimately provide your personal information to fraudsters.

- Always make sure that when you receive the letter from the bank that is consisting of your debit card or credit card, the seal should be there it should not tamper if it is, then contact your bank immediately.
- After receiving the card if it is not tempered, open the letter and make sure you change the pin immediately. Always make a strong PIN so that no one can easily trace it.
- While you are entering your pin, be careful that no one is watching you and always be careful with the CCTV's present over there.
- One of the most necessary things that one should remember is always to link your bank account to your phone number and also subscribe for the mobile notifications so that if any of the activity occurs, you will get the notification then and there only.
- Always make sure that you have disabled your international transactions unless you are not traveling abroad.

To prevent oneself from getting trapped one must ensure that he/she is vigilant enough then only one can protect themselves. Always take proactive measures.

13.11   CASE STUDIES

13.11.1   *ICICI–PUNE BANK FRAUD CASE*

The case was basically a credit card scam and three persons were found accused in this case. They committed a crime in which they used to gain the details of credit card of a particular individual and then used those details for booking various airline tickets. During the investigation conducted by the city cybercrime investigation cell, it was found that around 100 people became the victims of this crime (Walker-Munro, 2021). An officer of ICICI Prudential life insurance, i.e., Mr. Pravesh Chauhan along with his customers, filed an FIR against these suspects. The complaint was lodged on behalf of one of the customers of Mr. Pravesh Chauhan. After doing the investigation, the cyber cell got a lead in the case and arrested Mr. Sanjeet Mahavir Singh Lukkad, Dharmendra Bhika Kale, and Mr. Ahmed Sikandar Shaikh. Lukkad used in a personal Institute, kale was one of his friends and Mr. Sheikh was an employee in one of the branches of a renowned Indian Banking company. It was found that one of the buyers received an SMS for the acquisition of price tags even after the credit card was under the control of the buyer. The client was quite attentive and understood that something was wrong. The complaint this suspicious activity to the bank. During the investigation, police also found the involvement of many other banks. It was also known that the tickets were booked through online means. Police investigated the details of the register and obtained the full contact details of the Institute. The thorough investigation of police revealed that the main points or the credits were obtained from the Indian banking company. Now the three accused work on a particular pattern, i.e., Sheik worked in the credit card department and used to access the information of Mastercard from his client or customers and then gave this information to Kale. Kale then used to pass this information to one of his friends Lukkad. Then Lukkad used this to buy tickets and then sell these to his customers for money. The investigation was done for eight days where cyber cell got all the leads that they wanted, and the culprits were arrested.

13.11.2   *PUNE CITIBANK MPHASIS CALL CENTER FRAUD*

It's an example of procurement engineering in action. Over the Internet, US $3,50,000 was transferred for profit from the city bank accounts of four US clients to bogus accounts in Pune. Employees at one center exploited the bravado of American clients to obtain their PIN codes under the guise of assisting customers in difficult situations. The numbers were then utilized to perpetrate fraud (Cascavilla, Tamburri, & Van, 2021). When call centers in India are afraid of losing business, they practice maximum security. Decision-making center employees are checked once upon entering and leaving so that they cannot transcribe the numbers and therefore have not written them down. They had to remember these numbers, go directly to an internet restaurant and gain access to clients' Citibank accounts. Customers also reported that money from their accounts was transferred to accounts in Pune and that perpetrators were being hunted down regardless. The police were able to establish the decision-making center's honesty and blocked all accounts where the money had been transferred.

13.12   CONCLUSION

From this study, which was conducted based on existing data, it was found that how in this era with digitalization the primitive fraud and financial crime have shaped themselves into financial cybercrime. We also came across the ways one can trap people and there are many ways how we can protect ourselves from these types of crimes. There are so many ways by which an individual can come across your cyberspace. Every single day comes with a new loophole and vulnerabilities in our system. To eradicate these offenders from abusing these vulnerabilities, one must be aware of technologies that can be used. If one has become the victim of the crime, he/she must know it then only the offenders can be caught. One has to be careful to protect themselves from these kinds of cyber-attacks. Awareness among people needs to be spread as much as possible to avoid such crimes.

KEYWORDS

- **anti-money laundering**
- **artificial intelligence**
- **automated teller machine**
- **computer-related crimes**
- **digitalization**
- **finance**
- **financial cybercrime**
- **offender**

REFERENCES

Cascavilla, G., Tamburri, D. A., & Van, D. H. W. J., (2021). Cybercrime threat intelligence: A systematic multi-vocal literature review. In: *Computers and Security* (Vol. 105). https://doi.org/10.1016/j.cose.2021.102258.

Dashora, K., (2011). Cybercrime in the society: Problems and preventions. *Journal of Alternative Perspectives in the Social Sciences, 3*(1), 240–259.

Financial Cybercrime: How the System is Failing the Victims. (n.d.). Retrieved from: https://fineksus.com/financial-cybercrime-how-the-system-is-failing-the-victims/ (accessed on 10 January 2022).

Financial Cybercrime: How the System is Failing the Victims. (n.d.). Retrieved from: https://fineksus.com/financial-cybercrime-how-the-system-is-failing-the-victims/ (accessed on 10 January 2022).

Harshita, S. R., (2019). Cybercrime in banking sector. *International Journal of Research-Granthaalayah, 7*(1), 152, 153. https://doi.org/10.5281/zenodo.2550185.

Hasham, S., Joshi, S., & Mikkelsen, D., (2019). *Financial Cybercrime and Fraud* | McKinsey.

Hasham, S., Joshi, S., & Mikkelsen, D., (2019). *Financial Cybercrime and Fraud* (pp. 1–3). | McKinsey.

Hoonakker, P. L. T., Carayon, P., & Bornø, N., (2009). *Spamming, Spoofing and Phishing E-Mail Security: A Survey Among End-Users.* https://www.researchgate.net/publication/268412871_ Spamming_spoofing_and_phishing_E-mail_security_A_survey_among_end-users (accessed January 30, 2023).

Kaur, S., (2017). E-banking & cybercrime. *IOSR Journal of Business and Management (IOSR-JBM), 19*(11). Ver. IV pp. 60–63. e-ISSN: 2278-487X, p-ISSN: 2319-7668.

Nkotagu, G. H., (2011). Internet fraud: Information for teachers and students. *Journal of International Students, 1*(2), 73. ISSN-2162-3104.

Sarmah, A., Sarmah, R., & Baruah, A. J., (2017). A brief study on cybercrime and cyber laws of India. *International Research Journal of Engineering and Technology (IRJET), 4*(06), e-ISSN: 2395 -0056.

Sharma, U. M., Ghisingh, S., & Ramdinmawii, E., (2014). A study on the cyber - crime and cyber criminals: A global problem, 10.20894/IJWT.104.004.001.003. *International Journal of Web Technology, 177*, 3.

The Institute of Company Secretaries of India, (2016). *Cybercrime Law and Practice* (1st edn., Vol. 1). The Institute of Company Secretaries of India. https://www.icsi.edu/media/webmodules/publications/Cyber_Crime_Law_and_Practice.pdf (accessed on 10 January 2022).

Vadza, K. C., (2013). Cybercrime & its categories. *Indian Journal of Applied Research, 3*(5), 130. ISSN – 2249-555X.

Walker-Munro, B., (2021). A case for the use of cyber-systemics to combat financial crime in Australia. *Kybernetes, 50*(11). https://doi.org/10.1108/K-09-2020-0581.

CHAPTER 14

Enhanced Security Based on Face Detection in Payment Gateway

M. R. EBENEZAR JEBARANI, S. LAKSHMI, and P. KAVIPRIYA

Department of ECE, Sathyabama Institute of Science and Technology, Chennai, Tamil Nadu, India

ABSTRACT

In this modern world, all are marching towards digital mode of communication. The security plays a vital role. This is a new initiative to provide secure cash transaction, so that to avoid unauthorized access. This mode of transaction can be used by everyone whether they are knowing the technology or not. This is the easiest way as well as more secure way of transaction.

The combination of machine learning and cryptographic techniques was initiated to perform the payment. In this approach a new innovative methodology is used for secure payment gateway.

14.1 INTRODUCTION

In the digital era, the public face so many problems because of online shopping, fund transfer, etc. This advancement leads to cybercrime, and that insists on the importance of cyber security. To enhance the security of the payment gateway, transaction is insisted by giving two-step authentication, i.e., before inputting the UPI pin in the transactions, face detection and proxy detection are performed. It also provides double security, not only check the face, but also ensures if a person tries to make a proxy transaction and it

Advancements in Cybercrime Investigation and Digital Forensics. A. Harisha, Amarnath Mishra, & Chandra Singh (Eds.)
© 2024 Apple Academic Press, Inc. Co-published with CRC Press (Taylor & Francis)

prohibits from the entire process. In this, an enhanced face net algorithm is used for authenticating the right person. The main operation of this system is proxy detection, which is implemented using image processing technique. This provides more security to the system. To overcome the limitations and security holes of the existing techniques, a new initiative is developed to have two-step authentication in transaction process. Face verification will be implemented in the system by incorporating 128 feature points of the face utilizing triplet loss functions and the face net deep learning algorithm. Support vector machines (SVMs) will be used to detect proxies (SVM). SVM will be used to classify each person's facial features. If an individual doing the operation is the correct person, the transaction will be executed, and the balance will be presented after the successful transaction is completed. If there is any mismatch it will exit from that portal. The novelty of the work is to help the public by avoiding unauthorized access by hackers to their personal account by asking the one-time password and leads to cybercrime. Elderly people and illiterate are not aware of the process and innocently reveal their passwords to the anonymous calls. This system assures a secured payment gateway process that will be able to help the public in particular.

Through the use of debit or credit cards, banks need infrastructure investment as well as large deposits. As a result, the bank must devise a low-cost infrastructure to address the problem. The effects of numerous recent technological advancements can be seen in various sectors such as education, government, and so on. To protect its capital, a bank must invest in expensive infrastructure. To deal with these problems, the bank must constantly think of fresh and innovative solutions. As a result, created biometric initiatives to address this problem. Cryptography can be used to encrypt data and transactions in a variety of ways. Biometric technology has many advantages, but it also has drawbacks. Biometric authentication has recently gained popularity in this area. Fingerprint, eye recognition, facial tracking, and iris technology are all examples of biometrics. Machine learning (ML) can be used to enforce these biometric security measures. ML is a kind of data processing technique. Its methodology is to automate the structure of a systematic model. Its major framework is that it can learn from data, recognize various examples and also make decisions with minimal or no human inputs.

14.1.1 MACHINE LEARNING (ML) TECHNOLOGY

Machine learning (ML) is a data-analytics technology that creates models automatically. A computerized reasoning theory states that frameworks should learn from data, recognize examples, and make conclusions with minimum human participation. The iterative nature of ML is important because models can evolve with the introduction of new information. They use information from previous estimates to provide dependable, repeatable choices, and outcomes.

14.1.2   DEEP LEARNING TECHNOLOGY

Deep learning is a branch of ML that uses artificial neural networks (ANNs) to create algorithms that are inspired by the structure and properties of the brain. It's a feature that simulates the human mind's data processing and the development of decision-making styles. Deep learning has developed alongside the virtual era, resulting in a vast influx of data of all forms and from all corners of the globe.

14.2   LITERATURE REVIEW

Several options for facial recognition applications have been implemented over the last decade. The methods and algorithms used in one piece of software may be vastly different from those used in another. When facial recognition software extracts the components of the face, it uses the findings to try to find the name. Normalization, followed by facial recognition: Other algorithms are used to normalize fixed-face images. The words 'enter' and 'face' are synonymous in an equation. This is an innovative approach to facial recognition that is likely to gain widespread acceptance. To figure out who it is, it use a 3D scanner to pick out features of the face, such as the jawline, nostrils, and chin are not hampered, and recognition from several perspectives is possible; additionally, recognition is made simpler with this approach. Researchers also created a new method for distinguishing offenders based on facial characteristics such as pores and skin texture analysis. After a normal skin scan, a statistical representation of the exact lines and spots can be obtained. An overview of the facial recognition systems that have been introduced for security purposes is given.

Hemery et al. (2008) discuss the benefits and advantages of face authentication as a biometric technology for banking applications. First, to understand the operation it is must to go over some fundamental financial concepts. They proposed a method to replace PIN code authentication with biometric data. The authentication of biometrics is then thoroughly clarified.

They present a face recognition method that reveals itself to be a biometric solution candidate. In a real-world application, it shows the benefits and drawbacks of this approach. Several reports on biometric authentication have been released. The majority of the time, it is a matter of developing an algorithm that is mainly focused on image processing and pattern recognition methods. Before it can be used in a real-world transaction, many problems, such as security and material considerations, must be addressed. The aim of biometrics is to reduce fraud and simplify authentication. Standard methods of authentication are used. Across the board, biometric algorithms generated excellent results. Using biometric technology for bank transfers has both benefits and drawbacks. For face detection, no proxy detection is used in this chapter.

Agostinho et al.'s (2019) paper, according to the Indonesian Institute of Statistics, society's level is rising in 2019. Based on data, the bank conducted a group to simplify transaction payment in the market. The bank organized a group to simplify market transaction payment based on data. The bank just completed the transaction using a debit or credit card, but banks require additional infrastructure expenditure, which is too expensive. The bank now requires a new low-cost infrastructure solution as a result of this. This software allows for online transactions. To improve communication security and transaction information, this study employs biometric encryption, or decryption transaction authorization, and QR Code Scan transaction permits. The Biometric Cloud Authentication Platform's implementation tests reveal that AES 256 agents may be utilized for face biometric encryption and decryption. Scan the QR code to finish a transaction permit. Face verification transaction permits have a 95% accuracy rate for a sample of 10 persons, and the transaction process takes 53.21 seconds per transaction for a sample of 100 transactions. The level of society in Indonesia is rising in 2019, according to the Indonesian Institute of Statistics.

The facial-recognition age as non-intrusive and low-cost. Face recognition has many benefits over other payment systems. Each person's face plays a distinct role. Iris and fingerprint recognition, on the other hand, require additional hardware, whereas facial recognition does not because smartphones already have cameras. In 2013, Bhatia conducted research on different forms of biometrics, writing, "Biometrics is an emerging era, and has been commonly used in forensics, safe entry, and jail surveillance," and weighing the benefits and drawbacks of various authentication technologies (Ahmad, Surya, & Fairuz, 2016; Chinchu, Anisha, & Mahesh, 2017).

As a result, individuals are on the lookout for innovative ways to safeguard themselves. The human body, according to Parmar and Mehta, is unique and difficult to duplicate, copy, or lose. During the purchasing process, mobile phones and credit cards are no longer required. They just want to "check their face" in a shop to finish their order. One of the key advantages of biometric bills is their versatility and ease of usage, which eliminates the need for coinage or credit cards in locations like retail and public transportation, while mobile serves all consumer needs.

14.3   EXISTING SYSTEM

The existing system consists of QR Code Authentication and Biometric Authentications. Existing system was implemented using face encryption and decryption, finger prints and other bio metric authentications to avoid hackers while transaction. The existing system which implements Face encryption uses Azure API Face Verification. Azure Face Verification recognizes, analyzes, and recognizes human faces in photos. Facial recognition software is critical for identifying a person's face. This program addresses a wide range of issues, including natural user interfaces, security, visual content analysis and management, robots, and mobile apps. This face service offers a variety of facial analysis options (Chellapa, Wilson, & Sirohey, 1995; Dileepkumar & Yeonseungryu, 2008).

14.3.1   FACTORS INFLUENCING PAYMENT THROUGH FACIAL RECOGNITION

To ensure the transaction's safety and trustworthiness, Suh and Han adopt a security mechanism in which the machine manipulates personal records and preserves their secrecy, integrity, validity, and non-replicability. There are various researchers who are interested in the secured payment gateway systems. The most important factors are security, which was easily affect the public is the usage of mobile-based transaction. Similarly, many academics consider security to be the most important research topic when investigating cellular payments and utilize the TAM-based model is to examine, why people use new technologies. One of the most important influencing variables is the nature of the brand-new technology itself. New technologies also bring with it some associated dangers, such as privacy, non-public information and the transaction, that raise the perceived risk of mobile payment services. Shah and his colleagues, as a result, Ashrafi and Ng argue that safety and security risk awareness are critical challenges in reality while using the digital enabled transaction systems, and that they stymie the usage of innovative mobile-payment instruments. As a result, the mobile payment system must be tackled in order to ensure security for the generation to be effectively employed (Dileep, Yeonseung, & Dongseop, 2008; Hung-Yuan, Chun-Cheng, & Shou-Jyun, 2017). Customers, on the other hand, have recently been aware of a few concerns about the security of identification, non-public records, payments, and other technical data when using IT technology. Many researchers have looked into the issue of confidentiality protection. Issues of privacy and security are additional variables that influences the implementation of mobile-payment providers. As per Johnson et al. (2015) research, customers are very much concerned about the security of their personal data, which diminishes their interest for being familiarized and utilized. Customer worries about the data breaches and hence diminish enthusiasm to adopt to the digital era. As a result, it is dealers' important responsibility to alleviate buyers' anxieties and boost up the protection of the private or personal information. Face-recognition payment deployment and acceptance among customers require careful consideration of privacy considerations. Confidentiality, secrecy, maintaining the security and online identification control are always a big challenge for customers, as they are with other new technology dealt (Maria, n.d.; Turk & Pentland, 1991; O'Regan & Norfolk, 2020).

14.3.2 VISIBILITY

If the new services or products have a large number of clients, as per Moore and Benbasat visibility is defined as the degree of innovativeness the company can view and also the acknowledgment of implementing various new innovative services or goods by ability customers. Russell et al. (2018) says that the volume to which people employ new advances to serve others is sometimes referred to as visibility. As per Johnson et al. (2015) research, relative advantage, perceived ease of usage, comparative advantage, perceived safety and visibility and all these influences whether or not people use mobile payment services. To put it another way, if witness humans use new innovative technology or facilities in the societies, it may cause social stress and, as a result, have an impact on prospective awareness of new technology or products. Potential customers of novel technologies may be affected by the societal framework, especially if there are numerous customers present. Customers with the ability to accept new technology or offers may be under pressure from their social surroundings. Yang & Choi (2001) stated that the usage of technological acceptance model as a hypothetical model of the social impact on the acceptability of record structure (Hee-Dong & In Choi, 2001). The innovations revealed that by increasing the audience's societal influence boosted visibility, and that this effect was judged good. Immoderate intake or state intake, which occurs in a class or role of intake, is a sub concept of product visibility. The connection between availability, entertainment, and the societal impact of mini-domestic transaction for payment was investigated by Lee and Lee using the technological acceptance model. It was considered that how the visibility influences societal perception, which in turn influences in what way purchasers evaluate the utilization of a particular product. "Innovation, application, universality, reliability, and visibility" of digital charge schemes are crucial characteristics to consider while adopting them, according to users. Most notably, facial-recognition became immediately obvious as an efficient method of safeguarding public safety De-Cormis. Furthermore, there are numerous stimulating openings for business to use face identification or recognition to boost up the efficiency. This in turn leads to the comfort of dealing with customers and also to provide really personalized services such as De-Cormis. Similarly, another researcher says De-Cormis that was owned by KFC and other related fast-food franchises, had combined together to form a team with the mobile fee company Alipay to implement the facial popularity fee authentication features (Nikita & Vibha, 2017f; Phil & Timothy, 2013).

14.3.3 EFFORT TO BE EXPECTED

The expected attempt is defined as the degree to which humans remember how to use the upcoming new informative technology in a very simple methodology. This leads to the initiation to use the records oriented technological interface is to be simple to view and employ. It also insists how to implement a plenty of tips to help once all are immediately linked to the predicted try. Many studies undergone previously have documented those efforts have had a significant impact on the translation of cause to practice. The introduction of a new technology necessitates a calculated effort. When technology designers like Orlikowski ignore elements relating to ease of use, the adoption of a upgraded technology might be limited or even may be limited or even unsuccessful. According to Davis et al. (1991) expected effort has a significant impact on the user recognition of data generation. Expected efforts for analyzing the performance, effort, and societal recognition that plays a vital role in the decision-making technique used in a mobile-payment system. The Unified Technology Acceptance and Use Theory (UTAUT) is used by Koenig Lewis et al. to expand the outline by incorporating variables such as reported gratification, societal impact, understanding, and apparent hazard. Expectations about performance and behavioral goals are both influenced by expected attempts says Tiago (2016). De Luna et al. developed TAM, to explore the features that was influenced by the buyers in adopting to mobile payment systems. It was discovered that perceived safety, perceived compatibility, non-public innovation, and the character movement were in and around among them. According to the earlier research, predicted effort has a strong and favorable association with the cause of implementing a mobile-payment system (Rafael & Richard, 2018; Chellapa, Wilson, & Sirohey, 1995).

14.3.4   SOCIAL IMAGE

The volume to which a customer believes his or her condition or photograph is advanced within the organizational community to which the consumer belongs is referred to as social image. Moore and Benbasat are two names that come to mind. Venkatesh and Davis established the TAM2 technology acceptance model, and the social photograph is an part of a professed utility idea inside it.

In their study, Jaradat and Faqih looked at the key effect factors (Faqih & Jaradat, 2015). Customers' familiarization in the usage of mobile payment offerings can be determined related to what motivated to do, with the following factors having a significant impact: perceived utility, perceived simplicity of usage, subjective models, photograph, quality of the output, self-efficacy, perceived outside control and interest. In other words, societal pictures impact perceived ease of use and the incentive to apply records production in studies on the inspiration for continuous use of contemporary technological data offerings. Furthermore, the data reveal that societal impact might influence incentive to use cellular-fee, which is consistent with prior research suggesting that customers' opinions are influenced by their social context (Liebana-Cabanillas et al., 2019). Koenig-Lewis (2015) found that societal consequence, generational information, and the perceived risk all had an impact on customers' readiness to receive simple payments. Mostly in the economic situations, public are not only anxious just by their material attainments, but they are aware of preserving societal image up to certain level. As a result, the motivation of this observation leads how to keep in mind the perspectives of many contributors to the society while implementing new technology services or goods (Rafael, Richard, & Steven, n.d.; Richard, Diego, Santiago, Paul, & Diego, 2017).

14.3.5 OPENNESS (OPENNESS TO EXPERIENCE)

Costa and McCrae were the first to offer openness, which they well-defined as the inquisitiveness and an exposed mind to tales and beliefs. According to Ziegler et al. (2015), those who are with a high level of openness like to relish new concepts and are more approachable to novel info. The term openness related to people's desire to learn new things and hear new stories. People that are open to change are more likely to experiment with new technology. In our example, the generation of facial-popularity-ratio generation is a novel generation. Customer features have a significant impact on the propensity to embrace innovative skills and products, and most of the students have conducted much research on various areas related to this.

Nowadays, various attempts have been made, with the advancement of the records generation, innovative technology, or innovative technology-oriented services, to observe the role of character persona in behavioral and usage patterns. The researchers Moore and McElroy discovered the changes in buyer personas have a significant impact on the desire to use Facebook. Customer behavior is spurred by way of persona developments, impact on "Role of persona in laptop-related comprehensive learning." Openness has a huge influence on apparent practicality and simplicity of usage. Most of the researchers used a value-mindset conduct model to investigate individual values as predictors of motivation to practice self-provider generation in trading. They discovered that personal characteristics had a significant impact on motivation to utilize self-provider generation (Selvia, Ichsan, & Gitarja, 2017; Varthika & Deepak, 2015).

14.4 SECURED PAYMENT GATEWAY

To address the shortcomings and security gaps, we proposed a two-step verification mechanism in the transaction process. The use of image recognition techniques for proxy detection is a benefit of our proposed approach. As a consequence, the machine becomes more robust. Face Verification will be used in the proposed process by embedding with 128 attribute points of the face using the Face Net Deep Learning Algorithm. Proxies can be detected using SVMs. The triplet loss function will evaluate 128 feature points on each person's face. SVM may be used to categorize each individual's facial features. The kernel approach allows SVMs to execute effective non-linear classification by implicitly mapping their inputs into high-dimensional feature spaces. The support-vector clustering algorithm uses support vector statistics provided by the SVMs algorithm to categorize unlabeled data and verify the facial image. After the face is checked with a proxy, the account will be processed, and the balance will be shown (Figure 14.1).

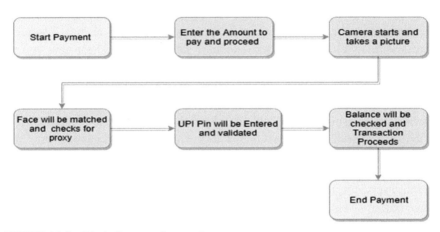

FIGURE 14.1   Block diagram of secured payment gateway.

14.4.1   *DEEP LEARNING-FACE DETECTION AND RECOGNITION*

Face net employs a deep convolutional neural network (CNN), as shown in Figure 14.2. The squared L2 distance between the embeddings linked to face similarity should be used to train the network. The photographs used in training must be resized, altered, and usually cropped closely around the face area. Face Net is distinct from other systems in that it learns the mapping from the image. Rather than using a bottleneck layer for recognition or verification, it builds embeddings.

14.5   IMPLEMENTATION

14.5.1   *IMAGE PREPROCESSING – FACENET ALGORITHM*

FaceNet is a single face recognition and clustering embedding algorithm. In a number of benchmark face recognition bases, including the labeled face in the wild (LFW) and YouTube face databases, it has obtained state-of-the-art success. FaceNet algorithm suggested a way of generating high-quality image face mapping using in-depth learning architectures like ZF-Net and inception network. A major component of the architecture is end-to-to-end learning. Its middleware uses ZilogbyNet or Inception. These additional 1*1 convolutions are to cut down on the number of parameters as well as processing. This system uses deep learning to perform a linear two-layer normalization L2 on the image f(x). These embeddings were reduced by the same proportion until they were passed into the cost feature to measure the loss. The aim of this loss function is to have the squared distance between two images is nearly the same, regardless of pose or state. Triplet failure is implemented as a result of triplet digits rather than single digits. In our model, triplet loss establishes a margin between face features of identity. In triplet loss, an image's embedding is defined by f(x), such as x. This embedding is a 128-dimensional vector that is normalized into positive and negative pairs. In triplet selection, to ensure quicker learning, triplets must take the challenges. Generate triplets based on previous checkpoints and calculate the minimum and maximum on a subset of data for each step.

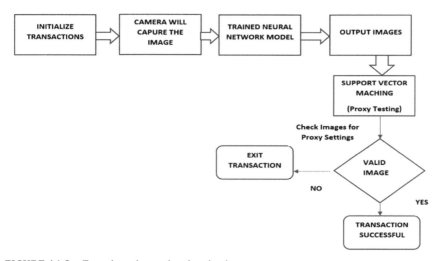

FIGURE 14.2   Face detection and authentication process.

14.5.2 *ONE SHOT LEARNING*

One shot learning is a classification task in which one can take a large number of examples and apply a single rule to each one. This classified image encountered in face recognition, such as those with different facial expressions, accessories, and hair light, as well as those with specific faces and no hairstyles. The method for one-shot face recognition employs a high-dimensional function called a face embedding, which is computed in a straightforward manner. When learning novel concepts was needed for one-time applications, the Siamese network was used. FaceNet used the triplet loss feature after practicing on a group of Siamese networks, which became state-of-the-art in facial benchmarking competitions.

14.6 RESULTS AND DISCUSSIONS

14.6.1 OUT MODULES

FaceNet application in dart, which will run on the flutter platform. ML algorithms are written in python and are converted in dart files using starflut package. These samples are protypes after integrating with the bank software's that can be modified. For the face recognition approval portal by using python SQL server database. After approving the payment will get processed further.

Figure 14.3 shows the app home. The new app was identified by the name as Face Pay. Like saving our credit/debit card, here face can be registered and once everything over can be deleted.

14.6.2 MODULE 2

While signing up in app and training the dataset done by registering customer's face (Figure 14.4). First step to register the customer face.

14.6.3 MODULE 3

Figure 14.5 shows the face is detected in camera range. If the rectangle is green face detected is valid else invalid.

14.6.4 MODULE 4

Figure 14.6 shows that face is detected and verified that is it was matched with the data already available in the trained data set. So, this identifies the person with name and then it proceeds for payment.

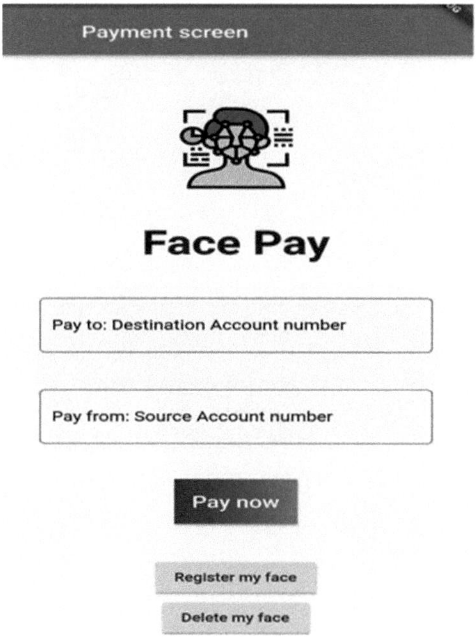

Payment screen

Face Pay

Pay to: Destination Account number

Pay from: Source Account number

Pay now

Register my face

Delete my face

FIGURE 14.3   App home working screenshot.

Name

Account number

Sign Up!

FIGURE 14.4   Sign up window.

FIGURE 14.5   Face detection by camera.

FIGURE 14.6   Face detection and verification.

14.6.5 MODULE 5

Figure 14.7 shows the payment was successfully completed. After the verification and authentication, the payment completed.

FIGURE 14.7 Payment completed.

14.6.6 MODULE 6

Figure 14.8 shows the face validation database using SQLite and flask modules in python. Here the stored database of the face was verified and validated.

14.7 DISCUSSIONS

In this segment, the pros and cons of implementing this algorithm were discussed.

14.7.1   ADVANTAGES

- Only with live face payment will be completed;
- No need to share the OTP for friends and family;
- Proxy Detector is included, if showing any photo in front the prediction system, it automatically detect the proxy and placement will not be completed;
- Secure payment is established.

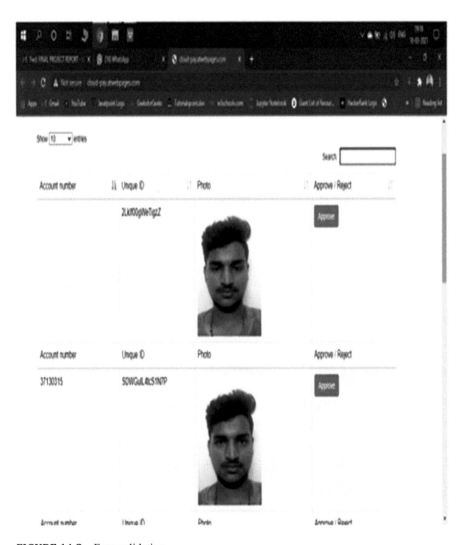

FIGURE 14.8   Face validation.

14.7.2   DISADVANTAGES

- Lighting may be the disadvantage of the system;
- Implementation in the real time, to change entire process may be difficult;
- Training time may take longer for the large dataset.

14.8   CONCLUSIONS AND FUTURE SCOPE

In this work, an authentication for online payment was created, which is a mobile application-based authentication using biometrics that allows users to complete financial transactions. The data used for authentication is face identification or verification which is used to grant transaction authorization. The method entails using an AES algorithm that uses 256-bit for the biometric encryption and decryption of face in conjunction with QR (quick response) biometric authentication for online payment. The transaction licenses used to counterfeit biometric face data include QR Code and also face authentication cards. The face data information has been stored on the cloud server so that it is easier to control and succeed. The Android OS with a minimal portion of API (application programmable interface) and to build mobile app authentication Level 22 Lollipop is used. As a result, it was determined that there are two key implementation processes: registration and transaction. Here, double authentication is used because of this operation and it can help to prevent cybercrime and provide a safe payment transaction.

Mobile application-based online transaction authentication are utilized for commercial transactions and consecutively online access. For transactional purposes, dual authentication was implemented. The first uses a QR code, and the second uses facial recognition. The 256-bit AES method is used to encrypt and decrypt biometric facial authentication data. This application had a 95% accuracy rate. Each transaction takes about 53.21 seconds to process. Incorporating blockchain technology for secure and encrypted data storage could improve this strategy in the future. This ensures that the transaction will not be tampered with. In the current situation, the person must proceed to the counter and place his or her face in front of the computer or any other electronic device where the database is stored. This work could be expanded by including it in a shopping mall where payments for items purchased are made automatically. In the exit, the face may be checked and authenticated, and the bill payment can be completed automatically. Compared to traditional approaches, this strategy saves a significant amount of time. Unauthorized access is impossible with this facial recognition system because it is incredibly secure.

KEYWORDS

- **application programmable interface**
- **convolutional neural network**
- **FaceNet deep learning algorithm**
- **QR code**
- **quick response**
- **support vector machines**

REFERENCES

Agostinho, M. X., Sritrusta, S., Amang, S., & Hasan, B., (2019). Mobile platform biometric cloud authentication. *INTEK Jurnal Penelitian., 6*(2), 75–84.

Ahmad, A., Surya, M. N., & Fairuz, A., (2016). Implementation of cryptography algorithm for biometric payment. *E-Proceeding of Engineering, 3*(1).

Baptiste, H., Julien, M., Marc, P., & Christophe, R., (2008). Face authentication for banking. *First International Conference on Advances in Computer-Human Interaction (ACHI)* (pp. 137–142). Sainte luce, France.

Chellapa, R., Wilson, C., & Sirohey, S., (1995). Human and machine recognition of faces: A survey. *Proc. of the IEEE, 83*(5), 705–741.

Chinchu, S., Anisha, M., & Mahesh, B. S., (2017). A novel method for real time face spoof recognition for single and multi-user authentication. In: *2017 International Conference on Intelligent Computing, Instrumentation and Control Technologies (ICICICT)*.

Chung, H.-Y., Hou, C.-C., & Liang, S.-J. (2017). "Face detection and posture recognition in a real time tracking system," *IEEE International Systems Engineering Symposium (ISSE)*, Vienna, Austria, 2017, pp. 1-6, doi: 10.1109/SysEng.2017.8088265.

Davis, F. D., Kottemann, J. E., & Remus, W. E. (1991). "What-if analysis and the illusion of control," *Proceedings of the Twenty-Fourth Annual Hawaii International Conference on System Sciences*, Kauai, HI, USA, vol.3, pp. 452–460, doi: 10.1109/HICSS.1991.184174.

Dileep, K., Yeonseung, R., & Dongseop, K., (2008) A survey on biometric fingerprint: The cardless payment system. In: *2008 Second International Symposium on Biometrics and Security Technologies*.

Dileepkumar, & Yeonseungryu, (2008). A brief introduction of biometric and fingerprint payment technology. In: *2008 Second International Conference on Future Generation Communications and Networking Symposia*.

Faqih, K. M., & Jaradat, M. I. R. M., (2015). Assessing the moderating effect of gender differences and individualism-collectivism at individual-level on the adoption of mobile commerce technology: TAM3 perspective. *Journal of Retailing and Consumer Services, 22*, 37–52.

Hee-Dong, Y., & In Choi, (2001). Revisiting technology acceptance model with social influence factors. *PACIS 2001 Proceedings* (pp. 509–523).

Huang, R.; Tlili, A., Yang, J., Chang, T.-W., Wang, H., Zhuang, R., & Liu, D. (2020). Handbook on Facilitating Flexible Learning During Educational Disruption: The Chinese Experience in Maintaining Undisrupted Learning in COVID-19 Outbreak; Smart Learning Institute of Beijing Normal University: Beijing, China.

Johnson, M. H., Atsushi, S. & Przemyslaw, T., (2015). The two-process theory of face processing: modifications based on two decades of data from infants and adults. *Neuroscience & Biobehavioural Reviews, 50,* 169–179, ISSN 0149-7634.Phil, T., & Timothy, F. C., (2013). *Mobile Biometrics: Combined Face and Voice Verification for the Mobile Platform.* IEEE, Books.

Maria, M. P., & Petrou, C., (2010). Image Processing: The Fundamentals, (2nd edn.). ISBN: 978-0-470-74586-1 May 2010. Wiley Publications.

Nikita, B., & Vibha, P., (2018). Face recognition system for access control using principal component analysis. In: *2017 International Conference on Intelligent Communication and Computational Techniques (ICCT)*. Manipal University Jaipur.

O'Regan, M., & Norfolk, L., (2020). Biometric technologies at music festivals: An extended technology acceptance model. *Journal of Convention & Event Tourism, 22*. 10.1080/15470148.2020.1811184.

Rafael, C. G., & Richard, E. W., (2018). *Digital Image Processing* (4th edn.). Pearson Education Inc.

Rafael, C. G., Richard, E. W., & Steven, L. E., (2010). *Digital Image Processing Using MATLAB* (2nd edn.). Tata Mcgraw hill education Pvt. Ltd.,

Richard, M., Diego, N., Santiago, R., Paul, R., & Diego, P., (2017). Face detection and classification using eigenfaces and principle component analysis: Preliminary results. In: *2017 IEEE International Conference on Information System and Computer Science (INCISCS).*

Selvia, R., Ichsan, T., & Gitarja, S., (2017). *AES Algorithm Implementation (Advanced Encryption Standard) 256 Bit and Compression Using the Huffman Algorithm in the Voice Recorder Application.* SENTER 2017.

Turk, M., & Pentland, A., (1991). Eigenfaces for recognition. *Journal of Cognitive Neuroscience, 3*(1), 71–86.

Varthika, M., & Deepak, P., (2015). A fascinating territory approaching edge detection using feasibility of eigenface to identify an individual. In: *2015 2nd International Conference on Advances in Computing and Communication Engineering.*

Ziegler, M., Cengia, A., Mussel, P., & Gerstorf, D. (2015). Openness as a buffer against cognitive decline: the openness-fluid-crystallized-intelligence (OFCI) model applied to late adulthood. Psychol. *Aging 30*, 573–588. doi: 10.1037/a0039493.

CHAPTER 15

Artificial Intelligence Applied to Computer Forensics

MANISHA VERMA

Assistant Professor, Department of Computer Science and Engineering, Hindustan College of Science & Technology, Mathura, Uttar Pradesh, India

ABSTRACT

Most personal information has moved to computer formats in recent years, including books, films, photographs, medical data, and genetic material. Devices are widely accustomed convert PC information into laptops, tablets, good phones, wearable systems, which have become an integral part of our daily lives. As a result of these changes, we tend to might tend to become simpler. PC rhetorical investigation permits lost or deleted things for pretend digital tools. Still, the prevailing human and executing fund aren't sufficient for managing cybercrime. Sadly, many modern methods of digital research require a high level of user engagement with people, which slows down the process. The scientific subject of ML has dominated the field of AI. It uses programming to understand human performance. The aggregate of ML and automation of digital analysis is a winning combination. In various stages of research, this approach provides many opportunities to assist applied scientific researchers. The purpose of this section is to provide the results of research into computerized intelligence investigations, which identifies gaps, challenges, and potential problems in the field. And while we tend to argue that the present facts of facts, especially as they are used by law enforcement, are not sufficient for various criminal investigations. The increase in crime, which is why the complexity of cybercrime, as well as the

Advancements in Cybercrime Investigation and Digital Forensics. A. Harisha, Amarnath Mishra, & Chandra Singh (Eds.)
© 2024 Apple Academic Press, Inc. Co-published with CRC Press (Taylor & Francis)

limitations of procedures and human resources in dealing with cybercrime, poses a growing need for computer investigators to use computer forensics and cybernated inspection to produce the outcome. Then it is easy to deal with the difficulties, more market resources should be used, and current forensic instruments should be developed in addition to their strengths and challenges. We tend to insist that complex procedures are necessary and should be applied ahead of time. The ability to quickly detect a huge amount of data in the search for relevant evidence during a criminal investigation is essential to the performance of computer testing. For this reason, efficient use of resources is required, although there are limits to the strategies used in the past. We show how AI can be useful using a multiagent system and case-based thinking (CBR) can be applied to computer forensics (Steve, 2006). This strategy utilizes highly specialized agents that incorporate information into the technical area of the observers. Their purpose is to investigate, research, and correlate the information included in the interrogation evidence and, it will provide the most interesting facts about analyst, which will lesser number of items to be examined individually. This chapter explores and explores the benefits of applying AI principles and methods to computer intelligence forensics and intelligent forensics, as well as the need for such tools and methods (Bruschi & Monga, 2004). It is suggested that through the use of modern digital investment methods, research, the opportunity to address the challenges and high-level sectors in which cybercrime occurs.

15.1   INTRODUCTION TO ARTIFICIAL INTELLIGENCE (AI) IN DIGITAL FORENSICS

Digital forensics is an emerging field that assists organizations in detecting and preventing breaches of privacy. It is relatively new compared to many traditional data protection environments, such as authentication and access control, and was born with the need to identify and exploit the electronic system in a way that will be acceptable to the legal system (Beebe & Clark, 2005). There are several categories of digital forensics, such as computer, network, and embedded forensics. Each strives to understand their technological platform so that they can use the evidence they have obtained. Extracting evidence from hard drives and dynamic media, for example, computer forensics has developed tools, methods, and processes. Understanding the nature of file systems has taken great effort to ensure that all art objects are accessible, and that information is valued. The state of the basic construction of embedded forensics, such as mobile devices or game consoles, requires the use of special tools and techniques to produce important art in a meaningful way. To date, law enforcement and the identification of traditional crimes have been key indicators of the use of computer forensics. This eventually led to cybercrime, but it remains largely ahead of law enforcement and their need to analyze systems in a legally authorized manner to investigate those responsibly. Because this trend has not changed, businesses are increasingly focusing on the importance of developing computer forensics skills.

Artificial intelligence (AI) is used in computer forensics to maintain an accurate series of storage while identifying, collecting, and analyzing digital data while maintaining integrity. The ever-increasing number of malicious cyber acts involving the use of digital devices and services is one of the major reasons for the use and acceptance of computer forensics.

Performing dangerous cyber operations using AI has become more widespread in government and corporate systems over the years and relies on information and communication technology. Harmful cyber-attacks include hacking, data leaks, data breaches, infringement of information, viruses, and viruses.

There is rising worry that the digital investigation technologies, techniques, and procedures are incompatible with the technology used by criminals to conduct crimes. So, many important issues are facing cyber researchers are that it is not always clear where the digital evidence will be at the beginning of the investigation (Gogolin, 2010). Even any proof is obtained effectively, so it will be challenging to determine whether the available evidence will be useful in the inspection. Cybernated proof will be instrumental in exploring the variety of violations, including child abuse, document fraud, tax fraud, and even terrorism. The steady increase in computer storage capacity, as well as its widespread presence in all aspects of human life, has led to an increase in the demand for such tests, as well as the amount of information that will be tested. In addition to the current issue, current forensic technology is not enough when it comes to analyzing comprehensive data and combining results. Because of this, the work of forensic computer specialists requires excessive time. Because many forensic tools do not have the functionality of a dispersed process, the process resources needed to try such tests are also troubling. We aim to provide a tool to assist specialists during special forensic tests to obtain better results than those currently produced by the technology, taking into account three factors:

- The reduction of routine and repetitive analyzes, and the amount of evidence to be examined by an expert;
- the relationship of the evidence; and
- allocation of procedures.

15.1.1   MACHINE LEARNING (ML) AND ARTIFICIAL INTELLIGENCE (AI)

It is necessary to look at what AI is, how strategies can help determine computer forensic issues, and how these methods differentiate themselves.

15.1.1.1   ARTIFICIAL INTELLIGENCE (AI)

AI is the study of developing good things or the ability of machines (such as visual perception, NLP, etc.), to perform human activities. The main point is that AI is not the same as machine learning (ML) or intelligent things. AI can be considered anything that can make human activities and make them easier. The use of AI technology is growing during the day, and its widespread discovery will greatly increase the number of destructive actions. The term "smart agent" refers to computing (AI) systems. Intelligent agents are used to communicating with their surroundings. The agent uses this technology to locate areas through its senses, after which it will take steps to affect the government through its sensors. How the senses are used to collect information and how it translates into actuators are key elements of AI technology; this could be how inter-agency activities could produce these results. The final objective of AI is to develop a machine that acts equally to human being. This task will use training approaches or training data or different algorithms, designed to build a diagram of the neural network studies for understanding the human being mind. AI technology offers many benefits and has a promising future. However, this technology will inevitably be used to commit some serious crime that could harm the public.

15.1.1.2   MACHINE LEARNING (ML)

ML refers to the ability of a computer program or algorithm to obtain large amounts of data and to make predictions or to draw conclusions as a result. Ml examines past data and future data. The ML algorithm can be learned in two ways: supervised and supervised (Jiang, Liu, & Chen, 2019).

ML is one of the AI methods that use a program that can learn from previous data alone, that data doesn't work for AI chore like mimicking any person's behavior, but it should reduce human activity and time spent on complex and even simple tasks. ML may be thought of as a program that learns by example and information rather than by programs. ML is defined as a process that learns continuously and makes decisions based on information rather than planning. ML is a new technology being developed to create new computer jobs and is being used commercially and scientifically. ML supports a variety of independent solutions in medical science, robotics, engineering, and other fields. Computer Forensic Science is a branch of computer forensic science.

There are three types of surveillance: surveillance, surveillance, and consolidation. In which Strategies will use the traditional or trained algorithms are monitored that are not ML. Learning reinforcements, both supervised and unsupervised, are complicated.

15.1.1.2.1   Supervised Techniques

Model of the trained algorithm is surveillance reading. Trained dataset model will be clear for implementation purpose. That data is presented show up labeled examples. The data is sometimes referred to as "trained data." Individually, various learning algorithms can be fed to those pairings such as labels. As a result, algorithms can anticipate a label for each model. It will also deliver critics and provides correct answer. At the last stage, the any pattern will be trained using a large number of previously trained data (inputs). So, it is a quick, straightforward process. Algorithms are designed to learn over time to get closer to the existence of relationships between samples and labels. Qualified supervised education will see brand new data that has never been seen before and predict its excellent label. The most widely used and simplest method is to use surveillance. The most popular way to learn a machine is to read surveillance.

15.1.1.2.2   Unsupervised Techniques

The uncontrolled reading format is not well organized. Training details have no purpose. That method (unsupervised) must find its interpretation of the facts provided. It has no labels on it. Because algorithms are tired of dealing with big data, a tool is provided to help them understand data structures. The purpose of the program in this case is to find out how to collect and organize data in the same procedure and how one can do that. Somehow, unattended reading attracts a lot of attention. This type of technology can benefit industries in the sense that we have a terabyte of unlabeled facts and editing that detail will be useful to business and benefit of editing it with little or no human work.

15.1.1.2.3 Reinforcement Techniques

Learning reinforcement is different from targeted and unattended machine reading. The presence and absence of labels may be related to the relationship between supervised and unsupervised. Strengthening learning, moreover, learns from mistakes. When you use reinforcing learning trained algorithms for any pattern, then it will get some critics then it will learn from element and take action and getting improvement. Good and bad behavior can be strengthened by algorithms. The algorithmic law will learn over time to make fewer mistakes than before.

15.1.1.3 FORENSIC EXAMINATION

Strategic computer program research involves a series of steps aimed at storing, collecting, and analyzing evidence obtained from a computerized approach that is a way to represent, facts for criminal acts. Such as, we may look at the intrusion of an Internet server, which could lead to the removal of visible information or illegal access to personal data, among other things.

As a result, human resources and equipment are often used in a variety of ways. Experts are not always able to predict which evidence will prove to be invaluable in investigating crime in real forensic computer tests. Think of a coffee shop or other situation where a large number of computers appear to have the same IP address. When you combine evidence of fraud with multidisciplinary businesses and consumers, the same problem occurs more frequently. Preliminary analysis of suspicious machines will reduce the number of assembly machines in these cases, and reduce the time required to complete the spy tests. The problem is the lack of smart tools to assist pre-analysis counselors, which results in the collection of bulk equipment to be analyzed, many of which will not affect the results of the investigation and will only increase the time the process required to complete the tests. This study looks at the need for smart tools and increases the use of mechanical resources during forensic testing. Given the large amount of data collected and forwarded to the intelligence laboratory for testing, we believe that it is not possible for intelligence coordinators to perform only analytical content analysis, let alone cross-analysis, given the lesser time and provided trained data or information available. In the best scenario, specialist should test pattern or model for trained facts.

Main ideal situation, that is not uncommon, as computers are heavily connected to networks when exchanging knowledge or facts. Additionally, the high energy range of magnetic disk or magnetic tape adds to the confusion (Qureshi, 2008). Due to the lack of resources to assist with counter-investigations, these removed computers and media were tested separately. As a result of the above-mentioned problems, a significant amount of evidence may be lost during this forensic examination. This can be an obstacle to not only the maintenance of the forensic computer but also to the response to network incidents. There are some important in network forensic or computer forensic that consider the objective of the research assignment, such as:

- How to collect and integrate expert data in a wide range of practical approaches;
- Manually ensure that these are required data or facts from the trained dataset will be collected.

   Finally, we usually use the outline in sequence. It provides a multi-level platform used to conduct digital research. The main objective of this structure for purpose-oriented sections, which can be used in several releases, with more details added as needed. The development of the special intelligent frameworks used in our product is guided by these objectives based on objectives. We often realize that objectives are based on cases, meaning that various ways of investigating various test data. That will allows us to use case-based thinking (CBR) is looking at actions of our learning agents.

- The file path agent has a group of folders in its database that are wonderfully used by many programs that may be interested in research, such as peer-to-peer, current electronic communication approach.
- File titles (the first eight bytes of a file) are checked to see file extension will match or not. This agent can detect if anyone changes the extension of file and to conceal real purpose. It also follows the most commonly used startups and file names, similar to those used by digital cameras.
- A timer analyzes the creation, access, and conversion dates for events such as system and software installation, backups, application usage, and other actions that may be related to the investigation.
- Windows Register Agent checks the Windows register and releases useful knowledge such as date of app installation, location setup, magnetic disk or tape details, etc.

System design makes it easy to install new agents. We will be able to expand our customer base in the future by wrapping up the functionality of many command-line tools available behind a clever specialized agent. The distributed platform has shown promising results, showing that the system can fund a large number of equipment before the cost of communications and communication disrupts distribution benefits. This dispersed feature allows the United States to develop a new and more cost-effective analysis of computer resources in the form of a single digital computer in a short period. In the age of computer simplicity, these could be a dead market within a spy laboratory. Because it currently has well-known diagnostic software, we hope to create agents that perform several analyzes such as finding and recognizing or searching keywords in a widely distributed way. Although integration options are limited, they demonstrate the ability to identify important evidence that cannot be reported if the information is not reviewed on its outside of the analysis phase. Integration features should be continuously expanded to hide the scope of the scenarios. With the addition of new agents, this can be improved. We expect to compare other CBR methods with learning methods when performing other tests. The combination of tools and productivity gains that will bring computer forensic experts and investigators dealing with the ever-increasing amount of digital evidence already demonstrates the power of the tool and the gains that will bring forensic computer experts and researchers facing a growing volume of digital evidence (Andrew et al., 2008).

15.2   TRENDS IN COMPUTER FORENSICS USING ARTIFICIAL INTELLIGENCE (AI)

Trends for computer operators using AI as the large range of electronic devices continues to increase, so that is per year increment the cybercrime and electronic filing cases involving the use of electronics. The annual statistics released by the FBI (Federal Bureau of Investigation) show the increment in multiple variation forensic investigations, a lot of information examined, and a lot of information evaluated each year (Garfinkel, 2010). Over the past 20 years, the global use of mobile devices has increased and has become an integral part of our daily lives. The term "smart device" refers to a variety of gadgets, including mobile phones, phones, tablets, GPS, and so on. Because of their processing power, high storage capacity, and low cost, those smart devices have gained a lot of grips. As a result, they will store a wealth of financial and personal information for users. These devices are an integral part of our lifestyle because they store personal and sensitive information about users. These gadgets, on the other hand, are vulnerable to attackers and are frequently used in criminal activities such as data theft, attacks, security risks, reconstruction of accidents, and so on. With the development of new technologies, such as computer hardware and forensic science, the number of digital crimes will increase. As a result, we have become victims of cybercrime and cyber-attacks.

"The use of analysis and experimentation, the collection, verification, identification, analysis, interpretation, and recording of processes, and additionally a presentation of computer-based proof to facilitate or advance the reconstruction of criminal activities, or the work of criminals," said computer experts. The wisdom of implants in computer forensics is becoming increasingly important in today's world. Its investigative processes make it easy to retrieve important data from a stolen device. These days entrepreneurs rely heavily on computer equipment and the Internet. It is also important to capture the required evidence from these devices. To support or refute any reason an investigator may have in connection with the incident, computer evidence must be obtained from the system. It is important to understand that there are ways to obtain computer evidence that investigators may be interested in. However, current human resources and various market resources are not enough to fully investigate devices, there are computer or digital crimes (Mitchell, 2010). In addition, current computer-based investigations and practices require extensive human interaction, which reduces the rate at which computer crimes are committed. We cover the current state of AI forensics (AF) in the in-depth computer research in this chapter. We usually provide the latest research in this chapter, as well as explanations of those methods that are importing and exporting in AI forensics (Figure 15.1).

1. **Learning that Intervenes:** Performance is based on a large number of specific facts. The details obtained are novel. It is important to get general ideas from a limited set of examples, just as it is important to get broad ideas from a limited set of examples. Experience is a word used to describe examples. The basis here would be to look at similarities in various examples. In-depth reading is aided by the methods used in these.

2. **Extract Thinking is a Way to Learn:** The details of the outpouring thinking come from well-structured mindsets. Through well-established methods, consultation is based on knowledge. Details are not new. The first knowledge, on the other hand, is born. New knowledge cannot undermine previously acquired knowledge or its basis in mathematical thinking.

3. **Big Data:** There are many samples or pattern of big data or issue finding the truth? Basically, "big data" usually refers to the challenges of processing huge amount of data sets that are collected at a

limited amount of data and require complex and constantly evolving processes (Jacobs, 2009). So, Many website—was considered a major real life fact issues. Then analysis problem or issues must pass a 'test' of volume, speed, and/or variability which will be considered a major data challenge. This means that the database to be processed is simply too large (varying in size) to be processed efficiently and effectively (volume), that the removal of meaningful data from the database takes too long (velocity), and/or that the database contains a variety of advanced information structures, such as logs or for programming, images, transactions, etc. (various).

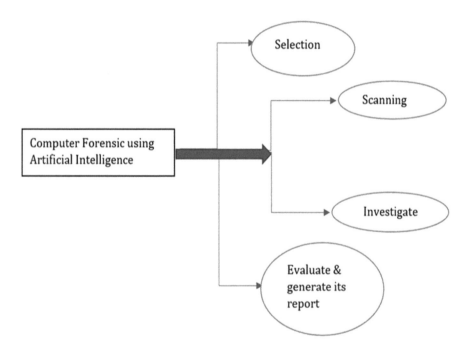

FIGURE 15.1   Computer forensic using artificial intelligence process.

The most important purpose in all cases is that data processing sets the technology on edge. We tend to suggest that the examples given here are more "big data" challenges than "big data" barriers. While digital research can address problems in dealing with large amounts of data, problems with volume, speed, and/or selection are very different for different data categories. Estimates of real time fact issues to analysis with computer forensics, that cannot be compared to other major facts complications such as CardioDX data translation. The magnitude of the challenge of finding the truth on a computer is huge, difficult to manage, but unresolved. We tend to believe that the argument is true until the forensics community is faced with a dilemma: that involves large databases, that it takes too long to produce useful results, and/or contains a variety of data formats the current tools for finding facts and strategies can handle. We are continuing to look at how the investigation will be expanded to include a variety of approaches that will provide more insight into the case and potentially contribute to the investigation. Computer Intelligence and Intelligent Forensics focus on complex digital research We believe that the computer forensics community should expand its tool and develop more effective data processing methods, especially extracting "intelligence" from sources to provide useful insight and understanding of the case under investigation. The Information Security Center and Cryptographic center has spent 15 months researching the major behaviors and profiles of cybercriminals who commit cybercrime, cyber fraud, and online auction site fraud. In addition, forensic intelligence is a well-researched field; Ribaux, for example, has produced extensive literature on the use of intelligence in the analysis of crime. It is important to distinguish between intelligence and evidence in criminal investigations. On view or real time proof is defined as "a set of markets or data that determines whether an idea or statement is accurate" and intellect is defined as "the capability to gather and use understanding and ability." that refers to "gain power" and is the definition of computer or system intelligence; the second refers to the acquisition "intentionally." While the second definition focuses on military and political intelligence numbers, it will be used in most cases and will be extended to various areas, such as business and legal. Spying, in the form of gifts, is relevant information that may or may not be evidence of a blatant or previously unknown crime. It has mandatory brands, for example, it has to have value, and for that to happen, it has to — as Ribaux says put it on time, correctly, and usable. The term "computational intelligence" seems to have several meanings. According to various scholars, the ability to under-stand, analyze, and apply technologies gives business managers a valuable opportunity. As a result, he is referring to computer intelligence as a way to make a profit and compete. Stanhope's definition is slightly different, and he suggests that computer intelligence is defined as: "capturing, managing, and analyzing facts to produce a complete perspective of the cybernated client understanding that enables computation, development, and implementation to sell strategies." of customers, and Stanhope's definition is focused on busi-nesses and customers. Many types of data collected provide input into his or her computer technology decoder in this way ranking, observation, online mail, advert, agreement, and internet community that are few examples of online information and business entries used and stored before analysis.

1. **Computer Forensic Espionage:** Integrating the two words yields a computer definition of forensic intelligence: in which information is gathered through forensic analysis and includes the number of investigative data obtained from law enforcement agencies or other entities through the use of conventional methods and investigations. There are many examples in the field of forensics, such as UK National Deoxyribonucleic Acid Data (NDNAD).

2. **Intelligent Forensics is a Company that Focuses on Forensic Science:** Intelligent forensics is a disciplinary mechanism for resolving technological advances and allocating resources in a highly efficient manner. The term "smart forensics" refers to a set of tools and processes that can be used to solve a variety of problems. We tend to believe that cognitive forensics will be used to investigate particularly complex situations. Smart forensics will be used in pairs before an incident occurs, and it is effective again after it has occurred. The effective use of smart forensics aims to detect threats before they become a problem. The effective use of scientific techniques is used for general examination to deliver the additional insight that can assist in a comprehensive assessment of data sources. We will continue to provide a comprehensive overview of the importance of these strategies in computer-based research. To deal with the complexities of large amounts of computer-based information, there are several ingenious solutions available. Solutions focus on reducing the dimensions of the research, for example, using exploitation to eliminate static or unstructured facts origin or to achieve the investigative various ways to use the scientific techniques. Extensive procedures and methods are used by intelligent forensics. In the past, computer forensics relied on questions to get information. While clever forensics will to carry on with that type of strategy, additionally, scientific methods that will be used for converting facts into questions, information into questions.

15.3   THE IMPORTANCE OF AI APPLIED TO FORENSIC COMPUTERS

Computer platforms, possibly to build systems, continue to acquire and better system execution to adapt to the changing conditions within the computer or network forensics area. According to American Medical Association (2014) "Many different techniques will be used for diagnostic features, collecting during this phase the possibilities of using computing in computer or network forensics recognized by the main motive to exploring process in which AI techniques give better consequence in computer or digital forensics research" (American Medical Association, 2014). The use of intelligence in computer forensics investigate in different stages of the life cycle of the investigation– the collection of electronic information, and provide protection of electronic information, inspection of computer base clues (Mitchell, 2010). That categories has the ability to update the data of computer forensics investigators are essential to the success of any inspection. However, it is hoped that the use of computers in computer forensic investigations will provide useful tools for the investigator to handle complex issues and, most importantly, will address issues related to speed and volume (the amount of information investigated instead of backlogs that may be a separate issue) of digital investigation cases. That method previously removed inactive files and "static" program files from digital search to some extent using hashing algorithms (Golden, Richard, & Roussev, 2006). If the view is made that the information used by the digital investigator in the investigation remains formal, it would be common to create a representation of information (computer forensic data to be consulted).

There are many challenges in applying computer science to computer forensics that is explaining the use of computer science algorithms within computer forensics is a thought process. There are two aspects to this: legal and IT. Legal offenses include acts that violate the law of a given authority, such as underage drinking or driving. Comprehensive computing has unusual computer device features, for example, a field that contains data in an unusual disk space; unusually built-in data packets; data outside normal limits, either in excessive data distribution or stored on static data storage; personal data of the relationship pointing to an unexpected relationship. The discovery of such differences incorporates a whole range of AI methods. According to Bruschi & Monga (2004), "Knowledge-based programs are designed to capture the legal expert's understanding of legal principles and to be able to demonstrate unusual behavior" (Bruschi & Monga, 2004). Artificial neural networks (ANNs) can be trained in the ethical/appropriate categories are also suitable for modeling the performance of various users to easily show unusual usage patterns to the current user for data processing and machine techniques will detect patterns of behavior.

In forensic computer investigations, it comes from AI to generate large amounts of information, analyze data to detect any criminal activities and risks, and separate data to detect criminal activity and behavior. Intelligent intelligence programs cannot make true learning skills and become real. Computer forensic through ML that the latest trends in capturing AI capabilities as capabilities to lead security solutions. Trained model act analytics that core is part of modeling, printing, forecasting, production, advertising, business knowledge and has recently implemented process. It will detect criminal behavior, MLF (forensic learning) uses wireless networks using cloud computing. The trained model objectives are therefore to give new facts and skills and to deliver structured knowledge structure, and provide continuous improvement in its operations. From AI algorithms may accustom to analyzing large amounts of information to identify risks, classify data and identify criminal behavior. AI algorithms enable researchers to search for widespread data sets that are embedded in the computer cloud. According to Sachan, Yang, Xu, Benavides, & Li (2020), summarize, AI algorithms consist of ideal software for identify model that typically analyzes a lot of information that tends to predict something morally (Sachan et al., 2020). AI algorithms seek to be informed of historical concepts that are accustomed to predict future performance and gain the ability to accept patterns of criminal activity using AI algorithms, to obtain information on the history of when and where these crimes will occur. Malicious activity from a set of extracted data will result in hacking, hiding, or hacking. This function is obtained by formalizing and analyzing servers, suspicious tool, wireless tool, the web, various types of information detection, linking, segregation, and forecasting illegal acts. Striker have developed sophisticated and sophisticated methods of attacking systems over time. The administrator will not be ready to receive this attack at any time. Further, human technology, expertise has certain limitations, and this ends up in the fact that the industry lacks the fast pace of events, long delays in detecting and preventing malware, trojan horse and it takes the latest technology to eliminate these malwares.

15.4   AI IN NETWORK FORENSICS

A network forensic is a main section of computer forensic. Basically, it is a fragment of network or digital activity. So, in that case, lots of researchers or experts to see advancement way to get authenticated data that consist of clues or evidence of a crime, that's why we use AI techniques in Network forensic that provide confidentiality of information in respect of law-breaking or digital offense.

There are two main techniques that is ANN and support vector machine (SVM) is used in Network forensic on base of AI. Both techniques provide integrity, confidentiality, and classified data regarding any digital crime. SVM is a more powerful technique to investigate of any digital offense comparison to the ANN.

In last few years, day by day increasing the digital offense so we use the AI techniques in Network forensic for investigate it. So, its main focus is on internal and external network attacks. Internal attacks mean unauthorized access to the network, and losing your secure data without giving any information. External attacks main focus on encrypting and destroying data, like network vulnerabilities, worms, and phishing, Man in the middle attack, ransomware attack (Figure 15.2).

15.5 PROSPECTS FOR THE FUTURE

The trends of the past years show that the nature of cybercrime and the use of computer crime in computer investigations are different and growth scale to be able to deal with cybercrime managers—identifying, compiling, recovering, analyzing, and documenting, there is a need to think of simple and effective computerized procedures and procedures. The event of the latest technologies and areas of digital offense, like advancement technology to get clue on base of offense and the proliferation of social networking platform, and the widespread usage of mobile security technology mean that there is a need to think about tools and techniques for detecting computer forensics (Lai, Chow, Fan, & Chan, 2013). In this chapter, we conclude address present and future digital challenges of digital offense or network forensics; there is a need to increase the use of available resources and to overcome the skills and challenges of the tools currently in use. We can say there are a large number of smart tactics can be used in computer surveys to magnify the investigation in terms of time and effectiveness. So this chapter outlines contingency available through the use of computer science principles and techniques in computer forensics and clever forensics that due to the use of smart methods in computer research has the potential to address the challenges of large and complex sectors where cybercrimes occur. From research, it has been shown that there are many challenges that legal professionals can face when experimenting. Firstly there is a huge increase in clarity within the data due to cheaper storage devices like hard drives, CDs, USB sticks and unlocked. This makes it almost impossible for people to do forensic in the short term of their time. As a result, it is almost impossible for forensic experts to perform accurate data analysis for each machine and perform analysis in each machine's process. That reduces the

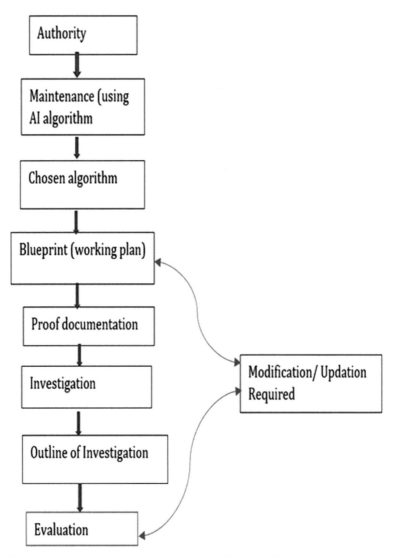

FIGURE 15.2   Network forensic process using artificial intelligence.

efficiency of human services. During this consultation period, most of the information should be sent to the laboratory for forensic purposes promptly with the available resources. In real-time computer investigations, it is very difficult to see in the early stages what evidence is most important and relevant to a criminal investigation, for example, when we look at a cybercafé or a computer network where multiple computers share the same IP address. On the other hand, smart tools are a big part of Forensic. However, these tools also reflect the issue of investigation within the pre-analysis phase. For that reason, the lack of data collection from distributed equipment should be assessed. Many tools are available to help resolve this issue and even extend the investigation time that need to develop intelligent methods and tools so that an automatic investigation of the suspect's equipment or dangerous activity can be analyzed and determined promptly. Information may be stored anywhere for malicious purposes. Therefore, AI Forensic strategies are easy sources for storing, testing, and using this data thanks to highly productive expectations and risky tasks. According to Mitchell F. "Where AI Forensic methods can perform Meta-analyzes of meta-information from a variety of sources and can simplify complex tasks into understandable and manageable data formats in the shortest possible time" (Mitchell, 2010). AI Forensic can provide a well-built repository that will contain well-defined data for computer-based investigations with known effects and consequence.

Solutions for ML network forensics should include the following features:

- Have data availability to support modeling;
- Address difficulty and strategies that have been put to good use;
- Explain the process of consultation;
- Structure systematically arranges the data structure;
- Have a systematic performance test;
- Include current builds, tools, and applications.

Finally, the intelligence of implants in computer forensic allows the investigation process to be carried out in a short period with high reliability.

AI is imitated behind human ingenuity and is intended to aid the hands-on process of decision-making.

KEYWORDS

- **artificial intelligence**
- **big data**
- **computational intelligence**
- **decision-making**
- **forensics**
- **machine learning**
- **meta-information**

REFERENCES

American Medical Association, (2014). *Improving Care-Priorities to Improve Electronic Health Record Usability.* © 2014 American Medical Association. CCJ:14-0462:PDF:9/14.

Andrew, C., Andrew, C., Lodovico, M., Golden, G. R., & Vassil, R., (2008). *Digital Investigation: Face Automated Digital Evidence Discovery and Correlation, 5*(Supplement 1), S65–S75. Elsevier.

Beebe, N., & Clark, J. G., (2005). A hierarchical, objectives-based framework for the digital investigations process. *Digital Investigation, 2*(2), 147–167.

Bruschi, D., & Monga, M., (2004). How to reuse knowledge about forensic investigations. *Digital Forensic Research Conference.* United States.

Garfinkel, S. L., (2010). Digital forensics research: The next 10 years. *Digital Investigation, 7*, S64–S73.

Gogolin, G., (2010). *The Digital Crime Tsunami: In Digital Investigation* (Vol. 7, pp. 3–8). Elsevier: Amsterdam, Holland.

Golden, G., Richard, III., & Roussev, V., (2006). Next-generation digital forensics. *Communications of the ACM, 49*(2), 76–80.

Jacobs, A., (2009). The pathologies of big data. *Commun. ACM, 52*, 36–44.

Jiang, L., Liu, S., & Chen, C., (2019). Recent research advances on interactive machine learning. *Journal of Visualization, 22*(2), 401–417.

Lai, P., Chow, K. P., Fan, X. X., & Chan, V., (2013). *An Empirical Study Profiling Internet Pirates: In Advances in Digital Forensics IX* (pp. 257–272). USA, Springer: New York.

Mitchell, F., (2010). The use of artificial intelligence in digital forensics: An introduction. *Digital Evidence and Electronic Signature Law Review, 7*, 35.

Qureshi, A., (2008). Plugging into energy market diversity. In: *Proceedings of the 7th ACM Workshop on Hot Topics in Networks.*

Sachan, S., Yang, J. B., Xu, D. L., Benavides, D. E., & Li, Y., (2020). *An Explainable AI Decision-Support-System to Automate Loan Underwriting: Expert Systems with Applications, 144*, 113100.

Steve, M., (2006). Unique file identification in the national software reference library. *Digital Investigation, 3*(3), 138–150.

CHAPTER 16

Open Standard Authorization Protocol: OAuth 2.0 Defenses and Working Using Digital Signatures

A. HARISHA,[1] LIKHITH SALIAN,[2] ADISH YERMAL,[3] and
C. G. AJAY SHASTRY[4]

[1]*Assistant Professor, Department of Computer Science and Engineering, Sahyadri College of Engineering and Management, Managluru, Karnataka, India*

[2]*Student, Computer Science and Engineering, Sahyadri College of Engineering and Management, Managluru, Karnataka, India*

[3]*Student, Information Science and Engineering, Sahyadri College of Engineering and Management, Managluru, Karnataka, India*

[4]*PG-Student, Computer Science and Engineering, Sahyadri College of Engineering and Management, Managluru, Karnataka, India*

ABSTRACT

Only a few authorization protocols could be used to offer protected server resources to Client applications. The requesting entity must generally have a token in order to use authorization methods. The safety of the tokens is crucial because the token is in the hands of an authorized user, it plays a major function in authorization architecture. The user can use it for malfunctions if he is a person with malicious intent. This issue can be avoided if the access tokens are encrypted and digitally signed. The use of digital signature makes the concept of OAuth 2.0 more efficient in terms of security measures.

Advancements in Cybercrime Investigation and Digital Forensics. A. Harisha, Amarnath Mishra, & Chandra Singh (Eds.)
© 2024 Apple Academic Press, Inc. Co-published with CRC Press (Taylor & Francis)

16.1   INTRODUCTION

The evolution of computer technologies, as well as network coverage, has brought tremendous change in communication. This has led to the advances in the speed of the network in the system to very easy communications. People use many applications and other technologies for a faster mode of communication.

With these technological advancements, electronic gadgets became cheaper, and millions of users are using these gadgets and many web platforms for their communication using the applications which are created for these gadgets. The usage of electronic gadgets is growing exponentially in this decade because of their availability and cheaper cost. Since millions of users are connected to the internet, different types of online platforms and mobile applications are being created. We can do many things like sharing multimedia content, articles, and expressing our views on a particular topic using web platforms and mobile applications.

Applications normally require user data to create accounts on their platform. The user has to enter his personal details like name, gender, age, date of birth, and so on if the user wants to create an account. These applications usually require this user data to create an account, which would be a tedious process to enter this data every time the average user must all be established an account in the desired platforms. Moreover, better the user experience these web platforms provide an easier mode of creating an account.

16.2   USING PROTECTED RESOURCES FOR USER ACCOUNT CREATION

During the account creation process, the platform will ask the user for permission to get the necessary information from a previously created account of the same user on another platform. If the user grants permission for the new platform to obtain the necessary data from an existing account, the new platform will be able to access the data and create an account. This method simplifies the process of creating an account for a normal user. In most cases, these details are saved in a platform's database as a "protected resource."

These details may be required by the new application in order to create an account. If the new application is willing to get the required information from the protected resource, it must first log in to the protected resource's platform. It takes a user id and a password to log in. Sharing personal passwords and similarly giving out user id is not a smart idea. A user can approve the new application to access protected resources from the platform without a user id or password. The new platform will have access to the required information from protected resources after it has been authorized. For the flow of authorization and resource accession, the system requires some protocols. The Application Programming Interface (API), is a gateway between computers or computer applications and hence governs these protocols. It is a type of software interface that provides a service to other software modules.

OAuth 2.0 (Open Authentication) provides across the board, a single authorization technology numerous APIs on the web application. The FAANG companies utilize a technique that allows Enables users to share account information with third-party applications or websites. OAuth 1.0 and 2.0 are two protocols that offer this type of authorization. This protocol connects four entities. The client, the authorization server, the protected resource, and the user are all involved (Jones & Hardt, 2012).

- A client is a unit that requests access to a protected piece of data from a protected resource.
- The authorization server is in charge of allowing access to confidential information to the client.
- A protected resource is a piece of information that a client requests from the server on behalf of a user.

16.3 WHAT LED TO THE CREATION OF OAUTH?

We often see in a classy hotel, a person who is also a car owner comes with his car and usually hands over his car key to a valet. A valet may park the car in the right parking spot if he is a person with the right intention. If a valet has any malicious intent he might take the car out for a joy ride or misuse the car in every other way possible by him. In order to avoid this mishandling. There came into the existence of a valet key, which a car owner might hand it to a valet and valet can only perform a selected task and the access is limited, which means a valet can only run a car at a particular speed, he cannot open the bonnet of a car or refuel the car. This is referred to as "Delegated Authorization." OAuth operates in the same manner: a user provides an application authority to conduct activities on instead of the user; the client application may only do the authorized actions limited by the user.

With the establishment of the authorization layer and segregating the client's functions from the resource owner, the OAuth 2.0 Protocol solves these problems. The client uses OAuth to gain access to content provided by the content provider and hosted by the resource server and is given credentials that are different from those used by the resource owner. Instead of utilizing the resource owner's credentials, the client acquires an access token, which is a string containing information about the scope, lifespan, and other access factors. An authorization server issues access tokens to third-party clients with the resource owner's consent. The client employs the access token to get access to the resource server's protected resources (Boyd, 2012).

In the OAuth protocol flows, there are multiple crucial actors:

1. **Resource Server:** The resource server is another name for API server, it handles the Authenticated requests after the application has obtained an access token. The list of scopes linked with the access token must be known to the protected resource servers. If the scopes in the access token do not include the required scope to perform the designated operation, the server is responsible for refusing the request. The OAuth 2.0 specification does not specify scopes, and there is no central registration of scopes. The list of scopes is left to the discretion of the service. The server that hosts OAuth-protected user-owned resources. An API provider typically stores and secures data such as calendars, photographs, videos, or contacts (https://www.OAuth.com/OAuth2-servers/the-resource-server/).

2. **Resource Owner:** The resource owner is the user who takes part in authorizing an application that allows them to access their account. Access to the user's account by the application is limited to the scope of the authorization provided for accessing their own data resource, which is hosted on the resource server.

3. **Client:** An application that makes API calls on behalf of the resource owner and with the ability to act on protected resources. The Client id is a public identification for applications that make such requests. The client secret is a password that only the application and the authorization server know. It must be random to be securely encrypted, which means you should avoid adopting mainstream UUID libraries, which frequently take into consideration the server's timestamp or MAC address. Using such a cryptographically safe library to produce a 256-bit number and convert it to a hexadecimal representation is an excellent approach to building secure encryption.

4. **Authorization Server:** An authorization server seeks permission from the resource owner and supplies access tokens to clients, allowing them to access protected resources mediated by a resource

server. Smaller API suppliers might have similar application and URL space for authorization and resource servers as major API providers.

5. **Access Token:** Access Token required by the protected server to give access to the client. An authorization server generates access tokens, which the client keeps. Access tokens are used by apps to perform API calls on behalf of users. The access token represents a certain plea for authorization to access certain portions of end-user data. Though being in transit and storage, access tokens must be kept confidential. Only the application, the authorization server, and the resource server should ever see the access token. This could be used via the HTTPS interdependence since transmitting it over an unsecured channel would make it easy for a third-party application to cut off access (Kaur & Aggarwal, 2013).

Figure 16.1 represents the basic Client-API authorization model. This model consists of a Client who sends the requests to access a particular resource. Here, the client wants to access data from the protected resource which is possible only when the client gets permission from the resource server to use the authorization server, which later provides a token to access the protected resource server. This process is also known as delegated authorization, as the client is only able to gain access to essential credentials.

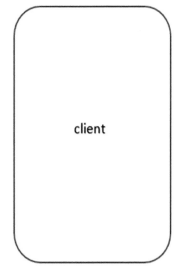 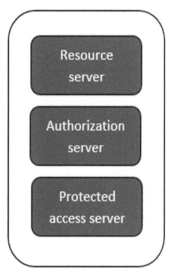

FIGURE 16.1   Client-API authorization model.

16.4   CONS AND SOLUTION FOR THE EXISTING PROTOCOL

The OAuth protocol has a number of lies. If a hacker or other unauthorized individual obtains this token, he or she may use it for malicious purposes. In the URL parameter and attributes field, an intermediary party may access the token, shared secret, and other client data. To address this issue, tokens may be signed with digital signatures and their duration can be extended. To make token security more secure. If we are able to use digital signatures to sign the tokens. The attacker will subsequently be unable to access and modify the token.

When Google first introduced the Google Calendar API, which allowed application developers to access and edit the consumer's Google Calendar, the flaws were mirrored. Giving the application his or her account login and password was the only option to grant delegate access, and Google's exclusive ClientLogin protocol would then be used by the application, because of these exclusive protocols like Client Login and standard protocols like HTTP Basic authentication, both small and large programmers have had to ask users for passwords in order to access their data. This wasn't limited to desktop programs; it was happening all over the Web, demanding certain certificates. Flickr, a photo-sharing website, was one such application. Flickr, an independent firm, was acquired by Yahoo! a few years after Google purchased Blogger. The prospect of Yahoo! requesting Google user passwords alarmed both companies, prompting the creation of new exclusive protocols to address the issue.

If an app sought access to user information, these new protocols, such as Google's AuthSub and Yahoo's BBAuth, would send users to an authorization page on the provider's site. Meanwhile, developers who were integrating numerous API providers had to learn and apply a variety of web-based authentication methods in their applications. Startups developing new APIs were not comfortable utilizing proprietary authentication systems or establishing their own authentication technology, which would have introduced many security issues. Rather, these start-ups and big API providers agreed that a uniform protocol named OAuth was needed to ensure consistency for these web-based authentication operations (Boyd, 2012).

16.5   LITERATURE SURVEY

The Architecture and control flow of the OAuth 2.0 protocol is described by Steinbach, Karypis, & Kumar (2000). The chapter focuses on detailed architecture and analyzes some of the key issues in security in the OAuth 2.0 protocol. It points out 4 types of vulnerabilities that are present in the implementation of the protocol. They include learning the user credential by the attacker with help of the wrong HTTP status code (307), Confusing the client by providing wrong information about the resource server, Using credentials the attacker may force to be logged in his name.

According to Laney (2001), the OAuth protocol has become an authorization standard, Organizations will utilize this protocol to gain access to protected resources on each other's websites. OAuth, on the other hand, doesn't ensure the privacy of protected data or the privacy of secured information. The OAuth protocol has a number of risks.

16.5.1 INAUTHENTIC SERVER AND DATA CONFIDENTIALITY

OAuth doesn't provide the authenticity of the server. During data exchange, SSL/TLS can be used to decline the chances of eavesdropping. Because OAuth creates digital signatures and secure requests using keys, tokens, and secrets, it's evident that a large part of the main goal of this protocol was to completely eliminate the requirement for the HTTPS requests. The standard seeks to provide relatively solid standards for ensuring the integrity of the request using its signatures; but, it cannot ensure the secrecy of the request, which might result in a variety of dangers. The Auth nonce attribute is a number generated at random to sign the client request. The OAuth timestamp describes remembering a duration. This can help to prevent repeat attacks.

An attacker can listen in on traffic and get access to seek parameters and properties like the OAuth signature, consumer key, OAuth token, signature technique (HMAC-SHA1), timestamp, and custom parameter data if they have network access. These values may aid the attacker in fully grasping the request and generating a packet in order to begin a brute force attack against the Server. Long, random tokens and shared secrets that are susceptible to these sorts of assaults can be used to mitigate the problem (Beltran, 2016).

Brute force attacks can be performed against the Server by the Attacker. The attacker can view the URL parameter and Attributes (such as tokens, signatures, timestamp, etc.), using these, attackers can attempt the brute force attack. Tokens plus shared secrets which are lengthy and random can be taken to solve this problem.

16.5.2   CLIENT SECRETS ARE STORED NON-CONFIDENTIAL

The goal of this protocol is to authenticate the Client (secret key and consumer key) and the User to the Server rather than vice versa. During handshakes, there is no protocol feature for checking the Server's authenticity. User queries may be sent to a potentially dangerous or malicious server via phishing or other exploits, where harmful or fraudulent payloads can be delivered. This could have a negative impact on the Users, as well as the credibility and bottom line of the client and server.

The system must safeguard: (a) server shared secrets; and (b) client consumer secrets. The OAuth signature will be computed using a shared secrets server. If the server fails to protect these two entities, the credentials will be compromised and give the attacker access to all of the secrets. As a result, it is the authorization entity's responsibility to keep client identifying and confidential information secure (Boyd, 2012).

16.5.3   OAUTH PROTOCOL IMPLEMENTATION WITH IMPROPER SESSION MANAGEMENT OF USER ACCOUNT

Even after leaving the client applications, users might remain logged in. Resource providers must log out the user automatically if the user forgets to log out after OAuth authorization. Few protected servers might auto-process the access request from a previously authorized client. This habit must be stopped.

16.5.4   SESSION FIXATION ATTACK USING OAUTH

The server might not know the integrity of the key. It may not have knowledge of whose key is being authorized.

- It examines the design of the OAuth 2.0 protocol and determines whether it is acceptable for use in the industry for the authorization of protected data. It investigates the OAuth 2.0 architecture's potential security flaws and assesses the level of security implemented on various websites K-means clustering algorithms check for the datasets and the group the datasets and makes the centroid and mean for and machine learning (ML) algorithms can then leverage the cluster IDs to enhance the analysis of large datasets. As a result, the clustering output serves as feature data for downstream ML systems (Szymański & Wegrzynowicz, 2011).

- It describes the OAuth 2.0 framework's whole architecture and control flow between various entities. It specifies the standards that must be followed when the OAuth protocol is implemented in websites (Steinbach, Karypis, & Kumar, 2000).
- It describes the difficulties that application developers confront when creating social networking systems. It describes what social bots are and how they work (Kaur, Sahiwal, & Kaur, 2012).

The architecture of OAuth 2.0 is investigated, and some serious security problems are discovered. A sort of attack that the author examines is one of the security vulnerabilities that the author examines. An assailant could do so. Partial Redirection URI Manipulation Attacks is the name given to this type of attack. The attacker can obtain the bearer token via this technique, and with this token, the attacker may be able to gain access to sensitive information and access to the user's protected data through the required identity provider (Jones & Hardt, 2012).

This explains how to use the OAuth 2.0 protocol in the construction of mobile apps, particularly for smart city initiatives. It explains how signatures in the authorization framework's design can be used to achieve privacy in applications (Liu & Xiong, 2011).

Analyzes grant types of OAuth 2.0 protocol and find 4 types of security attacks in the model. These attacks will compromise the integrity of the OAuth 2.0 protocol and provide an opportunity for the attacker to access protected information from the protected resource or identity provider. The discovered attacks are as follows: Attack to the token using HTTP redirect state 307, mixing of the identity provider, attacking by not logging out from the logged-in state, and creating a duplicate identity provider which appears as a real identity provider to the client. This explains possible solutions to all four types of problems. Gives the study of threat models and possible security considerations in the OAuth 2.0 protocol (Shostack, 2014).

16.5.5 DIGITAL CERTIFICATE AND THE CERTIFICATE AUTHORITY

For secure communication, a *Digital Certificate* is required for authentication. Users must be able to verify the identification of individuals and authenticate their own identity while interacting in order to communicate with them. Because the communication parties do not physically meet when communicating, identity authentication via a network is difficult. An unethical individual could use this to intercept messages or impersonate another person or entity. To maintain the essential level of confidence in the communication process, a method must be developed.

A digital certificate is a standard credential that may be used to validate an individual's identity. A certification authority (CA) is a trustworthy institution that issues certificates and is known as the certificate's issuer. It provides server certificates and client authentication certificates to clients and servers who request them.

This signature certificate since it contains a public key for digital signatures. If the CA is a root authority, the CA certificate might be alluded to as a root certificate. A site certificate is another name for a digital certificate. A Certificate Server is a trusted application that is frequently used. The certificate server is managed by a CA running on a secure computer. This application has access to all of its clients' public keys. Certificate servers send out communications called certificates that contain the public key of one of its clients. Each certificate is signed with the CA's private key, and a reputable CA will only issue a certificate after validating the identity of the certificate's subject to an individual or entity with whom the individual has a public key. The person or entity to whom a certificate is issued is referred to as the certificate's subject (Kaur, Sahiwal, & Kaur, 2012).

16.5.6 *THE PASSWORD ANTI-PATTERN*

OAuth is an amazing and important authorization protocol. This makes OAuth an interesting solution as we use services and platforms on a daily basis. The usage of the internet is rapidly growing.

We also provide access to some of these services and platforms so that we can securely link and move data between them. With the introduction of APIs and SDKs, application development has expanded rapidly to include not only the web but also mobile and desktop platforms. While these APIs are really useful and convenient (at times/rarely), providing this interoperable service in a way that protects the user's security becomes tricky. HTTP Basic Auth or other login-based session-related authorization routines are used by traditional HTTP services, although they normally require submitting your credentials through a third party rather than directly to the authentication/resource server. The confidential or user-specific features, services/APIs usually require authentication and authorization. While this simple login-based authentication method is simple, it poses major security risks.

One of the most major security issues with standard username/password authentication is that a user cannot easily remove access or permissions from a third-party application. In most cases, users are left with only one option: to reset their password.

Moreover, if a vulnerability or security bug is discovered in a third-party application and made public, the developers of a platform or service like Twitter can (temporarily) suspend/revoke/disable that third-party application (*OAuth Client*) to protect all of their users who may be using it until a security patch is released. Another security issue with simple authentication is the ability to limit resource and information access. There's no way to limit how much data the third-party program has access to. It's either everything or nothing.

Users want and deserve the option to grant selective access while also having the ability to revoke any unauthorized access at any time without having to enter their password. Developers want to give their users this functionality and give them access to and from a range of popular web services and platforms, but there are so many distinct authentication and authorization protocols that it's difficult. OAuth provides a standards-based solution to the table, which all community developers can and have accepted. It's the first step toward a unified, standardized specification for password-free authorization in web, desktop, and mobile apps (https://medium.com/security-operations/what-is-OAuth-and-why-should-i-use-it-5aa2f27ce387).

16.6 SYSTEM ARCHITECTURE AND DESIGN

The system's architecture and design phase will convert system requirements into a format suitable for implementation. Understanding the system architecture leads to the project being divided into submodules. If the project is broken into modules, designing it gets easier, as does identifying and correcting errors. We can simply assess the control flow and data flow between multiple entities by separating the project into sub-divisions. We can utilize this information to examine the sort of data structure and algorithms that will be employed in the system.

16.6.1 SYSTEM ARCHITECTURE

The system architecture specifies the numerous internal pieces of the proposed system as well as how these internal parts will communicate with one another to accomplish various activities.

Considering the various communication methods available, the developer can study the flow of data between different entities and analyze any form of data conversions and other structural changes that may be required during implementation. The architecture also clearly indicates which components of the system are visible to the end-user and which are run in the system's back end. The system's information front end can be effectively designed by displaying only the user-required feature.

Figure 16.2 represents the design of the system, when a client wishes to read a secured piece of information about a user from a secured resource server, the client contacts the resource holder, and with the resource holder's approval, the client examines the secured piece of information. The client contacts the authorization server after acquiring an authorization grant from the resource owner that includes the client's identity, secret, and current time. If the authorization server detects that the client is already registered with it, this generates a token and uses the client's identity and the secret to digitally sign it with the client's id and key. The client receives the token when it has been signed. This token will be passed to the protected data server by the client. The token will be decrypted by a secure data server. If the token is authentic, the resource can be made available to the client. The token will be digitally signed here, and the token will be large (Wu & Lee, 2013).

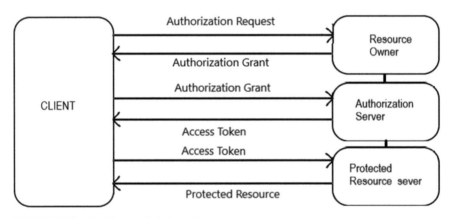

FIGURE 16.2   Architectural design of system.

16.6.2   *SIGNING YOUR OAUTH 2.0 REQUESTS*

When the API provider accepts or requires signatures, the MAC authentication specification defines how OAuth requests are signed by the client.

1. **Getting the Keys:** The client must first get a MAC key before using MAC authentication to sign requests. This may be accomplished through the OAuth authorization server. Each time the authorization server delivers an access token, the key is returned. This MAC key must be compatible with the HMAC-SHA-1 or HMAC-SHA-256 algorithms available. The MAC key might also be generated via an out-of-band operation, such as when a developer working on a project is testing a new feature. Their application is registered with an API provider. The keys must be distributed over a secure SSL/TLS connection and kept private.

2. **Making the API Request:** When integrating with OAuth-enabled APIs that need signatures, each API request must include a MAC signature in the Authorization header. To generate this signature, create a normalized request string (nonce, HTTP method, request URI, host, port, optional body hash, etc.), and perform a cryptographic signature. If an OAuth MAC signature is required, developers should strongly consider using a prebuilt library (Boyd, 2012).

16.6.3 *ACCESS TOKENS, CLIENT CREDENTIALS, CLIENT PROFILES, CLIENT AUTHENTICATION, AND AUTHORIZATION FLOWS*

OAuth was designed to manage API permissions for typical client-server web applications. There was no direction in the benchmark on how to handle authorization in mobile apps, desktop apps, JavaScript apps, browser extensions, or other scenarios. Regardless of the fact that each of these apps uses OAuth 1.0, the implementation approach is inconsistent and usually inadequate because the protocol was not built for these circumstances. OAuth 2.0 has been developed keeping a range of circumstances in mind.

16.6.3.1 *ACCESS TOKENS*

Access tokens are a form of credential that may be used to get access to resources that are restricted. As an access token, a string is utilized. A visual representation of a customer's permission. The client is usually unaware of the Strings. Tokens are unique access scopes and durations that are issued and implemented by the proprietor of the resource. Resource servers and authorization servers are the two sorts of servers. The token might be a unique identifier for locating authorization data, or it could be self-contained authorization data in a verifiable format (which means the token string is made up of some data and a signature). The access token functions as an abstraction layer, substituting several authorization constructs (such as username and password) with a single token that the resource server recognizes. This abstraction facilitates the issue of access tokens that are more restricted than the authorization grant that was used to get them, as well as removing the requirement for the resource server to understand a broad range of authentication methods. Access tokens can come in a number of forms, structures, and methods of use, depending on the resource server's security requirements (e.g., cryptographic properties) (Jones & Hardt, 2012).

16.6.3.2   REFRESH TOKEN

Refresh tokens are the credentials required to acquire access tokens. Clients receive refresh tokens from the authorization server, which can be used to obtain a new access token if the current one becomes invalid or expires, as well as additional access tokens with the same or narrower scope that must be approved by the resource owner. A refresh token can be issued by the authorization server. If the authorization server issues one, the refresh token is included in the access token. A refresh token is a string that represents the resource owner's authorization to the client. For the most part, To the client, the string is meaningless. The token is a unique identifier that is used to retrieve permission information. Refresh tokens, unlike access tokens, are only transmitted to authorization servers and are never forwarded to resource servers (Emil et al., 2001).

16.6.3.3   CLIENT CREDENTIALS

Client credentials could be used as an authorization permit when the permission scope is restricted to protected resources under the client's control or protected resources previously set up with the authorization server. Client credentials are often used as an authorization grant when the client is operating on its own behalf (the client is also the resource owner) or when seeking access to protected resources based on an authorization agreement with the authorization server.

16.6.3.4   CLIENT PROFILE

The OAuth 2.0 signifies many necessary client profiles:

1. **Server-Side Web Application:** The web server where an OAuth client application is running. A resource owner is able to access the web application, and the application utilizes a server-side programming language to make the required API calls. The user will not have any access to the OAuth client secret or any authorization server's access tokens.
2. **Web Browsers Running a Client-Side Program/Application:** An OAuth client that runs in the user's web browser and gives the users access to the application program and/or API queries. The application might be embedded into a web page as JavaScript, as a browser extension, or by using a plug-in technology like Flash. Certain API providers will not give client secrets to apps that utilize this profile because the OAuth credentials are untrustworthy and must be kept secret from the resource owner (Kaur & Aggarwal, 2013).
3. **Native Application:** An OAuth client that is practically identical to the client's program is used since the credentials cannot be trusted to be kept secret. Furthermore, the application is readily installed, and it may not have access to all of the web browser's capabilities.

16.6.3.5   CLIENT AUTHENTICATION

If the client characteristic is exclusive, both the client and the authorization server must agree on a client authentication process. The authorization server's security requirements. Any type of client authentication that fulfills the permission server's requirements is acceptable. Security requirements shall be confidential, and clients are frequently configured with the client credentials required to authenticate with the authorization system server (for example, a password or a public/private key pair). A client authentication mechanism may be established by the authorization server with the general public. The authorization server, on the other hand, must not rely on public client authentication in order to identify the client. In each request hence, the client may not employ more than one authentication mechanism.

16.6.3.6   AUTHORIZATION FLOWS

Each client profile must have its own protocol rush for obtaining the grant to access the resource owner's data. The basic OAuth 2.0 protocol defines four major "grant forms" for authorization, as well as a method for introducing innovative grant forms (Boukhdhir, Lachiheb, & Gouider, 2015).

In Figure 16.3, the communication between web API and Client. When the Client requests the resources from the API server it sends a request to the Resource Server and authorizes the client having the user credentials and requests for the resource, the API server validates the client and provides the client with the authorization code in compliance with the user permissions to the client, and the client use the redirection URL, authorization code and Access token to process the requests to the resource server.

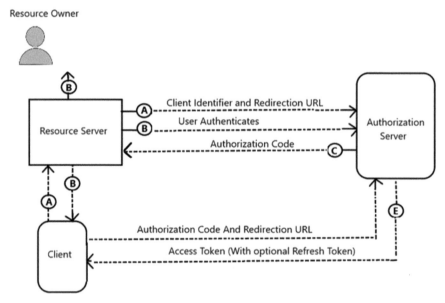

FIGURE 16.3   OAuth 2.0 client-server authorization flow.

1. **Resource Owner Password-Based Grant:** With this grant type, a resource owner's login and password may be exchanged for an OAuth access token. Mobile applications produced by API providers, except for extremely trustworthy customers, are permitted to utilize it. While the client may still see the user's password, it is no longer necessary for it to be stored on the device. Following the initial authentication,

just the OAuth token must be retained. Because the password is not stored, the user can withdraw access to the app without changing it, and the token is limited to a certain set of data, this grant type is more trustworthy than typical username/password authentication (https://www.OAuth.com/OAuth2-servers/the-resource-server/).

2. **Client Credentials:** When approval was "previously established with an authorization server," the client/user credentials grant type allows an application to receive an access token for resources held by the client. Apps that need to utilize APIs on their own behalf rather than on behalf of a particular user, such as storage services or databases, should use this grant type.

3. **Device Profile:** The device identification was set up to allow OAuth to be used on devices that do not have built-in web browsers or have constrained input options, such as a gaming console or an electronic photo frame. The user typically commences the flow on the device and is then requested to use a computer to visit a website and allow device access by providing an authorization number displayed on the device. The documentation on Facebook perfectly demonstrates this method.

4. **SAML Profile Holder Affirmation:** Using this profile, the SAML 2.0 assumption may be exchanged for an OAuth access token. This will be beneficial in business setups that already had SAML authorization servers in place to govern access to apps and data (Chen, Pei, Chen, Tian, Kotcher, & Tague, 2014).

16.6.4 PROTOCOL COMPLICATIONS

Although, with the implementation of the new OAuth Protocol 2.0, simplification and speed have improved.

Nevertheless, numerous constraints of OAuth 2.0 were added into existence:

1. **Unbounded Tokens:** For each protected resource request in OAuth 1.0, the client must provide two sets of credentials: token credentials and client credentials. Client credentials are no longer used in the latest version. Tokens are no longer restricted to a specific client type or instance.

2. **Bearer Tokens:** At the protocol level, OAuth 2.0 abolished all signatures and encryption. Instead, it entirely depends on the TLS (transport layer security) (Jones & Hardt, 2012).

3. **Expiring Tokens:** The tokens used by OAuth 2.0 may expire and therefore need to be renewed. The most substantial shift from 1.0 is that client developers must now integrate token state management. To allow tokens that are self-encoded and tokens that are encrypted that could be authenticated by the server without a database look-up are utilized, as is token expiration. Since self-encoded tokens can't be revoked, they should be limited in duration to reduce their vulnerability. With the inclusion of the token state management requirement, any benefit derived from removing the signature is now null and void (Kaur & Aggarwal, 2013).

4. **Grant Types:** In exchange for authorization grants, OAuth 2.0 Access tokens are collected. Grant is an ambiguous term that denotes end-user consent. It could be the user's true username and password, or a code retrieved when the user clicks the 'Approve' button on an access request. Grants were created with the intention of allowing for a variety of flows. 1.0 uses a single flow to support a wide range of client types. For several customer categories, 2.0 includes a significant level of competence.

5. **CSRF (Cross-Site Request Forgery) Attack:** These attacks are possible with OAuth 2.0. This is sometimes referred to as session riding or XSRF (Emil et al., 2001).

16.6.5   HIGH-LEVEL DESIGN

Architecture in which the system is designed, and the project procedure is discussed in detail. Overview of the solution, benefits supplied by the project, and system security. The envisioned system's processes are depicted in the high-level design. This level of design is feasible. Use case diagrams will be used to clarify the situation. A use case diagram is a graphical depiction of the interaction between the system user and the system overall.

16.6.5.1   GENERAL USE-CASE STRUCTURE

Using a use case diagram, we can understand the system design as well as the behaviors of the system's users. In general, UML is used to construct use case diagrams. The key components of a use case diagram are the use case scenario, the actor, and the use case. A use case refers to the number of tasks that a system user is capable of performing. The phrase "actor" refers to either the system's user or any other external system using the proposed system for its purposes. Users are split into 2 groups: major and secondary actors. The primary user is the person who will have the most direct interaction with the system. The secondary user is the one who will engage with the system in an ambiguous way. The use case diagram of the proposed system, which contains only one player by name „actor" and uses cases are 'Ask authorization from the owner,' 'Ask access token from resource server,' 'Get the token,' 'Using the access token to get the required data from protected server.' The actor is the client application that wants to read a secured piece of information from a secured place (Boyd, 2012).

Figure 16.4(a) illustrates the flow of process: how OAuth requests the resources the Client asks the permissions from the user and the Client requests the API server or Auth Server to validate the client and the user and provides an access token then the client requests the resource server to provide required resources.

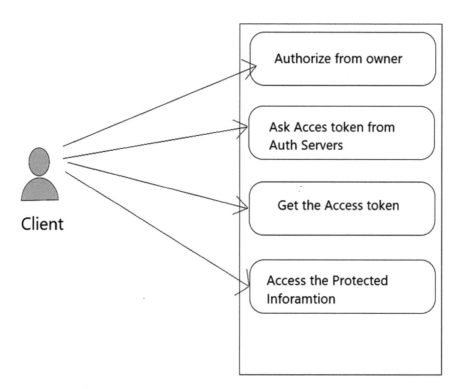

FIGURE 16.4(a)   Use case diagram.

16.6.5.2   *DATA-FLOW STRUCTURE*

A pictorial representation of data flow between different units of the system throughout its operation. It is beneficial to comprehend the overall operation of the system units and data transmission link between different sections of the system. It offers design and data flow to the system's developer so that the developer can grasp the mechanism depicted clearly. It can be compared to a flowchart, which is suitable for analyzing the data. A problem to be solved during the system design phase This includes important components like external entities, internal system operations, and databases are all examples of external entities. The external unit could be the system's user or any other system that makes use of the proposed system. It can either consume data from the system under development or supply data to the system under implementation. A process is a task that is used to process data. Start For permission, contact the owner of the resource.

Figure 16.4(b) describes obtaining a token, contact an OAuth server using the client secret, client ID, and current time. Make a token using the client's data and sign it with a digital signature. Obtain the token and use the token to contact the resource server. Get the data and then halt or save it or use a certain data set. An arrow mark represents the flow of data. The orientation of the arrow mark is critical. The direction of a data flow diagram is always in one direction of the data flow of the proposed project. In the proposed system, the client will first contact the resource owner for authorization. If the client is approved, he or she can request a token from the authorization entity. The client sends the server its id, secret, and current time. The server generates a token based on the information shared by the client and returns it to the client user. With this token, the client can access the relevant information from a secure location.

16.7   WORKING MODEL

The implementation can be done with the needed coding language by using the system's architecture and design diagrams. The project that is implemented must meet all of the user's needs and an implemented project must complete all of the tasks specified by the user. The entire functional requirements document contains the required functionalities. Implementation must be carried out in such a way that it achieves all of the application goals. The concept of implementation must be basic and straightforward to use. To build a system designer must perform the following actions flawlessly:

- Accurately arranging System units;
- Procedures that can manage the changes;

- Inspection of the framework and the necessities.

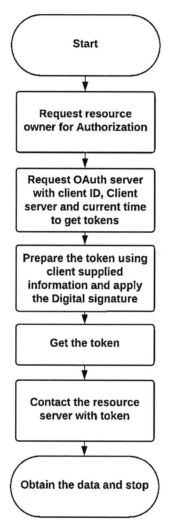

FIGURE 16.4(b) Data flow diagram.

16.7.1 FRAMEWORK USED FOR IMPLEMENTATION

XAMPP is a piece of software that may be used to create client and server applications. A developer can simply manage different databases in the system by using *XAMPP*. It is simple to create new databases and edit pre-existing databases using the *XAMPP* framework and its plug-ins, and *XAMPP* also provides a very clear view of the existing database. *XAMPP* supports

PHPMyAdmin, which is used to create website backends. Using *PHPMyAdmin*, a developer can create PHP scripts that are required for many types of database-related activities and to assist server-side programming. Programming at the server level is required to implement various security methods such as authentication and authorization, encryption, and digital signatures, among others. Websites are created using HTML and XML, however, they do not support programming, which is critical at the server end. PHP covers all types of programming needs, making it the programmers' default choice for developing server-end software during application development.

For system development, *PHPMyAdmin* can provide the following features:

- Create and update database tables;
- Create views to learn about the structure of tables;
- Extract a table from an existing database;
- It can be used to keep track of how the tables are changing;
- It supports a wide range of programming languages;
- Has the ability to create documents in pdf format;
- It can operate with databases that use the well-known MySQL extension;
- It includes ways for securing the existing database;
- It has mechanisms for backing up the current database.

16.7.2   LANGUAGE USED FOR IMPLEMENTATION

The project is implemented using PHP and HTML. HTML is mostly used to develop the user application's front end. PHP is a programming language that is used to create backend programs. PHP is utilized in this case to retrieve the contents of the database and compare those details for authentication and authorization. PHP is also used on the server to generate access tokens for users and to digitally sign those tokens for encryption.

16.7.3   ALGORITHM UTILIZED

To digitally sign user access tokens, the proposed system employs the RSA method. Tokens are created for the client, who can only access protected information if the user possesses the token.

  ➢ **Step 1:** When a client application wants to access information from a secure location, the algorithm is invoked. The algorithm is begun by the authorization server. The authorization server requires three parameters from the client. They are client id, client secret, and current time.

➤ **Step 2:** The algorithm begins by producing a token based on the parameters supplied by the client. The token will be encrypted using the public key. The RSA Algorithm is used to digitally sign it. Internally, the token contains three parameters, which are listed in the previous step.

➤ **Step 3:** The token is given to the client.

➤ **Step 4:** The client uses the tokens to request the required resources from the specified protected resource server.

➤ **Step 5:** At the server end, the token will be decrypted using the private key.

➤ **Step 6:** If the time does not exceed 30 seconds and all parameters are valid, the resource server returns the necessary information to the client.

Client parameters = (Client ID, Client Secret, Current time);
Generate token of client parameters; RSA (token); at Auth server end.
Decryption (token); at resource server end.

If information is valid, give the requested resource (Jones & Hardt, 2012).

The Web API using the Client Id, Client Secret and the current time generates a token using the RSA algorithm and provides it to the client to access the resources from the protected resource server (Figure 16.5(a)).

The token received by the client from the Web API server is then sent to the protected resource server and the token is decrypted and authorized and then the limited access to the server is given to the client (Figure 16.5(b)).

16.8 CONCLUSION

Protected resources from a protected place can be accessed by the required authorized client using tokens. Tokens are created on the basis of parameters provided by the client. The token can be secured using digital signatures to prevent intruders from accessing protected resources. OAuth 2.0 elevates the concept of social authentication by offering more security than previous methods. However, it can be more secure if all of the challenges presented by this protocol are fulfilled. Issues such as the CSRF attack should be considered while creating a new protocol.

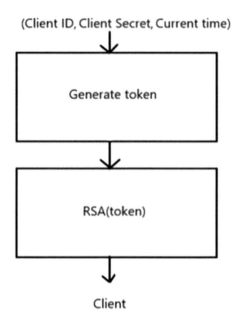

FIGURE 16.5(a)   Token generation.

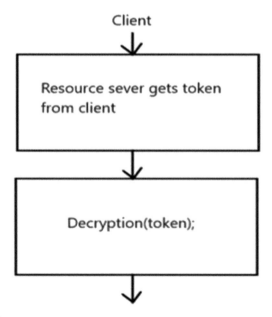

FIGURE 16.5(b)   Token signing.

ACKNOWLEDGMENT

This work is supported by VGST, Department of ITBT and ST, Government of Karnataka, funded project, The Center of Excellence Digital Forensics Intelligence GRD 853.

KEYWORDS

- **application programming interface**
- **certification authority**
- **client authentication**
- **cross-site request forgery**
- **digital signatures**
- **OAuth 2.0**
- **refresh token**
- **transport layer security**

REFERENCES

Beltran, V., (2016). Characterization of web single sign-on protocols. *IEEE Communications Magazine, 54*(7), 24–30.

Boukhdhir, A., Lachiheb, O., & Gouider, M. S., (2015). An improved MapReduce design of k-means for clustering very large datasets. In *2015 IEEE/ACS 12ᵗʰ International Conference of Computer Systems and Applications (AICCSA)* (pp. 1–6). IEEE. doi: 10.1109/AICCSA.2015.7507226.

Boyd, R., (2012). *Getting Started with OAuth 2.0.* O'Reilly Media, Inc.

Chen, E. Y., Pei, Y., Chen, S., Tian, Y., Kotcher, R., & Tague, P., (2014). OAuth demystified for mobile application developers. In: *Proceedings of the 2014 ACM SIGSAC Conference on Computer and Communications Security* (pp. 892–903).

Digital Certificate, https://docs.microsoft.com/en-us/windows/win32/seccrypto/digital-certificates (accessed on 10 January 2022).

Emil, K. F., Fu, K., Sit, E., Smith, K., & Feamster, N., (2001). Dos and don'ts of client authentication on the web. In: *Proceedings of the 10ᵗʰ USENIX Security Symposium.*

https://www.digitalocean.com/community/tutorials/an-introduction-to-oauth-2 (accessed on 10 January 2022).

Jones, Michael, & Dick Hardt. *The oauth 2.0 authorization framework: Bearer token usage.* No. rfc6750. 2012. https://www.rfc-editor.org/rfc/rfc6750 (accessed on 10 January 2022).

Kaufman, L., & Rousseeuw, P. J., (2009). *Finding Groups in Data: An Introduction to Cluster Analysis* (Vol. 344). John Wiley & Sons.

Kaur, G., & Aggarwal, D., (2013). A survey paper on social sign-on protocol OAuth 2.0. *J. Eng. Comput. Appl. Sci., 2*(6), 93–96.

Kaur, N., Sahiwal, J. K., & Kaur, N., (2012). Efficient k-means clustering algorithm using ranking method in data mining. *International Journal of Advanced Research in Computer Engineering & Technology, 1*(3), 85–91.

Laney, D., (2001). *3D Data Management: Controlling Data Volume, Velocity and Variety* (Vol. 6, No. 70, p. 1.). META Group Research Note.

Liu, F., & Xiong, L., (2011). Survey on text clustering algorithm-Research presents the situation of text clustering algorithm. In: *2011 IEEE 2nd International Conference on Software Engineering and Service Science* (pp. 196–199). IEEE.

Resource Server. Client: https://www.OAuth.com/OAuth2-servers/the-resource-server/.

Shostack, A., (2014). *Threat Modeling: Designing for Security.* John Wiley & Sons.

Steinbach, M., Karypis, G., & Kumar, V., (2000). *A Comparison of Document Clustering Techniques KDD Workshop on Text Mining.* Department of Computer Science/Army HPC Research Center.

Szymański, J., & Wegrzynowicz, K., (2011). 0-step K-means for clustering Wikipedia search results. In: *2011 International Symposium on Innovations in Intelligent Systems and Applications* (pp. 253–257). IEEE.

The Password Anti-Pattern. https://medium.com/security-operations/what-is-OAuth-and-why-should-i-use-it-5aa2f27ce387 (accessed on 10 January 2022).

Wu, M. Y., & Lee, T. H., (2013). Design and implementation of cloud API access control based on OAuth. In: *IEEE 2013 Tencon-Spring* (pp. 485–489). IEEE.

Yadav, A. K., Tomar, D., & Agarwal, S., (2013). Clustering of lung cancer data using foggy k-means. In: *2013 International Conference on Recent Trends in Information Technology (ICRTIT)* (pp. 13–18). IEEE.

CHAPTER 17

Child Pornography: The Filth of Society

ARPIT NIRVAN,[1] SWAROOP S. SONONE,[2] VINAY ASERI,[3]
PANDIT PRITAM,[3] RUSHIKESH CHOPADE,[3] and MAHIPAL SINGH SANKLA[4]

[1]*Department of Computer Science and Engineering,
I.T.S. Engineering College, Greater Noida, Uttar Pradesh, India*

[2]*Department of Forensic Science, Dr. Babasaheb Ambedkar Marathwada
University, Aurangabad, Maharashtra, India*

[3]*Department of Forensic Science, Vivekanand Global University, Jaipur,
Rajasthan, India*

[4]*Department of Forensic Science, University Centre for Research &
Development (UCRD), Chandigarh University, Mohali, Punjab, India*

ABSTRACT

The Internet plays an important role in the crime commission, as it lures
children to engage in sexual activity. Most of the children are exposed to
unknown predators including pedophiles because of communication over the
Internet, and it is found that more severe abuse was there on the younger ones.
Online sex crimes widely involve child pornography (CP) and sextortion,
which are rising as a business all over the world. Current Internet searches
specify that there is an increase in search for child pornographic material
over the Internet. The interline between online sex crimes and sexual violence
grows with online harassment, online solicitation, and sex trafficking, which
includes methods to abuse children sexually and mentally, methods may vary
from person to person, but last, the motive is the same for all that is sextortion. Worldwide, 80% of women and 43% of men reported experiencing some
form of sexual harassment or assault in their life. Excess circulation of ideas

Advancements in Cybercrime Investigation and Digital Forensics. A. Harisha, Amarnath Mishra, &
Chandra Singh (Eds.)
© 2024 Apple Academic Press, Inc. Co-published with CRC Press (Taylor & Francis)

related to sexual crimes is done in the form of magazines like "Playboy" and successful movies like "Deep Throat" in which real acts of sexual violence are presented to viewers. To combat this and other online sex crimes, countries collaborate internationally to amend the anti-pornography civil rights ordinance which can minimize these threats. The criminal justice systems of different countries fight against online sex crimes and CP by enforcing laws that will find out and identify the abuse and the predators. National, state, and local law enforcement can become much stronger if they get media attention, investigation assistance, and training programs.

17.1 INTRODUCTION

Children spend most of their time on social media over the Internet, and that invites troubles in the form of unknown predators like pedophiles. They can use children's facial images and private intimate data as pornographic content and make them the victims of child pornography (CP). Around 70 million reports of child sexual abuse material were acquired by United States authorities in 2018–2019 (Salter, 2021). After that, some of the reports from INTERPOL confirmed that child sexual exploitation activities increased at a very high speed during the pandemic (Salter, 2021). At a time when social media companies are earning profits in billions, child sexual abuse victims suffered from a lack of health care and help assistance to overcome the effects of social media (Salter & Whitten, 2021). Everyone is using social media in each part of this world, as a result, anyone can become a friend of another one without even meeting him/her in real life, which can make people suffer consequences same as sextortion.

Sextortion is practice all over the world as it is a pervasive cybercrime that also comes under the criminal acts of CP. In sextortion, sextortionist exploit and harass the victim by stealing their intimate data and threaten them to release their private images and videos. If they don't come up with their demands of money or additional sexual favors, predators either manipulate victims through social media to involve in sexting or she/he hacks the person system to acquire their intimate images and private data. Sexting had a direct impact on the mind of the victim as sextortionist force the victim to create and transmit sexual content of oneself. Sexting is a more sensitive criminal act because this can be done through any social media like E-mails, instant messaging, and social networking sites. Sexting also comes under the criminal definition of CP as this can make a victim a slave. And such kind of modern-day Slavery is known as sex trafficking.

Thousands of children and women become part of this modern-day slavery every year, and as a result, they were forced into prostitution practice and other sexual abuses. The demand for trafficking is very high among some countries because of some of the factors that are: the demand for commercial sex, sex industry expansion, destination countries, cultures. Countries oppose sex trafficking but fail to do so because there were immigration and asylum policies that support the trafficker's right to move from one country to another. These all-sexual abuse activities are terrifying and dehumanizing. Victims who fulfill the demands of sextortionist are forced to do sexual favors, sexual activities are pleasurable but done under harassment, and the threat becomes sexual assault and can have a direct impact on the mind of the victim. He/she feels a loss of control and goes into a stage of grief and develops a sense of shame (Jurecic, Spera, Wittes, & Poplin, 2018). Sexual matters are not openly discussed even in today's era and are considered to be a sensitive subject among society (Motsomi, Makanjee, Basera, & Nyasulu, 2016). Because of this nature of society victim feels fear, betrayed, angry, anxious, embarrassed, and guilty (Agrawal, 2020).

Society needs to change its mind concerning such kinds of sexual abuses. Mentally and sexually harassed people need to get proper health care and personal assistance to overcome all the effects of sexual exploitation that they had gone through. Victims need to realize that it's not their fault and they don't deserve this.

17.2 PORNOGRAPHY

Pornography is explained as the different types of sexually explicit materials, or we can call it pornographic materials that arouse and enhance sexual feelings which contain genital and sexual acts. The definition of pornography and sex varies from country to country as each country define their boundaries complexity to determine pornography.

For the first time, pornography was widely seen in England in the late 1800s. Since then, pornography branches tend to grow at a rapid speed worldwide. The Danish government was the first government that legalized the production of all forms of pornography in 1969 (O'Donnell & Milner, 2012). The 1990s era was considered to be the evolution era of digital recording of sexually explicit materials.

According to records, we can say that about 40 million users regularly come in contact with sexually explicit material, the total revenue that this pornography industry generates is about US \$97 billion worldwide (Ropelato, 2007). As given in Figure 17.1, every second \$3,07,564 (INR 228912.20) is being spent on pornography. When an individual tries to gain sexual knowledge or he/she searches online for sexual resources are instead find sexually explicit materials. Outcomes of these materials are considered to be positive as well as negative in nature depends on "How we use it."

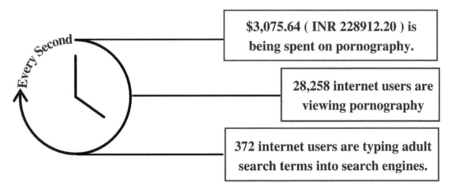

FIGURE 17.1   Factual data for pornography.

As of positive aspect, it is found that pornography can work as a source for users to get sexual knowledge. Users can use sexually explicit materials as a source of sexual information, and they can even get to know about how one's body reacts to different erotic conditions. Moreover, people who consume pornographic materials have better sexual knowledge and improved sex skills than people who do not. Pornography used by partners in a couple is accepted to be positive.

In contradiction to the positive aspect, some other studies believe that pornography focuses more on "Strange," "Kinky" and "Fetish" behavior not on the sexual health of the user. Pornographic content even led to aggression as it provides unreal sexual acts expectations (Malamuth, Addison, & Koss, 2000). Unreal like the long duration of sex, fake emotions, aggressive looks, and slavish nature of women are considered to be unrealistic and dehumanizing. Male dominance over women and exerting power also come under sexual aggression.

Males and females both appear to be similar in their motivation to consume pornography. Negative emotion is observed to be the vital factor for the motivation to use pornography. Under negative emotions, certain sub factors motivate the user to use pornography. Sub factors mainly include habitual motivation, mood management motivation, fantasy management, and relational management (Paul & Shim, 2008).

Consumption of pornography or sexually explicit materials provides just a little period of inner pleasure and happiness. Pornography never accords lifetime pleasure and satisfaction. Even on a societal level, the consumption of pornography is observed to be a major concern about public health like more hook-ups, prostitution, lesser use of condoms, extramarital affairs, etc. (Braithwaite, Givens, Brown, Fincham, & sexuality, 2015).

It is believed that people consume pornography because they do not get a better platform to gain sexual knowledge. Well-trained teachers, parents, and even the Internet can provide more realistic sex education (Gesser-Edelsburg & Arabia, 2018). At least one source of information is needed to teach accurate sexual health education and open communication needs to be encouraged regarding pornography issues who need it.

17.3 CHILD PORNOGRAPHY (CP)

Pornography creates a hedonistic spirit (Koesnoen, 1964), which makes CP a part of itself. CP is a cybercrime in which children are enticed to engage in sexual activity for the sexual gratification of unknown predators including pedophiles because it involves the production and distribution of child sexually explicit materials. CP comes under the criminal acts of fornication (sexual intercourse between people not married to each other) that violates the dignity or we can say virtue of a person (Marpaung, 1996).

As shown in Figure 17.2, CP came into light in the late 1800s. The 1970s was the starting period of CP that witnessed many children's sexually explicit films and more than 250 child sexually explicit magazines. The first child sex magazine named "Bambina Sex" was published in 1971 with nude pictures. After that, in 1984, CP was raised as a serious social issue by media, and then in 1986 CP was finally included in the criminal code with the legalizations (O'Donnell & Milner, 2012).

Before the 19th century, CP was expensive as well as difficult to obtain. At that time magazines imported to the US costs $6 and $12 and domestic magazines and photos cost $20 and $50, respectively (Lanning & Burgess, 1984). Digital CP begins in the 19th century with the introduction of digital recording to the international market. Even later on Internet advancement also changes the criminal code of children sexually explicit materials and sexual abuse. The problem of CP is believed to get even worse with the advancement of modern innovations and technologies that promotes children's sexually explicit materials.

Predators or pedophiles, those who have immense interest in CP collect child pornographic materials in the form of magazines, photos, videos, and other sexual materials. Child Pornographic materials are categorized as erotica materials, nudity materials, and sexually explicit materials (Shackel, 1999).

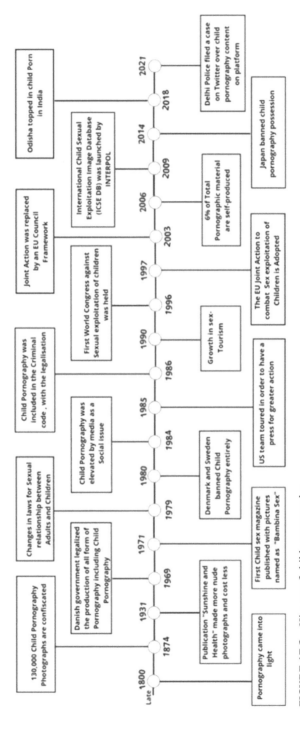

FIGURE 17.2 History of child pornography.

Erotica materials do not contain sexual acts or nudity, but pictures of children's swimsuits, underwear, and upper wear. Such type of erotica materials is not likely to come under the definition of CP. Nudity materials as the name itself say contain nudity like children's nude poses photographs. Nudity materials come under the definition of CP, but they were known as "Softcore" pornography in the adult industry. Sexually explicit materials focus particularly on the sexual acts and on the genital or anal areas of the child's body. Sexually explicit materials come under the definition of CP and are known as the "Hardcore" Pornography in the adult industry.

Lack of Social knowledge about cyber-criminal law and because of the innocent behavior, Children were abused by predators to acquire sexual benefits from them in multiple forms. In today's world, smartphones have created their cyberspace, or we can call it a pseudo world which has increased the rates of children's sexually explicit materials production and distribution as smartphones can take and share high-resolution images and can record better quality videos.

Most Internet CP cases are handled by the local police departments. Sometimes investigation that begins from one district police will almost cross-jurisdictional boundaries because there are two methods for investigation of child sexually explicit materials.

One is through nation-sanctioned Investigation that can collaborate various countries, take assistance from IT companies and non-profit organizations to punish the guilty.

Another is a personal vigilante operation performed by the people who want to assist jurisdiction law enforcement and punish the guilty ones. Child Protection acts need to be implemented in every country to protect children so that they can practice their rights and obligations (Gultom, 2014). Children's sexually explicit materials in cyberspace are not slowing down as a result more and more production, distribution, and consumption of these materials are expanding. To combat this distribution, national and international law enforcement agencies need to come together to place pressure on government and internet service providers (ISPs) to arrange much better judicial and data tools to combat CP.

17.3.1 SOCIAL MEDIA AND ONLINE SEX CRIMES

Around 93% of the initiated sex crimes began from social networking sites out of which 61% voluntarily self-create sexually explicit materials as shown in Figure 17.3. Sex crimes on social media come under the acts of voyeurism (the practice of gaining sexual pleasure by watching others when they are naked or engaged in any sexual activity). Online sex crimes include harassments include harassment in the form of sexting, cyberstalking, and cyberbullying, cyberstalking, and the production or circulation of sexually explicit images and videos without the consent of victims. Even circulation of sexual assault images and videos is a part of Online sex crimes.

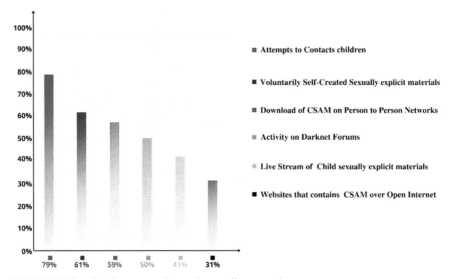

FIGURE 17.3   Graph representing various online sex crimes.

Online sex crimes come under exhibitionism, an emotional disorder that is characterized by sexually arousing fantasies. Predators with exhibitionism are known as exhibitionists, who reveal their genitals to get sexually excited and exploit the unwilling observer like woman or child of either sex over social media. Social media portrays a fictional and futuristic parallel universe called "Cyberspace." Availability of social media among teenagers and adolescents has increased the production of child sexually explicit material as social media provides wide exposure to their Pseudo world to share good quality images and high-definition videos.

Sharing genitals photographs and videos, are all are clear abuse. Sharing genitals photographs and videos, are all clear abuses. That's why many juris-dictions launched targeted criminal legalizations (Hanson, 2019), as they consider consexual sharing of sexual imaginary under CP. A perfect example of this consexual sharing is "Sexting," sharing of sexual images and videos in sexting over social media comes under the criminal acts of CP.

Social media platforms are widely operated by co-operating companies, such as YouTube, Facebook, and Twitter. In 2006, Twitter launched its social media platform which is now having 330 million active monthly users and 145 million active daily users. It is entitled as one of the most influential platforms among academics and socialism. Some of the pedophile groups spread hate speech and child sexually explicit materials over the platform, which had a direct impact on the share price and reputation of Twitter (Salter, Media, & Culture, 2018). While many self-identified victims retweet the same content to claim that it was against their rights to circulate the images and videos of their abuse, which is an online sex crime in itself. Twitter faced a lot of pressure and in 2020, a notable revised child sexually explicit policy was published by Twitter (Dodds, 2020) to combat online sex crimes in which the abusive and exploitive behavior of pedophiles is completely explained.

Another biggest social media platform with a particular focus on teenage users is TikTok. This online platform has age restrictions that can easily be bypassed by giving wrong information (Salter & Hanson, 2021). TikTok provides a platform where teenagers can upload their content in the form of short clips, even some children perform sexually suggestive or explicit songs which attracts exhibitionists and results in online sex crimes. This platform restricts or discourages privacy settings because privacy settings reduce the number of views and user-content interaction (Salter, 2021). TikTok updated its policies in 2021 after it gets banned in India in 2020. New policies include a change, that account of the user under the age of 12 are set to private by default (Peters, 2021).

Many other technology and innovations came forward to fight against online sex crimes over social media. PhotoDNA technology was introduced by professor Hany Farid in collaboration with Microsoft in 2008 in which child sexually explicit material gets automatically matched with the known CSAM database (Farid, 2017). This technology is free for all organizations to combat online sex crimes. Big companies like Google introduce this PhotoDNA technology after five years of its launch on their services. As PhotoDNA was not able to filter out child sexually explicit videos, a new PhotoDNA tool was developed in 2018 to detect child sexually explicit videos. Even Facebook and Google introduce new technologies in 2018 to detect explicit materials and to combat online sex crimes.

17.3.2 SEXTORTION

Sextortion is the union of sounds that combine the explanation of two words, i.e., "Sex" and "Extortion." In extortion, a sense of fear is created inside the victim's mind by offenders, as they threaten them to circulate their confidential private data. Sextortion is an image-based sexual abuse crime in which images were used to gain access over the victim and later on to harm them. The desired outcome of this cyber sextortion may vary differently depending on offenders, which can either be in the form of sexual favors or financial benefits. Increased access to social media and smartphones has changed how predators commit crimes. The methods and motives that were for other cybercrimes are all similar in cyber sextortion. Offenders can gain possession of victim's data by manipulating them or by hacking their systems (Liggett, 2019).

There are four factors, i.e., cyberspace, possession, and extortion in which cyber sextortion is different from other types of sexual crimes. One is cyberspace, as a predator and the victim never shared any physical space, crime is performed only and only in cyberspace. Another is, possession as offenders possess sexually explicit images of the victims by uneven means either by coercion or by using social engineering techniques. The last one is extortion, as by possessing sexually explicit images of victims over cyberspace predators force victims into acts, which can either be sexual, behavioral, financial (Açar, 2016).

As Sextortion requires a sense of power and control over the victim, offenders maintained it by creating a sense of fear inside the victim that obliges them to act according to predators. Sextortionists mentality differs from each other, and they were mostly direct in their actions. Most sextortionist use the tricks of manipulation and coercion while some believe in using social engineering techniques such as hacking to steal images from victims rather than wasting time on using grooming tricks to manipulate or create a relationship with someone.

Violent sextortionists mainly target minors as they get, manipulated easy with little use of grooming tricks, and threatening them is much easier than adults because they had a fear of punishment if they disclose anything to their parents (Mishna, McLuckie, & Saini, 2009). Whereas, pedophile sextortionists target females to gain power over them and control their life to pursue sexual benefits. Transnational sextortionists target successful men and women either by hacking or by using other social engineering tricks to acquire financial benefits.

After possessing sexually explicit images of the victim by any means, offenders may circulate them to their friends, known ones, or widely on Internet, without any permission of the victim (Bates, 2017), which can even cause psychological distress to the victim. He/she feels angry, betrayed, embarrassed, and even can go into a stage of pain and can develop a sense of humiliation. Very few of them discuss with their family and friends the threat they faced. Approximately, half of the victims tell someone about the incident due to grief and fear that they would be blamed for the situation (Zweig, Lachman, Yahner, Dank, & Adolescence, 2014). As per records, male victims are less likely to tell anyone about the incident than females (Patchin & Hinduja, 2020). It is found that cyber sextortion victims are two times more expediently to get depressed than non-victims and three times more expediently to commit suicide than non-victims (Goebert et al., 2011).

Recently, the majority of people (mainly men) are getting in touch with women on Instagram and get attracted to their fake profiles. Those women immediately ask them for nude video calls and sex chat. When victims agree with them, they record the whole video call and start blackmailing them for money and start threatening them to share the video with their relatives and friends.

Current legalizations are created only for Image-based sex crimes in which the circulation of sexually explicit images takes place. Laws that require image distribution will never assist victims of cyber sextortion, as images are not shared every time, they are only used to create a sense of fear inside victims' minds.

17.3.3 SEXTING

Texting has now become a core part of everyone's social life. People use and misuse the advancement of technology for sexual interaction and inspections. Texting which involves such types of interactions and inspections is known as "Sexting." Sexting is the production and circulation of self-made sexually explicit materials through social media over the open Internet. Ownership of personal smartphones and applications like Snapchat are meant to be one of the potential reasons for the rise in sexting activities (Van Ouytsel, Madigan, Ponnet, Walrave, & Temple, 2019). People use sexting as their first step to experience real-life sexual contact but besides this, they get involved in transactional sexting in some extreme cases, i.e., Sexting which is done in exchange for something else which can either be drugs, drinks, or money (Van Ouytsel, 2020).

Sexting becomes problematic when sex chats get published or forwarded without permission or consent. Sexts can be posted on any social media platform or can be forwarded to friends, office colleagues, and known ones of victims. Such a type of sexting is called non-consensual sexting, which is also known as revenge porn or image-based sexual abuse. There is no gender-related difference in sexting as both boys and girls equally send suggestive pictures to one another. Boys are more likely to take part in sexting at an earlier age than girls, Whereas, on the other hand, girls suffer more ample aggression than boys during sexting, Girls are exposed to more extensive intellectual upshots than boys (Roulston & Shelton, 2015). Girls mainly fall into because of the fear that they lose their relationship with the person they love (Lippman, Campbell, & Media, 2014).

Sexting is mainly done to achieve three types of purposes, i.e., sexual purpose, instrumental/aggravated purpose, and body image reinforcement. The first one is a sexual purpose, in which sexting is practiced to bring off sexually related aims, to flirt with someone as a token of love, to attain intimacy, to allure a partner, or to increase familiarity in a dating relationship. Another is instrumental/aggravated purpose, in which sexting is practiced to attain secondary aims like getting things in exchange for sexting like drugs, drinks, and money. The last one is body image reinforcement, in which suggestive images attain feedback about the creator's body type.

Recent research studies show that sexual cyberdating is directly linked to dating violence, as dating violence is observed more than intimate violence during dating (Morelli, Bianchi, Baiocco, Pezzuti, & Chirumbolo, 2016). In this type of violence, the coercive behavior of one other partner in a couple is misunderstood as a sign of seriousness and commitment. According to studies, dating violence during sexting is directly proportional to the duration of a relationship. As long-term relationship results in seriousness and commitments, which increase dating aggression and even other risk factors (Luthra & Gidycz, 2006).

Other risk factors like continuous sexting activities gradually transformed into crimes like sexual harassment and sexual bullying (Mitchell, Finkelhor, Jones, & Wolak, 2012), which had a direct impact on the physical and mental health of victims. Society only blames the one who is depicted in the sexually related materials when those materials are only shared in an intimate relationship (Hasinoff & Shepherd, 2014). Youth face more emotional regulation issues than those who are not engaged in sexting (Houck et al., 2014) they started experimented with vicious substances like alcohol, drugs, and marijuana (Dake, Price, Maziarz, & Ward, 2012).

As both laws and law enforcement differ from jurisdiction to jurisdiction, but they practice sexting to punish the guilty ones. Law enforcement collaborates with different IT companies and government sectors to launch new technologies that help to prohibit such kinds of Sexting activities.

17.3.4 SEX TRAFFICKING

The method of stealing someone's freedom to acquire profits is called trafficking, and when humans of all ages, nationalities, and gender are trafficked all over the world to gain profits in exchange for labor and sex is called human trafficking or sex trafficking. These types of criminal act not only harm human, social, physical, and psychological but also erodes the social, economic, and political facet of a nation where these types of crime occur.

According to the International Organization of Migration Data, crime rates are four to five times higher in destination countries like the Czech Republic, the UAE, Yugoslavia, the Netherlands, Greece, Germany, and Hungary. Such types of destination countries generate sex and labor demands of women and minors, which are accomplished by traffickers who trade victims by using uneven means. Approximately 20% of women are recruited through fake advertisements of foreign jobs and others through marriage agencies which help pimps to reach those women who want to travel and emigrate (Aronowitz & Research, 2001). When traffickers acquire control over victims by any kind of means, either by threatening or blackmailing them, then they exploit them to gain huge profits. Control over victims is easily possible as exploits face language barriers, limited knowledge, fear, and lack of money that prevent them from breaking out of the jail of sex trafficking.

Traffickers exploit victims both physically and emotionally to retain trafficking profitable as it is the third-largest income generated in organized crime after narcotics and arms sales. The global sex trade is the fastest increasing form of business, which costs worth $32 billion annually. In an estimation, 1 million people are trafficked across international borders annually, out of which 70% are women and the remaining 30% are minors and others. Only want to break out from sex traffic rings is a Police raid which, results in expulsion. The legalization of prostitution makes it difficult to prove pimps guilty are accountable for their crimes. Traffickers elude prosecution by speaking that woman had their own choices, they know what they are getting into, and prosecutors face difficulty in proving the difference between optional and forced prostitution.

Victims face moderate to severe physical, psychological, social, mental, and emotional abuses/tortures. In most cases rescued victims suffer from sexually transmitted infectious diseases like gonorrhea, syphilis, urinary tract infections, and AIDS, victims feel pelvic pain, vaginal/anal tearing, rectal trauma, and/or urinary difficulties (Deshpande, Nour, & Gynecology, 2013).

Policies and laws of countries need to change to protect human rights. The government needs to focus more on the recognition of victims who have been trafficked into prostitution. Aggressive intervention against traffickers involved in illegal prostitution needs to be practiced by the law enforcement agencies as trafficking harm not only human being but also the security of the nation.

17.3.5 ONLINE SOLICITATION

Online sexual solicitation is one of the online sex crimes where adolescents and adults request, force, or encourage victims to take part in sexual activities like sexual talks and sharing sexual information for money. According to WHO, online sexual solicitation can be called non-contact sexual abuse (World Health Organization), which can be wanted or unwanted by victims because it may or may not lead to offline abuse of the victim. The increase in the risk of online sexual solicitation is because of the extensive use of direct messaging, e-mails, chatrooms, online magazines, and blogs over the Internet (Mitchell, Finkelhor, & Wolak, 2007).

Factually, it is studied that adults who are known and trusted ones are more likely to engage in suck kinds of criminal acts of online sexual solicitation. They can easily have control over children by using different types of grooming techniques which include bribing, threatening, controlling, coaxing, and deception (Gámez-Guadix, De Santisteban, & Alcazar, 2018). All these grooming techniques depend on various factors like adolescents or child responses, predators' personalities, and how grooming is performed by the solicitor (Smith, Thompson, Davidson, & Gynecology, 2014). According to studies, there are four risky behavioral things of predators when they come in contact with any adolescent or child over the Internet. The first one is a solicitor will send naked/sexual photos or videos, another one is they search someone over social media platforms to talk about sex, the third one is they disclose their fake personal information like their addresses/phone numbers and the last one is they search online someone to have sex with (Noll, Shenk, Barnes, & Haralson, 2013).

Online sexual solicitation over victims can result in psychological distress and various other kinds of traumas, anxiety, fear, and poor low esteem. Even engaging in online sexual solicitation can cause other high-risk offline sexual behaviors which include multiple sexual partners and inconsistent use of condoms (Li, 2007).

Basic parenting techniques are the basic thing, that needs to be performed to overcome the effect of online sexual solicitation, in which victims can have open and comfortable communication with their parents about the situation they faced. The government needs to put compulsion on schools and colleges to teach sex education in which principles of healthy intimate relationships and principles of Internet literacy are mainly focused.

17.3.6   ONLINE HARASSMENT

Online harassment, a clear act of aggression is the process of violating people over social media platforms to cause mental, social, emotional, and psychological distress to a victim. Cases of online harassment were observed for the first in the year 2004 and since then it is emerging at a very rapid speed which results in consequences like distress, trauma, and depression, etc. Problems of online harassment increase with the development of the advancement of technologies [. Passing ill-mannered or disgusting statements to someone on social media platforms or using new technologies to harass or humiliate people over the internet are all come under examples of online harassment. Humiliating messages that predators use to violate a victim can be in the form of a direct message or form of open content over the Internet concerning a specific person. Content uploaded to an open-source platform can cause more harm than that of another form because in such cases it gets difficult for the victim to delete those nasty messages which have noisy language including bad words, word validation, and slang (Sood, Churchill, Antin, & Technology, 2012) at the time of uploading.

According to some of the studies, men are more likely to get nasty or hate comments than women. More exposure to online space is considered as one of the reasons why men are subject more to online harassment than women. Men share their opinions openly on the Internet, that is why they receive back more hate. But as per some of the other data, it is found that aggressive effects of harassment are more on women than men.

Practice to detect online harassment messages need to be adopted by government agencies and strict actions against the guilty need to be implemented by law enforcement agencies to reduce the consequences of harassment and humiliation. Manual detection of harassment messages is much more laborious that's why the solution to such types of problems can be addressed by specially designed software based on machine learning (ML) algorithms, which itself detect and remove the creepy content that spread hate over social media networks.

17.4 KEYS FOR PREVENTION OF ONLINE SEX HARASSMENTS (Table 17.1)

TABLE 17.1   Nine Keys for Prevention of Online Sex Harassments

1.	Elucidate belief with policies	• Elucidate online sex harassment.
		• Draft consequences.
		• Give examples of incidents related to sexual harassment.
		• Elucidate zero-tolerance perspective.
2.	Imply guidance and superior management	Superior management should:
		• Exhibit guidance and commitment to approaches.
		• Lay down an example by their etiquettes.
		• Enact to upshots.
3.	Settle happenings	• Inspect punctually.
		• Answer to complaints speedily and cautiously.
		• Never ignore any complaints.
		• Impose upshots if needed.
4.	Upskill and execute pieces of training	Training help people as:
		• Make them clued up about their actions.
		• People can easily detect incidents.
		• Clued up them about ill-suited behaviors encircling them.
		• To tackle vicious incidents.
5.	Demoralize poor behavior	• Never cheer up poor behavior.
		• Never validate poor behavior.
		• Penalize poor behavior.
6.	Keep track of workplace	• Never get frightened to query or intercede.
		• Must be aware of warning indications of ill-suited actions.
7.	Afford a system for complaining	• Have a conversation about your situation with any of the resources accessible.
		• Elucidate how to report.
		• Assure people that the incident will remain confidential as possible.
8.	Assist victims	• Assure victims that they will be not counterattack by a predator.
		• Assist people to feel secure reporting incidents.
		• Be on the side of the victim who comes ahead.
9.	Precautions for reducing risks	• Awareness of sexual harassments policies.

17.4.1 PREVENTIVE MEASURES (IN SCHOOL, COLLEGE, HOMES, AND HOSPITALS)

While surfing on Internet, never allow anyone to access your microphone and webcam to record your talks or capture your intimate activity. Leaked data can cause harm to the owner's reputation. Always be willing to ask for help, if you are getting suspicious messages or requests on your smartphone then block that sender immediately, report the behavior of the site administrator or consult an adult. If you have been victimized online, then tell someone. Be dubious if you find someone on any kind of dating app or in an online voice chat game and they ask you your mobile number or your social media handles. Always be aware of people's photos and their profile pictures as they can be fake which might be stolen from someone's account or downloaded from the Internet. Always be selective before uploading anything on social media handles over the internet as anyone can read out your information from your open to all social media handles.

Everyone should have to supports victims so that they can step out of their fear to share their stories with adults, friends, or known ones and ask them for help without any fear and uncertainty. In schools and colleges, safety skills related to the Internet and complete education of age-based restrictions on the Internet need to be given to the students as well as to their parents. Schools have to schedule guided programs that will teach students and their parents about methods to address online sex crimes and prevention against such types of criminal acts. Which will guide students on how to concede vicious/unhealthy situations and how to intercede efficiently as a bystander (Hong, Lu, Wu, Jimenez, & Milanaik, 2020). In middle schools, drama roleplay should be prepared by the cultural committee of the school to teach students how to respond to predators and pedophiles with whom they get in touch on the Internet.

In-Home, parents should try to understand the feelings of their child because in most cases children try to hide a thing from parents because of fear of punishment, they feel that everyone blames them for the situation. Parents should try to spend most of their time with their children, make them feel that they are not alone.

In Hospital or a clinic, a physician should provide victims a comfortable zone so that he/she can speak out openly about their situation without consequences. In addition, clinicians have to support their victimized patients as a friend and have to understand their feelings of fear, anxiety, distress, and shame. Emotional, behavioral support and counseling should have to be provided to both parents and patients by the physician. They have to instruct parents on how predators use uneven means to trap victims in the rings of sexual criminal acts.

17.4.2   EFFECTS ON HUMAN PSYCHE

Online sex crimes impact victims' physical, emotional, mental, social, and psychological health conditions. Impacts can vary from person to person depending on how moderate to severe torture and abuse they face during and after they get victimized by online sex crimes. The impact on physical conditions is due to the sexual assault that was performed by predators on women and minors, which can be in form of any body part damage. Explicit, who are forced to perform sexual activities can have sexually transmitted infectious diseases like HIV AIDS, syphilis, gonorrhea, urinary tract infections. They can even have other types of body part damages like vaginal/anal tearing, rectal trauma, or urinary difficulties because pimps use uneven brutal means to force victims to perform sexual activities.

Victims face trust issues after getting victimized as they get emotionally harassed by predators. Emotional impacts led to the loss of control and which afterward results in distress and trauma. Loss of control can be characterized into three impacts, i.e., past, present, and future impacts. In past impact, the victim feels a loss of control at the time of harassment when they are assaulted sexually and mentally, whereas in present impact victims feel a loss of control when they are getting investigated after getting rescued by police during their type of recovery from past crimes. And in future impact, victims feel a loss of control when they get revictimized again and again (Frazier, Berman, Steward, & Psychology, 2001). Sharing of the sexual content explicit over the Internet will directly impact the mental condition of victims like they feel depressed and experience loss in self-esteem, they even feel other kinds of mental health issues like feeling guilty, anxious, cheated, substance abused, etc. According to studies, online sex crime victimized victims are three times more likely to attempt suicide and two times more likely to get depressed than non-victims (Goebert et al., 2011).

Victims even get socially impacted by such types of online sex crimes because society blames only the one, who they saw in sexually explicit materials. Society believes that it was the victim who is completely responsible for his/her situation. Society associates sex crimes with public health concerns like paid sex behaviors, penetrative hookups, lowered condom use, and extramarital intercourse (Braithwaite et al., 2015). Sexual assaults on victims change their psychological behavior as in the feeling of shame and helplessness explicit try to remove their nude/seminude sexual materials from the Internet, but it is very challenging for them to remove any content from the Internet. Victims are found to exhibit more psychological problems than normal people, and because of these psychological change's victims fall into the use of substances like alcohol, drugs, cigarette to overcome feelings of anxiety, depression, shame, helplessness, and loss of self-esteem.

17.5 METHODOLOGY

Summing over 45 papers, the current study has initiated possible areas for further studies about characteristics, consequences, and preventive measures of CP and online sex crimes. The current study is the first of its type to study important evolution aspects of CP from the late 1800s to 2021. Findings from this chapter are therefore useful for practice, policy, and further research on online sex crimes, current work is set out to examine the effects of different types of vicious crimes on human health conditions. With a likeness and dissimilarity in mind, future study should focus more on how to cope up with predators mentality as a deep study need to be done on how they get into such type of crimes and how they stand out from normal people to heartless ones.

17.6 CONCLUSION

In conclusion, this study has two vicious looming threats CP and online sex crimes. Lack of social knowledge about cybercriminal law and because of their innocent behavior children was abused to attain sexual benefits. Various factors like cyberspace, possession, and extortion make online sex crimes like sextortion different from other kinds of sex crimes. Online sex crimes are performed mainly to attain sexual, instrumental benefits or to attain body type reinforcement. Online sex crimes impact victims' physical, emotional, social, mental, and psychological health conditions. To overcome all these health conditions, schools and colleges need to introduce sex education in their education system. Parents and physicians should provide victims a comfortable zone so that victims can speak up openly about their situation with consequences.

KEYWORDS

- **child pornography**
- **cyberspace**
- **online sex crimes**
- **photoDNA technology**
- **pornography**
- **sexting**
- **sextortion**
- **sex-trafficking**

REFERENCES

Acar, K. V. (2016). Sexual Extortion of Children in Cyberspace. *International Journal of Cyber Criminology*, *10*(2).

Agrawal, S. (2020). Online sextortion. *Indian Journal of Health*, *6*(1).

Aronowitz, A. A. (2001). Smuggling and trafficking in human beings: the phenomenon, the markets that drive it and the organizations that promote it. *European Journal on Criminal Policy and Research*, *9*(2), 163–195.

Bates, S. J. F. C., (2017). *Revenge Porn and Mental Health: A Qualitative Analysis of the Mental Health Effects of Revenge Porn on Female Survivors, 12*(1), 22–42.

Braithwaite, S. R., Givens, A., Brown, J., & Fincham, F. (2015). Is pornography consumption associated with condom use and intoxication during hookups? *Culture, Health & Sexuality*, *17*(10), 1155–1173.

Dake, J. A., Price, J. H., Maziarz, L., & Ward, B. (2012). Prevalence and correlates of sexting behavior in adolescents. *American Journal of Sexuality Education*, *7*(1), 1–15.

Deshpande, N. A., & Nour, N. M. (2013). Sex trafficking of women and girls. *Reviews in Obstetrics and Gynecology*, *6*(1), e22.

Dodds, L., (2020). *Twitter Accused of Aiding Child Abuse by Allowing 'Explosion' of Online Paedophile Communities*. Telegraph.

Farid, H. (2018). Reining in online abuses. *Technology & Innovation*, *19*(3), 593–599.

Frazier, P., Berman, M., Steward, J. J. A., & Psychology, P., (2001). *Perceived Control and Posttraumatic Stress: A Temporal Model, 10*(3), 207–223.

Gámez-Guadix, M., De Santisteban, P., & Alcazar, M. Á. (2018). The construction and psychometric properties of the questionnaire for online sexual solicitation and interaction of minors with adults. *Sexual Abuse*, *30*(8), 975–991.

Gesser-Edelsburg, A., & Abed Elhadi Arabia, M. (2018). Discourse on exposure to pornography content online between Arab adolescents and parents: qualitative study on its impact on sexual education and behavior. *Journal of Medical Internet Research*, *20*(10), e11667.

Goebert, D., Else, I., Matsu, C., Chung-Do, J., & Chang, J. Y. (2011). The impact of cyberbullying on substance use and mental health in a multiethnic sample. *Maternal and Child Health Journal, 15*(8), 1282–1286.

Flora, H. S., & Gultom, M. (2021). Legal protection against girl victims of trafficking for prostitution. *International Journal of Business, Economics and Law, 24*(5), 44–49.

Hanson, E. (2019). Losing track of morality: Understanding online forces and dynamics conducive to child sexual exploitation. *Child Sexual Exploitation: Why Theory Matters, 87.*

Hasinoff, A. A., & Shepherd, T. (2014). Sexting in context: Privacy norms and expectations. *International Journal of Communication, 8,* 24.

Hong, S., Lu, N., Wu, D., Jimenez, D. E., & Milanaik, R. L. (2020). Digital sextortion: Internet predators and pediatric interventions. *Current Opinion in Pediatrics, 32*(1), 192–197.

Houck, C. D., Barker, D., Rizzo, C., Hancock, E., Norton, A., & Brown, L. K. (2014). Sexting and sexual behavior in at-risk adolescents. *Pediatrics, 133*(2), e276–e282.

Jurecic, Q., Spera, C., Wittes, B., & Poplin, C. (2018). Sextortion: The problem and solutions. *Center for Technology at Brookings.* Retrieved from: https://www. brookings. edu/blog/ techtank/2016/05/11/sextortion-the-problem-andsolutions/on 05/01.

Koesnoen, A., (1964). *Criminal Structure in the Socialist State of Indonesia*: Bandung Well.

Lanning, K. V., & Burgess, A. W., (1984). *Child Pornography and Sex Rings* (Vol. 8). Federal Bureau of Investigation, US Department of Justice.

Li, Q. (2007). New bottle but old wine: A research of cyberbullying in schools. *Computers in Human Behavior, 23*(4), 1777–1791.

Liggett, R. (2019). Exploring online sextortion offenses: Ruses, demands, and motivations. *Sexual Assault Report, 22*(4), 58–62.

Lippman, J. R., & Campbell, S. W. (2014). Damned if you do, damned if you don't… if you're a girl: Relational and normative contexts of adolescent sexting in the United States. *Journal of Children and Media, 8*(4), 371–386.

Luthra, R., & Gidycz, C. A. (2006). Dating violence among college men and women: Evaluation of a theoretical model. *Journal of Interpersonal Violence, 21*(6), 717–731.

Malamuth, N. M., Addison, T., & Koss, M. (2000). Pornography and sexual aggression: Are there reliable effects and can we understand them? *Annual Review of Sex Research, 11*(1), 26–91.

Marpaung, Leden. (1996). Crimes against decency and their prevention issues. [Jakarta]: Sinar Graphic.

Mishna, F., McLuckie, A., & Saini, M. (2009). Real-world dangers in an online reality: A qualitative study examining online relationships and cyber abuse. *Social Work Research, 33*(2), 107–118.

Mitchell, K. J., Finkelhor, D., & Wolak, J. (2007). Youth Internet users at risk for the most serious online sexual solicitations. *American Journal of Preventive Medicine, 32*(6), 532–537.

Mitchell, K. J., Finkelhor, D., Jones, L. M., & Wolak, J. (2012). Prevalence and characteristics of youth sexting: A national study. *Pediatrics, 129*(1), 13–20.

Morelli, M., Bianchi, D., Baiocco, R., Pezzuti, L., & Chirumbolo, A. J. P., (2016). *Sexting, Psychological Distress and Dating Violence Among Adolescents and Young Adults*, 137–142.

Motsomi, K., Makanjee, C., Basera, T., & Nyasulu, P. (2016). Factors affecting effective communication about sexual and reproductive health issues between parents and adolescents in zandspruit informal settlement, Johannesburg, South Africa. *The Pan African Medical Journal, 25.*

Noll, J. G., Shenk, C. E., Barnes, J. E., & Haralson, K. J. (2013). Association of maltreatment with high-risk internet behaviors and offline encounters. *Pediatrics*, *131*(2), e510–e517.

O'Donnell, I., & Milner, C., (2012). *Child Pornography: Crime, Computers and Society*. Willan.

Patchin, J. W., & Hinduja, S. (2020). Sextortion among adolescents: Results from a national survey of US youth. *Sexual Abuse*, *32*(1), 30–54.

Paul, B., & Shim, J. W. (2008). Gender, sexual affect, and motivations for Internet pornography use. *International Journal of Sexual Health*, *20*(3), 187–199.

Peters, T., (2021). *TikTok Reveals New Privacy Settings for Kids: What Parents Should Know*. Today.

Ropelato, J. (2007). Pornography statistics 2007. *Top Ten Reviews*, *1*.

Roulston, K., & Shelton, S. A. J. Q. I., (2015). *Reconceptualizing Bias in Teaching Qualitative Research Methods, 21*(4), 332–342.

Salter, M. (2018). From geek masculinity to Gamergate: the technological rationality of online abuse. *Crime, Media, Culture*, *14*(2), 247–264.

Salter, M., & Hanson, E., (2021). I need you all to understand how pervasive this issue is: User efforts to regulate child sexual offending on social media. In: *The Emerald International Handbook of Technology Facilitated Violence and Abuse*. Emerald Publishing Limited.

Salter, M., & Whitten, T. (2022). A comparative content analysis of pre-internet and contemporary child sexual abuse material. *Deviant Behavior*, *43*(9), 1–15.

Salter, M., & Hanson, E. (2021). "I need you all to understand how pervasive this issue is": User efforts to regulate child sexual offending on social media. In *The Emerald International Handbook of Technology-Facilitated Violence and Abuse*. Emerald Publishing Limited.

Sextortion, Cyber Cell Delhi. Available at http://cybercelldelhi.in/sextortion.html (Accessed on 10/1/2022). (accessed on 10 January 2022).

Shackel, R. J. M. L. R., (1999). *Regulation of Child Pornography in the Electronic Age: The Role of International Law, 3*, 143.

Smith, P. K., Thompson, F., & Davidson, J. (2014). Cyber safety for adolescent girls: bullying, harassment, sexting, pornography, and solicitation. *Current Opinion in Obstetrics and Gynecology*, *26*(5), 360–365.

Sood, S. O., Churchill, E. F., & Antin, J. (2012). Automatic identification of personal insults on social news sites. *Journal of the American Society for Information Science and Technology*, *63*(2), 270–285.

Van Ouytsel, J. (2020). A decade of sexting research: Are we any wiser?. *JAMA Pediatrics*, *174*(2), 204.

Van Ouytsel, J., Madigan, S., Ponnet, K., Walrave, M., & Temple, J. R. (2019). Adolescent sexting: myths, facts, and advice. *NASN School Nurse*, *34*(6), 345–350.

Wolak, J., Finkelhor, D., & Mitchell, K. J. (2012). How often are teens arrested for sexting? Data from a national sample of police cases. *Pediatrics*, *129*(1), 4–12. doi: 10.1542/peds.2011-2242.

Zweig, J. M., Lachman, P., Yahner, J., & Dank, M. (2014). Correlates of cyber dating abuse among teens. *Journal of Youth and Adolescence*, *43*(8), 1306–1321.

CHAPTER 18

Digital Forensics and Cybersecurity Tools

VAISHALI,[1] SOURABH KUMAR SINGH,[2] and AMARNATH MISHRA[3]

[1]*MSc Forensic Science Student, Amity Institute of Forensic Sciences, Amity University, Noida, Uttar Pradesh, India*

[2]*Research Scholar, Amity Institute of Forensic Sciences, Amity University, Noida, Uttar Pradesh, India*

[3]*Professor (Forensic Science) & Director, Lloyd Institute of Forensic Science, Greater Noida, Uttar Pradesh (affiliated to the National Forensic Sciences University, Gandhinagar, Gujarat), India*

ABSTRACT

Digital forensic is one of the branches of forensic science which play an extremely significant role in society due to the prevalence of various digital forensic. The main aim of this is to acquire the courtroom evidence which is extracted from the various digital device that is used by perpetrators in various cybercrime and physical crimes. Various digital forensic tools such as FTK, Encase, X-way forensic, Sleuth Kit, SIFT, COFEE were analyzed using a computer system and USB stick. USB imaging was performed to extract the performed information. Investigations can often be used to operate, analyze, report, and direct investigations, often using tools, software or hardware, or a combination of both, and ultimately to identify specific aspects that are difficult for people to understand. These tools are often open-source or proprietary. Some tools come as software packages with hardware, and some may be applications. Used in computer forensics, these tools are programs or important applications designed to obtain evidence or information to be used as evidence. Of course, legal action is required. A major crime scene can be named as a network of computers and connections. These clever special

Advancements in Cybercrime Investigation and Digital Forensics. A. Harisha, Amarnath Mishra, & Chandra Singh (Eds.)
© 2024 Apple Academic Press, Inc. Co-published with CRC Press (Taylor & Francis)

effects have far-reaching implications for boot connections, usage, and steps such as Windows, UNIX, Linux, DOS, MAC, and more.

18.1 INTRODUCTION

Due to the widespread use of various digital forensic, one of the subfields of forensic science called "digital forensic" and it has a huge impact on society. The primary goal of this is to gather trial evidence that is taken from various digital devices used by those responsible for various types of physical crimes and cybercrimes (Powell & Haynes, 2020). Conventional crimes leave behind various clues such as fingerprints, DNA, footprints, and witnesses for the investigators to examine and investigate. Similarly, any kind of digital activity on electronic devices leaves a trail of data for cyber investigators to investigate and inspect the crimes to find the perpetrators. In all the cybercrime cases, it is very crucial to acquire the digital evidence, and they should be handled properly so that they can be further admissible in the court of law.

Digital forensics is the science of location. It extracts and analyzes data types from a variety of devices and then interprets them as legal evidence by experts. Digital evidence can be found on computer hard drives, cell phones, iPods, flash drives, digital cameras, CDs, DVDs, floppy disks, computer networks, and the Internet (John, 2005).

The main purpose of digital forensic tools is to protect against identity theft, money laundering, confidentiality [3], threats, and unauthorized access to confidential information, and to protect against sexual harassment, corruption, and other similar cybercrimes when digital data and confidential information are leaked.

18.1.1 DIGITAL FORENSIC

Digital forensics is a scientifically sound and substantiated process for preserving, collecting, corroborating, identifying, analyzing, interpreting, authenticating, foster the reconstruction of a crime or events that may be considered criminal. It's about using style. Helps prevent unauthorized actions that could interfere with planned work (Kruse & Heiser, 2001). "Protection, attestation, abstraction, instrument, and explanation of computer media for evidentiary and/or root cause examination."

Digital forensic is about evidence from the PC that is adequately dependable to stand up in court and be prevailing advanced criminological is the study of finding, disengaging, and anatomizing kind of information from the colorful widgets, which experts also, at that point decrypt to fill in as legal evidence.

Digital forensic can be identified with internet banking cheats, online offer swapping misrepresentation, source law burglary, MasterCard misrepresentation, duty avoidance, infection assaults, digital detriment, phishing assaults, dispatch commandeering, turndown of administration, hacking, murder cases, coordinated wrongdoing, fear-grounded tyrannize tasks, libel, porn, compulsion, converting, etc.

18.1.2 TYPES OF DIGITAL FORENSIC

All the mentioned branches are essential for digital forensics investigators to thoroughly investigate all branches (Figure 18.1).

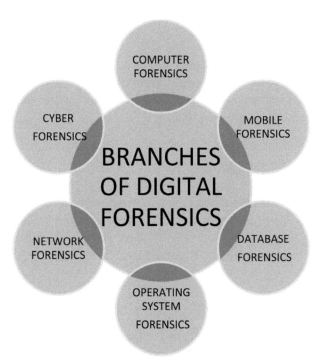

FIGURE 18.1 The various branches of digital forensics.

18.1.2.1   E-MAIL FORENSIC

The education includes documentation of the actual sender and recipient of the troubling e-mails, timestamp of the e-mail communication, purpose of mail, record of the whole e-mail operation. The various styles of crimes that square measure committed exploitation e-mails.

18.1.2.2   OPERATING SYSTEM (OS) FORENSICS

An OS is available on all PCs and so forth Since the presence of a working framework resembles essentially all over, have a criminology framework that oversees the prosperity and screens our activities and the cycles that go on in our PC to stay away from any kind of information misfortune or vindictive demonstrations that may occur in our framework. Like some other crime scene investigation activities, OS legal sciences additionally manages to check and look at the activities of an individual.

18.1.2.3   CYBER FORENSIC

Usually otherwise called computer crime scene investigation it is one of the fundamental parts of advanced criminology. Wrongdoing perpetrated utilizing PC and web which has a covered up and anticipated thought process of taking somebody's very own data.

18.1.2.4   MOBILE FORENSIC

Mobile forensics (MF) is a type of digital forensics that collects evidence from mobile devices. Also known as mobile device testing, which includes interactive elements such as government, people, finance, investigative platoon, process, and politics (Hazarika & Medhi, 2016; Ali et al., 2015; Punja, & Mislan, 2008).

18.1.2.5   COMPUTER FORENSIC

Assessment of virtual media via the clinical procedure of convalescing actual statistics for judicial review. The series and evaluation of statistics from several laptop resources, which includes laptop networks, telecommunication lines, laptop systems, and appropriate take a look at media (Ramadhani, Saragih, Rahim, & Siahaan, 2017; Sodhi et al., 2018; Al-Zain, & Al-Amri, 2018).

18.1.2.6 NETWORK FORENSIC

Communication networks allow computer data to be exchanged. Maximum digital offset like PCs, pads, and terminators connected via wired or wireless contacts on the network. Its goal is to limit online crime by finding evidence of cybercriminals in illegal activities (Chhabra & Singh, 2015).

18.1.2.7 DATABASE FORENSIC

Database Forensics is a branch of digital forensic wisdom related to the forensic investigation of databases and their associated metadata (Vishal & Meshram, 2012).

18.2 DIGITAL EVIDENCE

The definition proposed for Digital Evidence by the International Organization for Computer Evidence (IOCE) is – all data that is produced, put away, or sent utilizing electronic devices that might be depended upon in court during the trial is viewed as digital evidence. Even though it is ordinary for the expressions "electronic evidence" and "Digital evidence" to be utilized reciprocally, the last is, truth be told, just a subset of the more extensive class of "electronic proof" which, he sets, likewise remembers proof for the type of simple information, for example, video and sound tape accounts, photographic film, and copied captures of fixed-line telephone discussions. While this load of sorts of information likely could be "digitalized," they don't start in advanced structure.

18.3 THE DIGITAL FORENSIC INVESTIGATION MODELS

Digital forensics tools–forensic investigators gather information to help convict criminals through the process of locating, conserving, evaluating, and documenting digital evidence. Then it goes into the stream as shown in Figure 18.2.

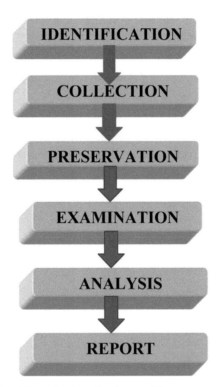

FIGURE 18.2   The digital forensics investigation models.

1. **Identification:** It's the technique for recognizing the violations and hence the connection information as proof for the examination.
2. **Collection:** Because computerized data is stored on the PC, an extended data set means writing the data to a set of equipment or some medium that contains the data.
3. **Preservation:** Preserving computerized proof early, is a basic advance toward expanding our odds of an effective examination, or episode reaction. It gives a straightforward comprehension of what happened to get the proof, and what the proof addresses.
4. **Examination:** It is best directed on a duplicate of the first proof. The first proof ought to be procured in a way that ensures and jam the respectability of the proof.
5. **Analysis:** During an investigation, agents typically recover evidence using a few different methods, and periodically initiate the recovery of erased material. This is often a way to destroy collected data. Find important information and links to help you keep track of who sent it.

6. **Reporting:** At the end of the survey, the data is often found in a structure that even non-specialists can accept. Reports may also contain survey data and other meta information.

18.3.1 TYPES OF ACQUISITION

The inconsistency of evidence is recognized, and digital evidence is collected in the order that secures maximum safety. Data is stored in a way that improves integrity and availability and reduces the impact on the system under investigation and investigation (Figure 18.3).

1. **Logical Acquisition:** This data extraction is mainly done to get basic device data. This import does not retrieve unallocated spatial data. In the absence of device root, retrieving information can be a huge advantage, but during this exposure only marked data will not affect deleted data.

2. **Physical Acquisition:** The actual way to get evidence of extraction is to access the device's RAM and isolate the information. Makes a bit-by-bit copy of the mobile device. First, it supports retrieving deleted records. These two types of data collection are often done on both rooted and non-rooted mobile devices for information and knowledge.

3. **Manual Acquisition:** Manually importing data is a simple way to import data from your Android smartphone. The investigator uses the mobile keyboard to query the contents of the mobile device. The main advantage is that it is as convenient as possible and does not require the training of an investigator to understand how to get the contents of the call. It does not retrieve all data on your Android smartphone, including deleted and hidden files.

18.3.2 EXTRACTION

It is the method of extracting data from e-mails, PDFs, PDF forms, text files, images, etc. It is a crucial part of a digital forensics investigation. Data extraction is the method of retrieving any deleted data, file, content, etc., from the electronic devices when they cannot be accessed, searched, or opened normally by the user. Being able to extract the deleted data could help the investigators or analysts to solve various civil or illegitimate cases.

"File carving is the method that is used in digital forensics to obtain the data from the disk drive or any other storage device without the assistance or permission of the file system that has created the file originally. It is the process which extracts the files at unallocated space and is also used to generate the data and execute the computer forensic investigation."

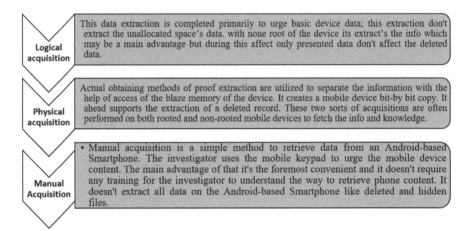

FIGURE 18.3   Types of acquisition.

18.4   DIGITAL FORENSIC TOOLS USED FOR DIFFERENT DEVICES

18.4.1   PC AND MEMORY FORENSIC TOOLS

PC explanatory instruments are acclimated to get data from a framework or a PC. They're acclimated secure libraries and diverse scrambled data from the PC. Though memory expository devices are acclimated procure and break down the PC's volatile memory (RAM). They're acclimated save evidence inside the memory that is lost once the framework is shutting. They straightforwardly analyze the OS and diverse running bundle inside the memory. The fluctuated stages whereby these devices work is Windows, UNIX, Linux, DOS, MAC:

1. **Windows:** It is a GUI progression of Sistema activity that is created and advertised more than Microsoft's subscription volume. UNIX: UNIX may be a multi-user PC work structure that performs a variety of tasks, and this exists in invariants. Unix exploited the C programming language.

2. **Linux:** A Unix-like PC OS based on a free and open-source package model. Customers change, create, and sell their OS versions for their benefit.
3. **DOS:** Partner OS that reflects and charges gadget memory devices and data.
4. **MAC OS:** It is primarily a GUI progression located within a business created by Apple INC. For the Macintosh computer line.

18.4.2 MOBILE FORENSIC TOOLS

Mobile logical instruments address every piece of equipment and bundle peripherals. Cell phones oblige a different fluctuate of connectors; the equipment gadgets support an assortment of different links and perform a steady job as a compose blocker in PC gadgets. Cell phones contain crucial individual data that might be utilized as confirmation. Data not inheritable from cell phones keep on being utilized as evidence in criminal, common, and even status cases.

1. **Android:** Humanoids by Google INC – Affiliate Degree Open Stock OS. Blackberry: The is a closed OS with limitations.
2. **iOS:** It is provided by Apple INC. Closed stock and limited. All Macs, iPhones, iPod bits, and iPads use iOS.
3. **Windows Phone:** This is Microsoft's inventory and proprietary product. Windows Phone devices are mainly manufactured by Nokia, HTC, Huawei, Samsung, and various companies.

18.4.3 NETWORK FORENSIC TOOLS

Network explanatory apparatuses comprise numerous screens that might be placed in totally various focuses inside the organization and utilized for appropriated network police work. These organization recognition devices coordinate data from the different screens and supply a whole and extensive read of the organized movement.

18.5 DIGITAL FORENSIC TOOLS

These tools have been developed by programmers and various software companies which help the investigators and digital forensic analysts to collect the digital evidence from the electronic devices at the crime scene. Digital devices can be desktops, laptops, USB drives, tablets, mobile phones, CCTVs, and many more. These tools can be both hardware and software. These can be the commercial ones that can be bought, or they are available online and can be used free of cost. There are pros and cons to every tool. Not all tools can perform every digital forensic process whether it is acquisition, extraction, or analysis. So, it's a good technique to have various tools available to perform multiple tasks from basic to advanced levels.

18.6   TYPES OF DIGITAL FORENSIC TOOLS

Whenever a crime is associated with a virtual world, it is called cybercrime (Lee, 2001). The process used to detect this crime is called digital forensics. In general, digital forensics involves the storage, retrieval, identification, analysis, and reporting of data (EnCase Tool). Tools we are working on are discussed in subsections (Table 18.1).

18.6.1   EnCase

EnCase version 7.12.01.18 is the tool which is distributed and maintained by Guidance Software. It is the tool that offers functionality and flexibility. Investigators have the flexibility to complete the investigation efficiently and quickly with this tool (http://www.arxsys.fr/features/).

- It gives the adaptability to any investigator to gain information from the wide variety of computerized gadgets and it likewise incorporates 25 kinds of cell phones like PDAs, tablets, iOS, and so on.
- It helps an investigator to complete the investigation in a forensically defined manner.
- It produces the extensive reports of the whole investigation and maintains the integrity of the evidence.
- It creates the exact duplicate files of original data which is further verified by hash and cyclic.

18.6.2   DIGITAL FORENSICS FRAMEWORK (DFF)

Crime scene investigation steps using the modified API. Commonly used by legal institutions, educational institutions, and private companies around the

TABLE 18.1 Caparisoning of Digital Forensic Tools

Attribute	EnCase	DFF	Win Hex	X-Ways Forensics	FTK	The Sleuth Kit	COFEE	SIFT	Window Scope
License	Proprietary	Open	Open	Proprietary	Open	Open	Proprietary	Open	Proprietary
Platform	Windows, MAC, Linux, DOS machine	Windows, Linux	Windows, Linux	Windows, DOS	Windows	Windows, MAC, Linux, DOS machine	Windows	Ubuntu	Windows
Disk imaging	Available	Available	Available	Available	Available	Available	Available	Available	Available
Language interface	Traditional Chinese	English	English	English	Chinese	English	English	English	English
Data recovery	Available	Available	Available	Available	Available	Available	Available	Available	Available
Password recovery	Yes	No	Yes	Yes	Yes	No	Yes	No	No
Techniques used for hash value calculated	MD5	MD5	MD5 and SHA1	MD5	MD5 and SHA1	MD5 and SHA1	MD5	MD5	MD5

world. DFF is available in three versions free, DFF Pro: €1,000 per year, and DFF Live: €1,300 with one-year support. DFF's runaway selection doesn't even offer expert help, reporting supervisors, robotic motors, client drilldowns, hash scanners, and Skype comparative studies, DFF Pro and DFF Live.

- can carry out cryptographic hash calculations;
- can extract EXIF meta information;
- all deprecated Microsoft Outlook symbols may be imported;
- investigate memory dumps;
- scripting and grouping capabilities;
- instinct alerts and web browsing of important data;
- information can be extracted naturally.

18.6.3　X-WAYS FORENSIC

X-Ways Forensics is closed-sourced in the marketplace (http://www.x-ways.net/forensics/). Integrated criminal PC programming with Win Hex, Disk Imager. The following components are:

- You have full access to platters, RAID arrays, and images that need to be seen more than 2 GB;
- Clipping elements must also be available in other documents;
- PhotoDNA hashing is used to distinguish famous photos;
- Multiple hash values can be defined together;
- Create a disk image;
- Password recovery to reduce the burden of documents found in packaging structures.

18.6.4　SCIENTIFIC TOOLKIT (FTK)

Access data group–FTK producers (http://accessdata.com/solutions/digitalforensics/forensic-toolkit-ftk). They are an important provider of training and certification for forensic science instruments. More than 1,30,000 regulatory agencies and law firms use FTK worldwide. Scans can be performed on PCs, PCs, network switches, and mobile phones. Screening and searching are faster than other available tools. The elements of FTK are:

- Can receive and store organization-wide information.
- It can receive data from 3,500 mobile phones.

- For the first time, you can differentiate between missing information, angry behavior, and information leaks. Different realms, FTK can recognize what did what, who changed the realm, and even see if there was a change in the first realm.
- Data can also be collected during static surveys.
- You can run FTK from a USB stick.
- There is an alternative to the master survey, where an audit is performed and compared to the observer in the last proof check.
- Password recovery.
- Plates were rendered using FTK Imager.
- Information can be viewed using hashing techniques and Boolean values.
- FTK internal observer allows researchers to view Excel, PowerPoint, and Word.

18.6.5 THE SLEUTH KIT

This is a library containing a set of order line devices (http://www.sleuthkit. org/autopsy/features.php). The primary function of the detective package is to help analyze information about the structure of the document. This helps to decompose the duplicate circular images of the framework understudy and recover the documents. We pursue the latest forensic science used by law enforcement, military, and corporate analysts and try to explain what's going on with our devices. It can also be used to recover recent media and documents from memory cards. The Sleuth Kit framework provides an open scene for working in the usage level module. The structure itself provides access to the document.

The Sleuth Kit's structure permits a client to effortlessly construct robotized, start to finish advanced criminological applications. The structure is utilized for a nittier gritty and complex examination of the circle picture. Assuming the client needs the subtleties of just the volume level help and record framework level help, then, at that point the first Sleuth Kit library can be utilized.

18.6.6 SANS INVESTIGATIVE FORENSIC TOOLKIT (SIFT)

SIFT (https://linuxhint.com/sans_investigative_forensics_toolkit/) changed into evolved with the aid of using a worldwide institution of experts. It is one of the maximums broadly used open-supply forensics tools. It changed into determined as an incident reaction computer and later opened to the public. The SIFT factors are:

- Real-time scanning capable;
- Fast reporting and investigation possible;
- Malware recognition capable;
- Cyclic imaging capable;
- Revocation framework;
- VMware appliance;
- Chief support also available observer design.

18.6.7   WIN HEX

Win Hex is essentially a multi-purpose location representation system editor that is very useful in the fields of computer rhetoric, information recovery, low-level processing, and IT security. A knowledgeable tool for every day and emergency use, viewing and editing all types of files, and recovering deleted files from burdensome drives or camera cards with corrupted file systems.

18.6.8   COMPUTER ONLINE FORENSIC EVIDENCE EXTRACTOR (COFEE)

Computer online forensic evidence extractor (COFEE) was created by Microsoft to remove evidence in Windows (https://www.semanticscholar.org/topic/Computer-Online-Forensic-Evidence-Extractor/1299045). Measurable people can perform real-time investigations by inserting COFEE into a flash drive or hard drive. Microsoft is currently working with Interpol and the National Center for Combating Crime (NW3C) on forensics. COFEE is not available to everyone except organizations that require compliance with the law.

18.6.9   WINDOWS SCOPE

Among the many tools (https://www.windowsscope.com/products/), Windows Scope performs the calculation. He has clients in 16 countries around the world. Graphical user interface (GUI) toolbox. Here's the fate of the Windows range:

- Able to perform a real-time investigation;
- Able to perform sporadic responses;

- Able to perform a circular rendering;
- Able to perform a memory dump check;
- Data must be recoverable;
- Alert the client if an attack occurs on the gadget.

18.7 CONCLUSION

The devices are programming-based, yet on occasion, equipment is likewise needed to procure proof. A portion of the products are unreservedly accessible, and some software is paid for. Unreservedly accessible programs are otherwise called Open-Source apparatuses. Broad aspects of this invention have been addressed in phases such as Windows, UNIX, Linux, DOS, MAC, etc., which can operate similarly, including boot association, use, and the like. These devices are not the only devices available for the goal of measurable research, and the list is much larger. These tools are not the only tools available for forensic purposes.

KEYWORDS

- **data acquisition**
- **data analysis**
- **data extraction tools**
- **digital forensics**
- **law enforcement**
- **mobile forensics**
- **SANS investigative forensic toolkit**

REFERENCES

Ali, A., Razak, S. A., Othman, S. H., & Mohammed, A., (2015). Towards adapting metamodeling approach for the mobile forensics investigation domain. In: *International Conference on Innovation in Science and Technology (IICIST)* (p. 5).

Al-Zain, M. A., & Al-Amri, J. F., (2018). Application of data steganographic method in video sequences using histogram shifting in the discrete wavelet transform. *International Journal of Applied Engineering Research, 13*(8), 6380–6387.

Barker, K., Askari, M., Banerjee, M., Ghazinour, K., Mackas, B., Majedi, M., Pun, S., & Williams, A., (2009). A data privacy taxonomy. In: *BNCOD 26: Proceedings of the 26ᵗʰ British National Conference on Databases* (pp. 42–54). Berlin, Heidelberg. Springer Verlag.

Chhabra, G. S., & Singh, P., (2015). Distributed network forensics framework: A systematic review. *International Journal of Computer Applications, 119*(19).

Computer Online Forensic Evidence Extractor. https://www.semanticscholar.org/topic/Computer-Online-Forensic-Evidence-Extractor/1299045 (accessed on 10 January 2022).

Digital Forensics Framework Tool, http://www.arxsys.fr/features/ (accessed on 10 January 2022).

EnCase Tool. https://www.guidancesoftware.com/encase-forensic (accessed on 10 January 2022).

Forensic Toolkit. http://accessdata.com/solutions/digitalforensics/forensic-toolkit-ftk (accessed on 10 January 2022).

Hazarika, B., & Medhi, S., (2016). Survey on real-time security mechanisms in network forensics. *International Journal of Computer Applications, 151*(2).

John, R. V., (2005). *Computer Forensics: Computer Crime Scene Investigation* (2ⁿᵈ edn., Vol. 1). Charles River Media River media.

Kruse, W. G., & Heiser, J. G., (2001). *Computer Forensics*. Incident Response Essentials. Addison-Wesley.

Lee, G., (2001). *EnCase: A Case Study in Computer-Forensic Technology*. Computer Magazine.

Powell, A., & Haynes, C., (2020). Social media data in digital forensics investigations. In: *Digital Forensic Education* (pp. 281–303). Springer, cham.

Punja, S. G., & Mislan, R. P., (2008). Mobile device analysis. *Small Scale Digital Device Forensics Journal, 2*(1), 1–16.

Ramadhani, S., Saragih, Y. M., Rahim, R., & Siahaan, A. P. U., (2017). Post-genesis digital forensics investigation. *Int. J. Sci. Res. Sci. Technol., 3*(6), 164–166.

SANS Investigative Forensic Toolkit. https://linuxhint.com/sans_investigative_forensics_toolkit/ (accessed on 10 January 2022).

Sodhi, G. K., et al., (2018). Preserving authenticity and integrity of distributed networks through novel message authentication code. *Indonesian Journal of Electrical Engineering and Computer Science, 12*(3), 1297–1304.

The Sleuth Kit. http://www.sleuthkit.org/autopsy/features.php (accessed on 10 January 2022).

Varsha, K. S., & Vanita, M., (2015). Comparative study and simulation of digital forensic tools. *International Conference on Advances in Science and Technology 2015* (*ICAST 2015).*

Vishal, R. A., & Meshram, B. B., (2012). Digital forensic tools. *IOSR Journal of Engineering, 2*(3), 392–398.

WindoesScope. https://www.windowsscope.com/products/ (accessed on 10 January 2022).

X-Ways Forensics Tool. http://www.x-ways.net/forensics/ (accessed on 10 January 2022).

Index